COSMOLOGY IN GAUGE FIELD THEORY
AND STRING THEORY

T0299724

Graduate Student Series in Physics
Other books in the series

GRADUATE STUDENT SERIES IN PHYSICS

Series Editor:
Professor Douglas F Brewer, MA, DPhil
Emeritus Professor of Experimental Physics, University of Sussex

COSMOLOGY IN GAUGE FIELD THEORY AND STRING THEORY

DAVID BAILIN

Department of Physics and Astronomy
University of Sussex

ALEXANDER LOVE

Department of Physics
Royal Holloway and Bedford New College
University of London

CRC Press
Taylor & Francis Group
Boca Raton London New York

CRC Press is an imprint of the
Taylor & Francis Group, an **informa** business

A TAYLOR & FRANCIS BOOK

British Library Cataloguing-in-Publication Data

A catalogue record for this book is available from the British Library.

ISBN 0 7503 0492 8

Library of Congress Cataloging-in-Publication Data are available

Commissioning Editor: John Navas
Production Editor: Simon Laurenson
Production Control: Leah Fielding
Cover Design: Victoria Le Billon
Marketing: Nicola Newey

Published by Institute of Physics Publishing, wholly owned by The Institute of Physics, London

Institute of Physics Publishing, Dirac House, Temple Back, Bristol BS1 6BE, UK

US Office: Institute of Physics Publishing, The Public Ledger Building, Suite 929, 150 South Independence Mall West, Philadelphia, PA 19106, USA

Typeset in LaTeX 2_ε by Text 2 Text Limited, Torquay, Devon

To Eva Bailin and the memory of William Bailin (1911–1994)
and
To Christine

Contents

Preface

The new particle physics of the past 30 years, including electroweak theory, quantum chromodynamics, grand unified theory, supersymmetry, supergravity and superstring theory, has greatly changed our view of what may have happened in the universe at temperatures greater than about 10^{15} K (100 GeV). Various phase transitions may be expected to have occurred as gauge symmetries which were present at higher temperatures were spontaneously broken as the universe cooled. At these phase transitions topological defects, such as domain walls, cosmic strings and magnetic monopoles, may have been produced. Various types of relic particles are also expected. These may include neutrinos with small mass and axions associated with the solution of the strong CP problem in quantum chromodynamics. If supersymmetry exists, there should also be relic supersymmetric partners of particles, some of which could be dark matter candidates. If the supersymmetry is local (supergravity) these will include the gravitino, the spin-$\frac{3}{2}$ partner of the graviton. Insight may also be gained into the observed baryon number of the universe from mechanisms for baryogenesis which arise in the context of grand unified theory and electroweak theory. Supersymmetry and supergravity theories may have scope to provide the particle physics underlying the inflationary universe scenario that resolves such puzzles as the extreme homogeneity and flatness of the observed universe. Superstring theory also gives insight into the statistical thermodynamics of black holes. In the context of superstring theory, bold speculations have been made as to a period of evolution of the universe prior to the big bang ('pre-big-bang' and 'ekpyrotic universe' cosmology).

These matters, amongst others, are the subject of this book. The book gives a flavour of the new cosmology that has developed from these recent advances in particle physics. The aim has been to discuss those aspects of cosmology that are most relevant to particle physics. From some of these it may be possible to uncover new particle physics that is not readily discernible elsewhere. This is a particularly timely enterprise, since, as has been noted by many authors, the recent data from WMAP and future data expected from Planck mean that cosmology may at last be regarded as precision science just as particle physics has been for many years.

We are grateful to our colleagues Nuno Antunes, Mar Bastero-Gil, Ed Copeland, Beatriz de Carlos, Mark Hindmarsh, George Kraniotis, Andrew Liddle, André Lukas and Paul Saffin for the particle and cosmological physics that we have learned from them. Special thanks also to Malcolm Fairbairn for helping us with the diagrams. Finally, we wish to thank our wives for their invaluable encouragement throughout the writing of this book.

We intend to maintain an updated erratum page for the book at http://www.pact.cpes.sussex.ac.uk/~mpfg9/cosmobook.htm.

David Bailin and Alexander Love
June, 2004

Chapter 1

The standard model of cosmology

1.1 Introduction

The principal concern of this book is the way in which recent particle physics, including electroweak theory, quantum chromodynamics, grand unified theory, supersymmetry, supergravity and superstring theory, has changed our standpoint on the history of the universe when its temperature was greater than 10^{15} K. This will be studied in the context of the Friedman–Robertson–Walker solution of the Einstein equations of general relativity. In this chapter, therefore, our first task is the derivation of the field equations relating the scale factor $R(t)$ that appears in the metric to the energy density ρ and the pressure p that characterize the (assumed homogeneous and isotropic) energy–momentum tensor. This is done in the following two sections. In section 1.4 we show how, for a given equation of state, energy–momentum conservation determines the scale dependence of the energy density and pressure. The standard solutions for the time dependence of the scale factor in a radiation-dominated universe, in a matter-dominated universe, and in a cosmological constant-dominated universe are presented in section 1.5; we give an estimate of the age of the universe in the matter-dominated case in section 1.6. In section 1.7, we present the evidence that there is, in fact, a non-zero cosmological constant and discuss why its size is so difficult to explain. The discussion of phase transitions and of relics that is given in later chapters also requires a description of the thermodynamics of the universe. So in the following two sections we describe the equilibrium thermodynamics of the expanding universe and derive the time dependence of the temperature in the various epochs. In section 1.10, we discuss briefly the 'recombination' of protons and electrons that left the presently observed cosmic microwave background radiation. Finally, the synthesis of the light elements that commenced towards the end of the first three minutes is discussed in section 1.11. The consistency of the predicted abundances with those inferred from the measured abundances determines the so-called baryon asymmetry of the universe, whose origin is discussed at length in chapter 4.

1.2   The Robertson–Walker metric

The standard description of the hot big bang assumes a universe which is homogeneous and isotropic with a metric involving a single function $R(t)$, the 'scale factor' (or 'radius' of the universe). The appropriate metric is the Robertson–Walker metric

$$ds^2 = dt^2 - R^2(t) \left(\frac{dr^2}{1 - kr^2} + r^2 \, d\theta^2 + r^2 \sin^2 \theta \, d\phi^2 \right) \qquad (1.1)$$

where the (time and spherical polar) coordinates (t, r, θ, ϕ), called the 'comoving' coordinates, are the coordinates of an observer in free fall in the gravitational field of the universe. The parameter k takes the values $-1, 0, 1$ corresponding to a universe which has spatial curvature which is negative, zero or positive, respectively. (This can be seen from the curvature scalar derived from the second equality of (1.30) with a change in sign for Euclidean rather than Minkowski space.) Units have been chosen in which the speed of light c is 1.

An immediate use of this metric is to calculate the size of regions of the universe that have been in causal contact (in the sense that there has been the possibility of causal influence occurring between points within the region at some time between the big bang at $t = 0$ and time t). Causal influences cannot occur over distances greater than the (proper) distance $d_H(t)$ that light has been able to travel from the the big bang at $t = 0$ to the time t being studied. This distance is called the 'particle horizon'. Without loss of generality, consider emission of a light signal from coordinate (r, θ, ϕ) at $t = 0$ to coordinate $(0, \theta, \phi)$ at time t along the (radial) geodesic with θ and ϕ constant. (It may be checked that this is indeed a geodesic by using the coefficients of affine connection given in the next section (exercise 1).) For a light beam, $ds^2 = 0$ and we have

$$\frac{dt^2}{R^2(t)} = \frac{dr^2}{1 - kr^2}. \qquad (1.2)$$

Thus, the largest value of r at $t = 0$ to be in causal contact with $r = 0$ at time t is given implicitly by

$$\int_0^t \frac{dt'}{R(t')} = \int_0^r \frac{dr'}{\sqrt{1 - kr'^2}}. \qquad (1.3)$$

This equation determines the particle horizon. The proper distance to the particle horizon at time t is

$$\begin{aligned} d_H(t) &= R(t) \int_0^r \frac{dr'}{\sqrt{1 - kr'^2}} \\ &= R(t) \int_0^t \frac{dt'}{R(t')}. \end{aligned} \qquad (1.4)$$

We shall discuss the time dependence of the scale factor $R(t)$ in the next section. Equation (1.4) then allows us to calculate the particle horizon. For example, when

$$R(t) \propto t^{2/3} \tag{1.5}$$

as is the case for a matter-dominated universe, we get

$$d_H(t) = 3t \tag{1.6}$$

and for a radiation-dominated universe in which

$$R(t) \propto t^{1/2} \tag{1.7}$$

we get

$$d_H(t) = 2t. \tag{1.8}$$

For an inflationary universe, such as will be discussed in chapter 7,

$$R(t) \propto e^{Ht} \tag{1.9}$$

with H approximately constant, and then

$$d_H(t) = \frac{1}{H}(e^{Ht} - 1). \tag{1.10}$$

The Robertson–Walker metric also allows us to calculate the redshifting of light from distant objects. Consider light, travelling on a radial geodesic, being received at $r = 0$ at (around) the present time $t = t_0$ from a distant galaxy at $r = r_1$. Suppose that two adjacent crests of a light wave are received at $t = t_0$ and $t = t_0 + \Delta t_0$ having been emitted from the distant galaxy at $t = t_1$ and $t = t_1 + \Delta t_1$. Equation (1.3) applies but with appropriate modifications to the limits of integration. Thus,

$$\int_{t_1}^{t_0} \frac{dt}{R(t)} = \int_0^{r_1} \frac{dr}{\sqrt{1 - kr^2}} \tag{1.11}$$

and

$$\int_{t_1 + \Delta t_1}^{t_0 + \Delta t_0} \frac{dt}{R(t)} = \int_0^{r_1} \frac{dr}{\sqrt{1 - kr^2}}. \tag{1.12}$$

Subtracting gives

$$\int_{t_1 + \Delta t_1}^{t_0 + \Delta t_0} \frac{dt}{R(t)} = \int_{t_1}^{t_0} \frac{dt}{R(t)} \tag{1.13}$$

so that

$$\int_{t_0}^{t_0 + \Delta t_0} \frac{dt}{R(t)} = \int_{t_1}^{t_1 + \Delta t_1} \frac{dt}{R(t)}. \tag{1.14}$$

Because the variation of $R(t)$ on the time scale of an electromagnetic wave period is very small, this equation may be approximated by

$$\frac{\Delta t_0}{R(t_0)} = \frac{\Delta t_1}{R(t_1)}. \tag{1.15}$$

But Δt_0 and Δt_1 are the times between adjacent crests; in other words, they are the periods of the waves. Thus, the waves have frequencies

$$\nu_0 = \frac{1}{\Delta t_0} \quad \text{and} \quad \nu_1 = \frac{1}{\Delta t_1} \tag{1.16}$$

respectively and, in units where $c = 1$, wavelengths

$$\lambda_0 = \Delta t_0 \quad \text{and} \quad \lambda_1 = \Delta t_1 \tag{1.17}$$

respectively. The redshift is usually defined by

$$z \equiv \frac{\lambda_0 - \lambda_1}{\lambda_1} \tag{1.18}$$

and, from (1.15), we conclude that

$$1 + z = \frac{R(t_0)}{R(t_1)}. \tag{1.19}$$

Equations (1.19) and (1.17), reinterpreted in terms of photons, mean that a photon emitted at time t_1 undergoes a redshifting of its wavelength as the universe expands, such that its wavelength at time t_0 is increased by a factor $R(t_0)/R(t_1)$. Since the momentum (or energy) of the photon is inversely proportional to its wavelength, the momentum (or energy) of the photon is reduced by a factor $R(t_1)/R(t_0)$ as a result of the expansion of the universe. This is often expressed as energy of photons being redshifted away.

When $|t_1 - t_0|$ is not too large, we can make the expansion

$$R(t_1) = R(t_0) + (t_1 - t_0)\dot{R}(t_0) + \tfrac{1}{2}(t_1 - t_0)^2 \ddot{R}(t_0) + \cdots$$
$$= R(t_0)(1 + H_0(t_1 - t_0) - \tfrac{1}{2}q_0 H_0^2 (t_1 - t_0)^2 + \cdots) \tag{1.20}$$

where

$$H_0 \equiv \frac{\dot{R}(t_0)}{R(t_0)} \tag{1.21}$$

is the present value of the Hubble parameter and q_0 is the present deceleration parameter

$$q_0 \equiv -\frac{\ddot{R}(t_0)}{R(t_0)H_0^2} = -\frac{\ddot{R}(t_0)R(t_0)}{\dot{R}(t_0)^2}. \tag{1.22}$$

The redshift may also be expanded in powers of $t_1 - t_0$:

$$1 + z = (1 + H_0(t_1 - t_0) - \tfrac{1}{2}q_0 H_0^2 (t_1 - t_0)^2 + \cdots)^{-1} \tag{1.23}$$

leading to

$$z = H_0(t_0 - t_1) + \left(1 + \frac{q_0}{2}\right) H_0^2(t_0 - t_1)^2 + \cdots. \qquad (1.24)$$

Since z is the physically measurable quantity, it is useful to invert (1.24). For small z

$$t_0 - t_1 = \frac{1}{H_0} \left[z - \left(1 + \frac{1}{2}q_0\right) z^2 + \cdots \right]. \qquad (1.25)$$

Then, after expanding $1/R(t)$ in (1.11) in powers of $t - t_0$, we may determine r_1 as a function of z. Expanding (1.11) gives

$$\frac{1}{R(t_0)} \left[(t_0 - t_1) + \frac{1}{2} H_0(t_0 - t_1)^2 + \cdots \right] = r_1 + O(r_1^3). \qquad (1.26)$$

Thus, in terms of the redshift,

$$r_1 = \frac{1}{R(t_0)H_0} \left[z - \frac{1}{2}(1 + q_0)z^2 + \cdots \right]. \qquad (1.27)$$

We shall use this result in section 1.7 to calculate the 'luminosity distance' of a (supernova) source as a function of the redshift.

1.3 Einstein equations for a Friedmann–Robertson–Walker universe

It is straightforward to calculate the coefficients of affine connection for the metric (1.1). The non-zero components are

$$\Gamma^0_{ij} = -\frac{\dot{R}}{R} g_{ij} \qquad \Gamma^i_{j0} = \frac{\dot{R}}{R} \delta_{ij} = \Gamma^i_{0j} \qquad (1.28)$$

$$\Gamma^i_{jk} = \frac{1}{2} g^{il} (\partial_k g_{lj} + \partial_j g_{lk} - \partial_l g_{jk}). \qquad (1.29)$$

Here x^i, $i = 1, 2, 3$, denotes the (spatial) coordinates (r, θ, ϕ). Equation (1.29) is just the coefficients of affine connection for the three-dimensional subspace (r, θ, ϕ). It is also straightforward to calculate the Ricci tensor $R_{\mu\nu}$ from the cofficients of affine connection (exercise 2). It has non-zero components

$$R_{00} = -3\frac{\ddot{R}}{R} \qquad \text{and} \qquad R_{ij} = -\left[\frac{\ddot{R}}{R} + 2\frac{\dot{R}^2}{R^2} + \frac{2k}{R^2}\right] g_{ij}. \qquad (1.30)$$

The corresponding curvature scalar is

$$\mathbb{R} \equiv g^{\mu\nu} R_{\mu\nu} = -6 \left[\frac{\ddot{R}}{R} + \frac{\dot{R}^2}{R^2} + \frac{k}{R^2}\right]. \qquad (1.31)$$

The Einstein equations for the Robertson–Walker metric, usually referred to as the Friedman–Robertson–Walker (FRW) universe, are

$$R_{\mu\nu} - \tfrac{1}{2}\mathbb{R}g_{\mu\nu} = 8\pi G_N T_{\mu\nu} + \Lambda g_{\mu\nu} \qquad (1.32)$$

where G_N is the Newtonian gravitational constant, $T_{\mu\nu}$ is the energy–momentum tensor and we are including a cosmological constant Λ. For a perfect fluid with energy density ρ and pressure p, the non-vanishing components are

$$T_{00} = \rho \qquad \text{and} \qquad T_{ij} = -p\delta_{ij}. \qquad (1.33)$$

The corresponding Einstein equations are, from the 00-component,

$$\left(\frac{\dot{R}}{R}\right)^2 + \frac{k}{R^2} = \frac{8\pi G_N}{3}\rho + \frac{\Lambda}{3} \qquad (1.34)$$

usually referred to as the 'Friedmann' equation, and, from the ij-components,

$$2\frac{\ddot{R}}{R} + \left(\frac{\dot{R}}{R}\right)^2 + \frac{k}{R^2} = -8\pi G_N p + \Lambda. \qquad (1.35)$$

Subtracting (1.35) from (1.33) gives the equation for $\ddot{R}$

$$\frac{\ddot{R}}{R} = -\frac{4\pi G_N}{3}(\rho + 3p) + \frac{\Lambda}{3}. \qquad (1.36)$$

In the case $\Lambda = 0$, this equation implies that $\ddot{R} < 0$ for all times[1]. Then, the present positive $\dot{R}$ implies that $\dot{R}$ was always positive and, therefore, that R was always increasing. Consequently, ignoring the effects of quantum gravity, there was a past time when R was zero—the moment of the 'big bang'.

Returning to the Friedmann equation (1.34) with zero cosmological constant, the universe is spatially flat when

$$\rho = \rho_c = \frac{3H^2}{8\pi G_N} = 3M_P^2 H^2 \qquad (1.37)$$

where H is the Hubble parameter,

$$H \equiv \frac{\dot{R}}{R} \qquad (1.38)$$

and M_P is the reduced Planck mass given by

$$M_P^2 = \frac{1}{8\pi G_N} = \frac{m_P^2}{8\pi} \qquad (1.39)$$

---

[1] A positive value of the acceleration $\ddot{R}$ can only arise if Λ is positive.

where m_P is the Planck mass, and

$$M_P \simeq 2.44 \times 10^{18} \text{ GeV} \qquad m_P \simeq 1.22 \times 10^{19} \text{ GeV}. \qquad (1.40)$$

Since the Hubble parameter varies with time, so does ρ_c. The density parameter Ω is defined as

$$\Omega \equiv \frac{\rho}{\rho_c} \qquad (1.41)$$

and measures the density as a fraction of the 'critical' density ρ_c. The current value of Ω, denoted by Ω_0, has a value [1]

$$\Omega_0 = 1.02 \pm 0.02. \qquad (1.42)$$

1.4 Scale factor dependence of the energy density

There is also conservation of the energy–momentum tensor to take into account:

$$D_\nu T^{\mu\nu} = 0 \qquad (1.43)$$

where

$$D_\lambda V^\mu = \partial_\lambda V^\mu + \Gamma^\mu_{\lambda\rho} V^\rho \qquad (1.44)$$

is the action of the covariant derivative D_λ on a contravariant index. The $\mu = 0$ component of (1.43) yields (exercise 3)

$$\dot{\rho} + 3(\rho + p)\frac{\dot{R}}{R} = 0. \qquad (1.45)$$

It is easy to see that this is just the first law of thermodynamics

$$dE + p \, dV = 0 \qquad (1.46)$$

for a comoving volume $V \propto R^3(t)$.

The energy density ρ may be related to the scale factor $R(t)$ once we have the equation of state. If this is of the form

$$p = w\rho \qquad (1.47)$$

then (1.45) leads to

$$\rho \propto R^{-3(1+w)}. \qquad (1.48)$$

In particular, for $w = \frac{1}{3}$, corresponding to radiation (massless matter)

$$\rho \propto R^{-4} \qquad \text{radiation} \qquad p = \tfrac{1}{3}\rho. \qquad (1.49)$$

For $w = 0$, corresponding to massive matter,

$$\rho \propto R^{-3} \qquad \text{matter} \qquad p = 0. \qquad (1.50)$$

Equation (1.50) may be understood as a constant number of massive particles occupying a volume expanding as $R^3(t)$ as the universe expands. Equation (1.49) may be understood as the number density of photons (or other massless particles) decreasing as $R^{-3}(t)$, as for massive matter but, in addition, the energy of each photon decreasing as $R^{-1}(t)$ because of the redshifting of the photon energy discussed in section 1.2. Another interesting case is $w = -1$, which gives

$$\rho = \text{constant} \qquad p = -\rho. \tag{1.51}$$

This may be interpreted as vacuum energy and allows us to incorporate the cosmological constant into the discussion without introducing it explicitly, if we wish.

1.5   Time dependence of the scale factor

It is easy to solve the Friedmann equation (1.34) in the case of zero cosmological constant and $k = 0$, a spatially flat universe. Both of these assumptions are always good approximations for sufficiently early times because, as discussed in section 1.4, $\rho \propto R^{-4}$ for radiation domination and $\rho \propto R^{-3}$ for matter domination. Consequently, for a 'big-bang' universe with $R \to 0$ as $t \to 0$, the $\frac{8}{3}\pi G_N \rho$ term in (1.34) becomes more important than the k/R^2 or $\Lambda/3$ terms. With the energy density ρ given by (1.48), the solution of (1.34) (provided $w \neq -1$) is

$$R(t) \propto t^{-\frac{3}{2}(1+w)}. \tag{1.52}$$

In particular,

$$R \propto t^{1/2} \quad \text{and} \quad H = \tfrac{1}{2}t^{-1} \quad \text{for radiation domination} \tag{1.53}$$

and

$$R \propto t^{2/3} \quad \text{and} \quad H = \tfrac{2}{3}t^{-1} \quad \text{for matter domination.} \tag{1.54}$$

However, if at some stage in the history of the universe the cosmological constant is (positive and) large enough to dominate over the energy density and curvature terms in (1.34), then the Friedmann equation has the solution

$$R(t) \propto e^{\sqrt{\frac{\Lambda}{3}}t}. \tag{1.55}$$

This is the de Sitter universe.

1.6   Age of the universe

We shall estimate the age of the universe in the case $\Lambda = 0$. We shall also assume a matter-dominated universe for the calculation. This is a reasonable approximation because, as can be seen from section 1.8, the universe was matter-dominated for most of its history. First, rewrite the Friedmann equation (1.34) in terms of the value ρ_0 of the energy density ρ today. From (1.50),

$$\frac{\rho}{\rho_0} = \left(\frac{R}{R_0}\right)^{-3}.$$
(1.56)

Thus, the Friedmann equation may be written as

$$\left(\frac{\dot{R}}{R_0}\right)^2 + \frac{k}{R_0^2} = \frac{8\pi G_N}{3}\rho_0\frac{R_0}{R}.$$
(1.57)

Next rewrite this in terms of the present value Ω_0 of the density parameter (1.41):

$$\Omega_0 = \frac{\rho_0}{(3/8\pi G_N)H_0^2}.$$
(1.58)

Then, at $t = t_0$, (1.57) gives

$$\frac{k}{R_0^2} = \frac{8\pi G_N}{3}\rho_0 - H_0^2 = H_0^2(\Omega_0 - 1)$$
(1.59)

where the last equality employs (1.58). Thus, the Friedmann equation may be written as

$$\left(\frac{\dot{R}}{R_0}\right)^2 + H_0^2(\Omega_0 - 1) = \Omega_0 H_0^2\frac{R_0}{R}.$$
(1.60)

This may be rewritten in terms of the variable

$$x \equiv \frac{R}{R_0}$$
(1.61)

as

$$\dot{x}^2 + H_0^2(\Omega_0 - 1) = \Omega_0 H_0^2 x^{-1}$$
(1.62)

with solution

$$t = \frac{1}{H_0}\int_0^x \frac{dx'}{\sqrt{\Omega_0(x'^{-1} - 1) + 1}}.$$
(1.63)

In particular, today, when $R = R_0$, x has the value 1 and the current age of the universe is

$$t_0 = \frac{1}{H_0}\int_0^1 \frac{dx}{\sqrt{\Omega_0(x^{-1} - 1) + 1}}.$$
(1.64)

We see that $t_0 \sim H_0^{-1}$ with the precise value depending on the value of Ω_0. For example, for an exactly flat universe (which is not consistent with observations) $\Omega_0 = 1$ and $t_0 = \frac{2}{3}H_0^{-1}$. It is usual to write H_0^{-1} in the form

$$H_0^{-1} \simeq h^{-1}9.78 \times 10^9 \text{ yr}$$
(1.65)

where the parameter h is measured to have the value

$$h = 0.72 \pm 0.05. \tag{1.66}$$

Thus, the present age of the universe is

$$t_0 \sim 10^{10} \text{ yr}. \tag{1.67}$$

1.7 The cosmological constant

In 1917, attempting to apply his general theory of relativity (GR) to cosmology, Einstein sought a static solution of the field equations for a universe filled with dust of constant density and zero pressure. The general static solution of (1.34) and (1.36) has

$$p = \frac{1}{3}\left(\frac{\Lambda}{4\pi G_N} - \rho\right) \tag{1.68}$$

and

$$\frac{k}{R^2} = \frac{8\pi G_N}{3}\rho + \frac{\Lambda}{3}. \tag{1.69}$$

With zero cosmological constant ($\Lambda = 0$), the only solution of these equations, apart from an empty, flat universe, requires that either the energy density ρ or the pressure p is negative. It was this unphysical result that led him to introduce the cosmological term. Then the solution for pressureless dust is

$$\rho = \frac{\Lambda}{4\pi G_N} \tag{1.70}$$

and

$$\frac{k}{R^2} = \Lambda. \tag{1.71}$$

Assuming that ρ is positive requires that Λ is positive, so that

$$k = +1 \tag{1.72}$$

and

$$R = \frac{1}{\sqrt{\Lambda}}. \tag{1.73}$$

Hence, the universe is closed and has the geometry of S^3 with volume V and mass M given by

$$V = 2\pi^2 R^3 = 2\pi^2 \Lambda^{-3/2} \qquad M = \frac{\pi}{2G_N\sqrt{\Lambda}}. \tag{1.74}$$

A non-zero cosmological constant also allows non-trivial static (de Sitter) solutions of the Einstein field equations with no matter ($\rho = 0 = p$) at all. It was, therefore, a considerable relief in the 1920s when the redshifts of distant galaxies were observed, the presumption of a static universe could be abandoned and there was no need for a cosmological constant.

However, anything that contributes to the energy density of the vacuum $\langle\rho\rangle$ acts just like a cosmological constant. This is because the Lorentz invariance of the vacuum requires that the energy–momentum tensor in the vacuum $\langle T_{\mu\nu}\rangle$ satisfies

$$\langle T_{\mu\nu}\rangle = \langle\rho\rangle g_{\mu\nu}. \tag{1.75}$$

Then, by inspection of (1.32), we see that the vacuum energy density contributes $8\pi G_N \langle\rho\rangle$ to the effective cosmological constant

$$\Lambda_{\text{eff}} = \Lambda + 8\pi G_N \langle\rho\rangle. \tag{1.76}$$

Equivalently, we may regard the cosmological constant as contributing $\Lambda/8\pi G_N$ to the effective vacuum energy density

$$\rho_{\text{vac}} = \langle\rho\rangle + \frac{\Lambda}{8\pi G_N} = \Lambda_{\text{eff}} M_P^2. \tag{1.77}$$

Thus, a cosmological constant is often referred to as 'dark energy', not to be confused with dark matter which contributes to the non-vacuum energy density (and has zero pressure).

A priori, in any quantum theory of gravitation, we should expect the scale of the vacuum energy density to be set by the Planck scale M_P. Since Λ has the dimensions of M^2, it follows that we should have expected that $\Lambda/M_P^2 \sim 1$. We shall see that, in reality, the scale of any such energy density must be much smaller. We noted in section 1.5 that the effect of the cosmological constant is negligible at sufficiently early times, because the energy density ρ scales as a negative power of R for radiation or matter domination. Thus, the most stringent bounds arise from cosmology when the expansion of the universe has diluted the matter energy density sufficiently. From the observation that the present universe is of at least of size H_0^{-1}, we may conclude that

$$|\Lambda_{\text{eff}}| \lesssim 3H_0^2 \tag{1.78}$$

where

$$H_0^{-1} \sim 10^{10} \text{ yr} \sim 10^{42} \text{ GeV}^{-1} \tag{1.79}$$

from (1.67). Then, in Planck units,

$$\frac{|\Lambda_{\text{eff}}|}{M_P^2} \lesssim 10^{-120}. \tag{1.80}$$

For many years, this tiny ratio was taken as evidence that the cosmological constant is indeed zero. However, during the past few years, evidence has accumulated that Λ is, in fact, non-zero.

   The first evidence suggesting this came from measurements of the redshifts of type Ia supernovae. Such supernovae arise as remnants of the explosion of white dwarfs which accrete matter from neighbouring stars. Eventually the white dwarf mass exceeds the Chandrasekhar limit and the supernova is born after the explosion. The intrinsic luminosity of such supernovae is considered to be a constant. That is, they are taken as standard candles and any variation in their apparent luminosity as measured on earth must be explicable in terms of their differing distances from the earth. In a Euclidean space, the apparent luminosity l of a source with intrinsic luminosity L at a distance D from the observer is given by

$$l = \frac{L}{4\pi D^2}. \tag{1.81}$$

We may, therefore, define the 'luminosity distance' D_L of a source from the observer by

$$D_L \equiv \sqrt{\frac{L}{4\pi l}}. \tag{1.82}$$

In GR we must be more careful. So consider the circular mirror, area A, of a telescope at the origin, normal to the line of sight to a source at r_1. Light emitted from the source at time t_1 and arriving at the mirror at time t_0 is bounded by a cone with solid angle

$$\omega = \frac{A}{4\pi R(t_0)^2 r_1^2} \tag{1.83}$$

as measured in the locally inertial frame at the source. The emitted photons have their energy redshifted by a factor

$$\frac{R(t_1)}{R(t_0)} = \frac{1}{1+z} \tag{1.84}$$

as explained in section 1.2, (see (1.18)). Also, photons emitted at time intervals of δt_1 reach the mirror at time intervals $\delta t_0 = \delta t_1 R(t_0)/R(t_1)$. Thus, the total power P received at the mirror is given by

$$P = L \left(\frac{R(t_1)}{R(t_0)} \right)^2 \omega \tag{1.85}$$

and the apparent luminosity by

$$l = \frac{P}{A}. \tag{1.86}$$

Then, using (1.27), the luminosity distance defined in (1.82) is

$$D_L = H_0^{-1}(1+z)\left[z - \frac{1}{2}(1+q_0)z^2 + \cdots\right] \tag{1.87}$$

$$= \frac{1}{H_0}\left[z + \frac{1}{2}(1-q_0)z^2 + \cdots\right]. \tag{1.88}$$

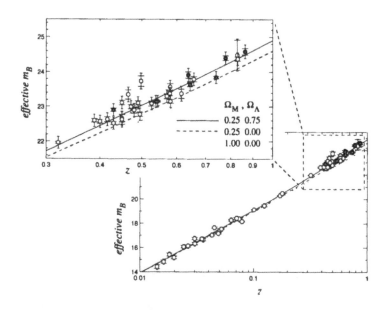

Figure 1.1. Hubble diagram giving the effective magnitude *versus* redshift for the supernovae in the primary low-extinction subset. The full line is the best-fit flat-universe cosmology from the low-extinction subset, the broken and dotted lines represent the indicated cosmologies.

Hence, for nearby supernovae the luminosity distance is proportional to the redshift of the source.

Astronomers measure the apparent magnitude m of the various supernovae sources. The difference $m - M$, where $M \sim -19.5$, is the (assumed constant) intrinsic magnitude of the source, is just the logarithm of the luminosity distance. So the apparent magnitude is predicted to be linear in $\ln z$ for small z. This is consistent with the data for $z \lesssim 0.1$, see figure 1.1 taken from [2]. For more distant supernovae the linear relationship between D_L and z is distorted by quadratic terms depending on the present deceleration parameter q_0 of the universe. The data for $0.7 \lesssim z \lesssim 1$ do display such a distortion, see figure 1.1 [2].

For an FRW universe, it follows from (1.36) and the definition (1.22) of q_0 that, in general, the deceleration may be written as

$$q_0 = \tfrac{1}{2} \sum_i (1 + 3w_i)\Omega_i \tag{1.89}$$

for a universe with components labelled by i having energy density ρ_i and pressure $p_i \equiv w_i \rho_i$; here $\Omega_i \equiv \rho_i/\rho_c$ where $\rho_c \equiv 3H_0^2/8\pi G_N$ is the critical density. In particular, for a universe with just (pressureless) matter and

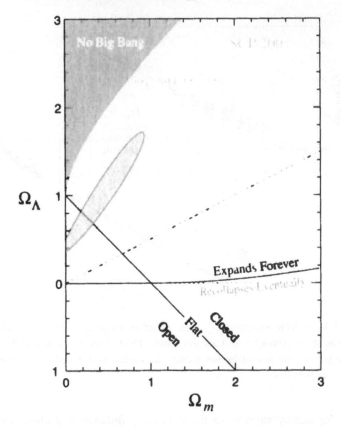

Figure 1.2. 68%, 90%, 95%, and 99% confidence regions for Ω_m and Ω_Λ.

a cosmological constant, we get

$$q_0 = \tfrac{1}{2}\Omega_m - \Omega_\Lambda \qquad (1.90)$$

where $\Omega_m \equiv \rho_m/\rho_c$ is the matter contribution and $\Omega_\Lambda \equiv \rho_{vac}/\rho_c = \Lambda_{eff}/3H_0^2$. As noted previously, a negative value of q_0, corresponding to an accelerating universe, can only arise with a positive cosmological constant. The data shown in figures 1.1 and 1.2 taken from [2] suggest that this is indeed the case.

The determination of Ω_m and Ω_Λ requires at least one further input. The recent data on the temperature anisotropies of the cosmic microwave background provide just such a constraint. Photons originating at the 'last scattering surface', when matter and radiation decouple (see section 1.10), having a redshift $z \sim 1300$, are seen now as the microwave background. Quantum fluctuations in the early universe give rise to fluctuations in the energy density of the radiation and these appear as temperature fluctuations in the microwave background (see section 7.7). These fluctuations may be analyzed by multipole moments, labelled by l, and are characterized by their power spectrum. The multipole number l_{peak} of the first peak in the power spectrum is determined by the total matter content of the universe. In fact, $l_{\text{peak}} \sim 220\Omega_0$, where $\Omega_0 \equiv \rho_0/\rho_c$ measures the total energy density ρ_0 relative to the critical density. The measured position of the first peak yields the value (1.42). Thus, for a universe with just matter and a cosmological constant, we get

$$\Omega_m + \Omega_\Lambda \sim 1. \tag{1.91}$$

When this result is combined with the supernova and other data, it is found that

$$\Omega_m \sim 0.3 \qquad \Omega_\Lambda \sim 0.7. \tag{1.92}$$

In Planck units, this means that

$$\frac{\Lambda_{\text{eff}}}{M_P^2} = \frac{\rho_{\text{vac}}}{M_P^4} = \Omega_\Lambda \frac{\rho_c}{M_P^4} \simeq 0.8 \times 10^{-120}. \tag{1.93}$$

There is currently no known explanation of this extremely small number. It corresponds to $\rho_{\text{vac}}^{1/4} \simeq 10^{-3}$ eV. It is generally believed that the particle physics vacuum is the minimum of an effective potential in which the electroweak gauge symmetry $SU(2)_L \times U(1)_Y$ is spontaneously broken (see section 2.5). The value of the effective potential at this minimum $\langle\rho\rangle$ has no effect on the particle physics. By adding a constant V_0 to the tree-level potential (2.93), it is easy to arrange that the potential, including any radiative and temperature-dependent corrections, has any desired value at the minimum. However, to do so requires the fine tuning of V_0 to ensure that the value (1.93) is obtained and it is this fine tuning that is regarded as unnatural and for which an explanation is sought. The obvious first approach to the problem is to seek a symmetry that requires $\Lambda = 0$ and then to explore mechanisms that break the symmetry only slightly. The only known symmetry that requires a vanishing cosmological constant is *global* supersymmetry. The (fermionic) supersmmetry generator Q satisfies the anticommutation relation

$$\{Q, \bar{Q}\} = 2\gamma^\mu P_\mu \tag{1.94}$$

where P_μ is the energy–momentum vector. It follows [3] that, for any state $|\psi\rangle$,

$$\langle\psi|P_0|\psi\rangle = \langle\psi|Q_\alpha Q_\alpha^* + Q_\alpha^* Q_\alpha|\psi\rangle \geq 0. \tag{1.95}$$

Thus, the energy of any non-vacuum state is positive and the vanishing of the vacuum energy defines a unique, supersymmetric vacuum state $|0\rangle$ that satisfies

$$\langle0|P_0|0\rangle = 0 \Leftrightarrow Q_\alpha|0\rangle = 0. \tag{1.96}$$

In a supersymmetric theory, all particles have supersymmetric partners (called 'sparticles') having opposite statistics. That is to say, the sparticle associated with a fermion is a boson and the sparticle associated with a boson is a fermion. The sparticles associated with the quarks and leptons, called respectively 'squarks' and 'sleptons', are (spin-0) scalar particles and, in a supersymmetric theory, they must have the same mass and quantum numbers as the original particles. This has the important consequence that the vanishing cosmological constant result is unaffected by quantum effects, because supersymmetry ensures that any quantum corrections arising from fermion loops, say, are cancelled by those that arise from the bosonic loops of the associated sparticle. It has yet to be demonstrated experimentally that supersymmetry has anything to do with reality. None of the sparticles associated with the known particles has ever be seen. (It is hoped that they will be discovered at the Large Hadron Collider (LHC).) Supersymmetry (susy), if present at all, is therefore a broken symmetry. It then follows from (1.95) that the vacuum energy is positive definite. The experimental limits on the sparticle masses require that

$$m_{\text{susy}} \gtrsim 100 \text{ GeV}. \tag{1.97}$$

If something like this bound were to set the scale for ρ_{vac}, then

$$\frac{\rho_{\text{vac}}}{M_P^4} \sim 10^{-68}. \tag{1.98}$$

Although small compared with the $O(1)$ expected in a generic quantum theory of gravity, this is still very much larger than the value (1.93) derived from the supernovae and Wilkinson Microwave Anisotropy Probe (WMAP) data. Thus, if this were the only contribution to the vacuum energy density, we should be confronted with an unmitigated disaster.

However, including gravity in any supersymmetric theory inevitably leads to a supergravity theory, in which supersymmetry is a local, rather than a global, symmetry. This is because in GR the momentum generator P_μ becomes a local field generating diffeomorphisms of spacetime. Then, in a supersymmetric theory incorporating GR, the supersymmetry generators too become local fields: this is why supergravity emerges as the low-energy limit of string theory. The form of the potential in a supergravity theory is given in section 2.8. The main point to note is that, as in the case of global supersymmetry, supersymmetric vacua are generally stationary points of this potential but that at such points the vacuum energy density is now generally *negative*. Non-supersymmetric (scalar) field configurations in which the energy density is zero do exist but (without fine tuning) these are not generally stationary points of the potential. Thus, supergravity does not solve the cosmological constant problem but it is no worse than in non-supersymmetric theories.

In the absence of any theoretical insight into the origin of the smallness of the cosmological constant, it is of interest to see whether 'anthropic' considerations can shed any light on the issue. Using the 'weak anthropic principle', we seek to determine which era or which part of the universe could support human life, so that physicists exist to pose such questions. A large positive cosmological constant leads to an exponentially expanding (de Sitter) universe, see (1.55).

This exponential expansion inhibits the formation of the gravitationally bound clumps of matter that are presumably a necessary precondition for life to evolve; once the clumps are formed, the cosmological constant has no further effect. Thus, the weak anthropic principle requires Λ_{eff} to be small enough to allow the formation of sufficiently large clumps of matter. Gravitational condensation began in our universe at a redshift z_c where $z_c \geq 4$. The energy density of matter at that time was greater than the present matter density ρ_m by a factor of $R^3(t_0)/R^3(t_c) = (1 + z_c)^3 \geq 125$. The cosmological constant has no effect so long as it is dominated by the matter density. Thus, provided $\rho_{vac} \lesssim 125\rho_m$, the vacuum energy density would not inhibit gravitational condensation. (A more careful treatment [4] gives a further factor of $\frac{1}{3}\pi^2$.) We conclude that if the anthropic principle accounts for the value of the (positive) cosmological constant, then we should expect $\rho_{vac} \sim (10 - 100)\rho_m$ because there is no anthropic reason for it to be smaller. This gives the prediction $\Omega_\Lambda \sim (10 - 100)\Omega_m$, at variance with the values (1.92) derived from the supernovae and WMAP data. Nevertheless, it implies a much smaller value ρ_{vac}/M_P^4 than that given in (1.98) which was derived from supersymmetry considerations.

In contrast, a negative cosmological constant does not affect gravitational clumping. We see from the Friedmann equation (1.34) that if Λ is negative, the expansion of the universe ceases (for a flat universe ($k = 0$)) when the matter density term is cancelled by the cosmological constant. We have already noted that the deceleration parameter q_0 given in (1.90) is positive for $\Lambda < 0$. It follows that after expansion has ceased, the universe begins to contract and, in fact, it collapses to a singularity in a finite time T. It is easy to show (exercise 4) that

$$T = \frac{2\pi}{\sqrt{3|\Lambda|}}. \tag{1.99}$$

Anthropic considerations would then require that this leaves sufficient time for life to evolve, say $T \gtrsim \frac{1}{2}H_0^{-1}$ where $H_0^{-1} = \sqrt{3/8\pi G_N \rho_m}$ is the Hubble time in our universe. This would give

$$\frac{\Omega_\Lambda}{\Omega_m} \lesssim \left(\frac{4\pi}{3}\right)^2. \tag{1.100}$$

Again, this would entail a much smaller value of ρ_{vac}/M_P^4 than was obtained from supersymmetry considerations. However, the supernovae data indicate a universal *acceleration* rather than a deceleration. Thus, Λ is positive and the previous bound is only of academic interest.

1.8 Equilibrium thermodynamics in the expanding universe

It makes sense to discuss equilibrium thermodynamics during most of the history of the universe because reaction rates were much faster than the time scale for the expansion of the universe which is characterized by the Hubble time H^{-1}. As discussed in section 2.2, the pressure p, entropy density s and energy density ρ due to a gas of ultrarelativistic particles (in which the temperature T is much greater than all masses) are given by

$$p = \frac{\pi^2}{90} N_* T^4 \tag{1.101}$$

$$s = \frac{2\pi^2}{45} N_* T^3 \tag{1.102}$$

$$\rho = \frac{\pi^2}{30} N_* T^4 \tag{1.103}$$

where

$$N_* = N_B + \tfrac{7}{8} N_F. \tag{1.104}$$

The numbers N_B and N_F of bosonic and fermionic degrees of freedom are defined after (2.19). The entropy S in a comoving volume $R^3(t)$

$$S = s R^3 \tag{1.105}$$

is expected to be conserved because a homogeneous universe has no temperature differences to generate heat transfer. (For an explicit proof of entropy conservation, see section 3.4 of Kolb and Turner or section 15.6 of Weinberg in the general references.) Thus, to the extent that the entropy density is dominated by the ultra-relativistic particles

$$RT = \text{constant} \tag{1.106}$$

while N_* is constant. Equation (1.106) is valid even for a matter-dominated universe because it is only the particles with mass m smaller than the temperature T that are present in thermal equilibrium with appreciable number densities and contributing to the entropy, although all particles contribute to the energy density. In reality, RT will show small discontinuous changes as the temperature drops below the mass of particular particle species. Subject to this caveat, equation (1.53) for the time dependence of the scale factor now implies the following connection between temperature and time for a radiation-dominated universe:

$$T \propto t^{-1/2} \qquad \text{for radiation domination.} \tag{1.107}$$

The constant of proportionality in this equation may be calculated from the Friedmann equation. When RT is a constant,

$$\left(\frac{\dot{R}}{R}\right)^2 = \left(\frac{\dot{T}}{T}\right)^2 \tag{1.108}$$

and, using (1.103), the Friedmann equation (1.34) may be rewritten as

$$\left(\frac{\dot{T}}{T}\right)^2 = \frac{8\pi G_N}{3} \frac{\pi^2}{30} N_* T^4 \qquad (1.109)$$

where we have neglected the cosmological constant and the curvature term, as in section 1.5. This has solution

$$t = \frac{1}{2}\left(\frac{90}{\pi^2 N_*}\right)^{1/2} M_P T^{-2} \qquad (1.110)$$

$$\simeq 1.51 M_P N_*^{-1/2} T^{-2}. \qquad (1.111)$$

If, for example, the appropriate N_* for T above 100 GeV is that of the $SU(3) \times SU(2) \times U(1)$ standard model or that of the supersymmetric standard model, then

$$N_* = \frac{427}{4} \quad \text{or} \quad \frac{915}{4} \qquad (1.112)$$

respectively.

Equations (1.54) and (1.106) imply the following connection between temperature and time for the matter-dominated, universe:

$$T \propto t^{-2/3} \qquad \text{for matter domination.} \qquad (1.113)$$

For a matter-dominated universe,

$$\rho(T) = 3M_P^2 H_0^2 \Omega_0 \left(\frac{T}{T_0}\right)^3 \qquad (1.114)$$

where we have used (1.56), (1.106), (1.58) and (1.40). Using (1.108), the Friedmann equation (1.34) may be rewritten as

$$\left(\frac{\dot{T}}{T}\right)^2 = H_0^2 \Omega_0 \left(\frac{T}{T_0}\right)^3 \qquad (1.115)$$

with solution

$$t = \frac{2}{3}(H_0 \Omega_0^{1/2})^{-1}\left(\frac{T}{T_0}\right)^{-3/2}. \qquad (1.116)$$

1.9   Transition from radiation to matter domination

As we have seen in (1.49) and (1.50), the energy density of radiation decreases as R^{-4} as the universe expands whereas the energy density of matter decreases as R^{-3}. Thus, radiation domination gives way to matter domination at some point in the expansion of the universe. For a matter-dominated universe, the energy density is given by (1.114) and for a radiation-dominated universe by (1.103). However, there is a subtlety in the interpretation of N_* which must be taken into account. We shall assume that the transition temperature is sufficiently low that the only relativistic particles are the photon and three neutrinos. Neutrinos drop out of thermal equilibrium below about 1 MeV when the (weak) interaction rate that keeps them in thermal equilibrium becomes less than the Hubble parameter. (See section 5.2.) When the temperature drops below the electron mass (about 0.5 MeV), electrons and positrons annihilate via $e^+e^- \rightarrow \gamma\gamma$, and the entropy of the electron–positron pairs is transferred to the photons. However, no entropy is transferred to the neutrinos, which are now decoupled. Before electron–positron annihilation, we have

$$N_* = 2 + \tfrac{7}{2} = \tfrac{11}{2} \tag{1.117}$$

but afterwards we should take

$$N_* = 2. \tag{1.118}$$

Note that we are *not* keeping any contribution from the neutrinos because they have now dropped out of thermal equilibrium and no longer contribute to the entropy. If $T_{\gamma i}$ and $T_{\gamma f}$ are the photon temperatures before and after electron–positron annihilation, conservation of entropy requires that

$$\tfrac{11}{2}T_{\gamma i}^3 = 2T_{\gamma f}^3 \tag{1.119}$$

so that

$$\frac{T_{\gamma f}}{T_{\gamma i}} = \left(\frac{11}{4}\right)^{1/3} \simeq 1.4. \tag{1.120}$$

However, the neutrinos do not share in this temperature increase. Thus, there is an effective N_* for the purpose of calculating the energy density of radiation

$$N_{*,\text{eff}} = 2 + \tfrac{7}{8} \times 6 \times (\tfrac{4}{11})^{4/3} \simeq 3.36. \tag{1.121}$$

Note that the relativistic neutrinos still have an energy density that varies as R^{-4} as the universe expands, because their number density varies as R^{-3} and they undergo redshifting of their energy as R^{-1}. Thus, the neutrino energy density also varies as T^4, where T is the photon temperature, i.e. the temperature in the usual sense. Then, from (1.103), in the radiation-dominated universe below the temperature at which e^+e^- annihilation occurs,

$$\rho(T) = \frac{\pi^2}{30} N_{*,\text{eff}} T^4. \tag{1.122}$$

If the matter energy density of (1.114) and the radiation energy density (1.103) are equal at a temperature T_{eq}, we find that

$$T_{\text{eq}} = \frac{90 M_P^2 H_0^2 \Omega_0}{\pi^2 N_{*,\text{eff}} T_0^3}. \tag{1.123}$$

Using (1.65), (1.121) and (1.40) gives

$$T_{\text{eq}} = 5.68 \Omega_0 h^2 \text{ eV}. \tag{1.124}$$

With h given by (1.66) and Ω_0 by (1.42), we have

$$T_{eq} \simeq 3 \, \text{eV}. \qquad (1.125)$$

This justifies the original assumption that the only relativistic particles are the photon and three neutrinos.

1.10 Cosmic microwave background radiation (CMBR)

During the radiation-dominated era, the photons were in thermal equilibrium with matter (at the same temperature) because of interaction with the charge of the electrons and protons. We are making the approximation here that all baryons in the universe at this time are in the form of protons. However, eventually the electrons and protons combine into neutral atoms. (This is referred to as 'recombination'.) Thereafter, photons decouple from matter and evolve at a temperature different from matter. In this way, black-body radiation at the recombination temperature develops into black-body radiation in the present universe at a lower temperature, because the temperature is proportional to the mean photon energy and the energy of the photons has redshifted with the expansion of the universe. Thus, for the photons,

$$T \sim R(t)^{-1}. \qquad (1.126)$$

The recombination temperature T_{rec} may be estimated to be

$$T_{rec} = 3575 \, \text{K} = 0.31 \, \text{eV}. \qquad (1.127)$$

Here, recombination has been defined as the point at which 90% of electrons have combined with protons. (See, for example, section 3.5 of Kolb and Turner in the general references.) We may calculate the time of recombination t_{rec} from (1.116), noting that the universe has been matter dominated from the time of recombination until the present by comparing (1.127) with (1.125). Using (1.42), (1.65) and (1.66), with $T_0 = 2.73$ K, gives

$$t_{rec} \simeq 1.89 \times 10^5 \, \text{yr}. \qquad (1.128)$$

1.11 Big-bang nucleosynthesis

In chapter 4 we shall discuss possible explanations of the 'observed baryon asymmetry of the universe':

$$\eta \equiv \frac{n_B}{n_\gamma} \simeq 6.4 \times 10^{-10} \qquad (1.129)$$

where n_B and n_γ are the present number densities of baryons and photons. Before doing this, it is important to understand the origin of this number whose explanation has been and remains a major topic of research. For this reason we shall outline here how this number emerges from measurements of the present abundances of the light elements, specifically deuterium (D), helium (^{3}He and ^{4}He) and lithium (^{7}Li). Light elements such as these and also tritium (T or ^{3}H) and beryllium (^{7}Be) were formed in a primordial nuclear reactor. We shall see that the process begins towards the end of the 'first three minutes', as the era was so memorably described by Weinberg [5]. The first step is the formation of the $A = 2$ nucleus deuterium via the process

$$np \to D\gamma \qquad (1.130)$$

conventionally written by nuclear physicists as p(n, γ)D. At earlier times, the process goes in both directions. However, since there are more than 10^9 photons for every nucleon in the universe at that time, any newly formed deuterium is dissociated before it gets a chance to capture a neutron or proton and begin building heavier nuclei. Thus, no appreciable deuterium density accumulates. This 'deuterium bottleneck' persists until there are too few sufficiently energetic photons to dissociate the deuterons before they can capture nucleons. The $A = 3$ nuclei ^{3}He and ^{3}H are then formed via

$$D(p, \gamma)\,^3\text{He} : \qquad pD \to \,^3\text{He}\,\gamma \qquad (1.131)$$
$$D(D, n)\,^3\text{He} : \qquad DD \to \,^3\text{He}\,n \qquad (1.132)$$
$$^3\text{He}(n, p)\,^3\text{H} : \qquad n\,^3\text{He} \to \,^3\text{H}\,p \qquad (1.133)$$

and ^{4}He via

$$T(D, n)\,^4\text{He} : \qquad DT \to \,^4\text{He}\,n \qquad (1.134)$$
$$^3\text{He}(D, p)\,^4\text{He} : \qquad D\,^3\text{He} \to \,^4\text{He}\,p. \qquad (1.135)$$

Since there are no stable $A = 5$ nuclei, the synthesis of heavier nuclei requires the ^{4}He nuclei to interact with D, ^{3}H or ^{3}He, all of which are positively charged. The Coulomb repulsion suppresses the reaction rates for such processes, thereby ensuring that virtually all of the neutrons available for primordial nucleosynthesis wind up in ^{4}He, the most tightly bound of the light nuclei. Subsequently, the processes T(^{4}He, γ) ^{7}Li, ^{7}Li(p, ^{4}He) ^{4}He, ^{3}He(^{4}He, γ) ^{7}Be and ^{7}Be(n, p) ^{7}Li form more ^{4}He and also small amounts of lithium and beryllium.

The first process p(n, γ)D is crucial, since an appreciable deuterium abundance must be built up before the others can proceed; the neutron and proton number densities are too low to allow the build-up of the other nuclear abundances by direct many-body processes. Clearly, the original abundance of neutrons and protons determines the light element abundances generated by these primordial processes. However, light elements are also created and destroyed in stars, supernovae and other astrophysical phenomena. Consequently, the light element abundances measured today differ significantly from those created in the first three minutes. Because of this, the primordial abundances can only be inferred from the observational data after corrections to allow for the effects of galactic chemical evolution. The discussion of these is beyond the scope of this book. Here we shall instead focus on the essential physics of the primordial processes. Following Bernstein *et al* [6] and Sarkar [7], we shall present a semi-analytical treatment that allows the ^{4}He abundance to be calculated quite accurately. The precise calculation of this and the other yields and, hence, of the present baryon asymmetry requires a detailed numerical analysis which also will not be presented here.

At sufficiently high temperatures (above a few MeV) neutrons and protons are in kinetic and chemical equilibrium with, as follows from (1.129), a very high value $(O(10^{11}))$ of the entropy per nucleon. During this era the equilibrium nuclear abundances are quite negligible. The first stage of nucleosynthesis is the freeze-out of the weak interaction processes

$$n\nu_e \leftrightarrow pe^- \qquad ne^+ \leftrightarrow p\bar{\nu}_e \qquad n \leftrightarrow pe^-\bar{\nu}_e \qquad (1.136)$$

that previously kept neutrons and protons in equilibrium. Kinetic equilibrium requires equality of the temperatures of the particles:

$$T_n = T_p = T_e = T_\nu = T \qquad (1.137)$$

and chemical equilibrium requires that the chemical potentials of the various species satisfy

$$\mu_n - \mu_p = \mu_{e^-} - \mu_{\nu_e} = \mu_{\bar{\nu}_e} - \mu_{e^+}. \qquad (1.138)$$

The total rate λ_{np} for converting neutrons to protons via these processes is the sum of the individual rates:

$$\lambda_{np} = \lambda(n\nu_e \rightarrow pe^-) + \lambda(ne^+ \rightarrow p\bar{\nu}_e) + \lambda(n \rightarrow pe^-\bar{\nu}_e) \qquad (1.139)$$

and the total rate λ_{pn} for the reactions that convert protons to neutrons is given by detailed balance

$$\lambda_{pn} = \lambda_{np}e^{-\Delta m/T} \qquad \text{where } \Delta m = m_n - m_p = 1.293 \text{ MeV.} \qquad (1.140)$$

(The difference arises because of the slightly different Boltzmann factors in the neutron and proton equilibrium densities, see (4.19).) Let us denote the fractional relative neutron abundance by $X_n \equiv n_n/n_N$, where n_n is the neutron number density and n_N is the total nucleon number density $n_N \equiv n_n + n_p$ where n_p is the proton number density. Then the fractional relative proton abundance is $X_p \equiv n_p/n_N = 1 - X_n$. The evolution of X_n is determined by the balance equation

$$\dot{X}_n = \lambda_{pn}(1 - X_n) - \lambda_{np}X_n. \qquad (1.141)$$

The equilibrium solution, found by setting $\dot{X}_n = 0$, is

$$X_n^{eq}(t) = \frac{\lambda_{pn}(t)}{\Lambda(t)} = \frac{1}{1 + e^{\Delta m/T(t)}} \qquad \text{where } \Lambda \equiv \lambda_{pn} + \lambda_{np}. \qquad (1.142)$$

Thus, we may rewrite (1.141) as

$$\dot{X}_n = -\Lambda(X_n - X_n^{eq}). \qquad (1.143)$$

This shows that X_n is always between its initial value and X_n^{eq}. At early times Λ is large compared to the rate of time variation of the individual rates and X_n quickly tracks its equilibrium value $X_n^{eq}(t)$. This persists until the scattering rate $\Lambda(t)$ decreases until it becomes comparable with the Hubble rate $H(t) \equiv \dot{R}/R = -\dot{T}/T$. At this point, because of the expansion of the universe, the nucleons become too dilute to maintain the chemical equilibrium, they decouple and the number densities become 'frozen' at the values they have at decoupling. Thus,

$$X_n(t_{dec}) \simeq X_n^{eq}(t_{dec}) = \frac{1}{1 + e^{\Delta m/T(t_{dec})}}. \qquad (1.144)$$

As explained in section 5.2, the decoupling (or freeze-out) occurs when the temperature is

$$T_{dec} \simeq 1 \text{ MeV}. \qquad (1.145)$$

It is a remarkable coincidence that these two numbers, T_{dec} and Δm (given in (1.140)), are of the same order. The former derives from the interplay between the weak and gravitational interactions, while the latter derives from the difference between the u and d quark masses, which is of unknown origin but presumably as a result of strong and electromagnetic effects. Because of this coincidence, a substantial fraction (of order 20%) of the neutrons survive and this, in turn, results in a significant amount of primordial helium formed in the early universe.

This calculation of the fractional relative abundance of neutrons when the weak interactions decouple is only a rough estimate. For a more accurate estimate, we must solve the balance equation (1.141). It is convenient to use the variable $y \equiv \Delta m/T$ instead of t. In this era, the temperature T is related to the time t by equation (1.111) with $N_* = 3.36$, as shown in (1.121). Using (1.140), the total decay rate is

$$\Lambda(y) \simeq \lambda_{np}(y)(1 + e^{-y}) \qquad (1.146)$$

neglecting the neutron decay rate compared to the scattering rates. Bernstein *et al* [6] have approximated $\lambda_{np}(y)$ by

$$\lambda_{np}(y) \simeq 2\lambda(n\nu_e \to pe^-) \simeq \frac{a}{\tau_n y^5}(12 + 6y + y^2) \qquad (1.147)$$

where $a \simeq 253$ and $\tau_n \simeq 887$ s. is the neutron lifetime. Then the solution is

$$X_n(y) = X_n^{eq}(y) + \int_0^y dy' \, e^{y'} [X_n^{eq}(y')]^2 I(y, y') \qquad (1.148)$$

where

$$I(y, y') = \exp\left[-\int_y^{y'} dy'' \frac{dt''}{dy''} \lambda_{np}(y'')(1 + e^{-y''})\right] \quad (1.149)$$

$$= \exp[K(y) - K(y')]. \quad (1.150)$$

Using (1.147) gives (exercise 4)

$$K(y) = \frac{b}{y^3}[4 + 3y + y^2 + (4 + y)e^{-y}] \quad (1.151)$$

where

$$b = a\sqrt{\frac{10}{N_*} \frac{3M_P}{\pi \tau_n (\Delta m)^2}}. \quad (1.152)$$

The required integral is easily evaluated numerically and gives

$$X_n(y \to \infty) \simeq 0.15. \quad (1.153)$$

This asymptotic value is essentially achieved when $y \simeq 5$ corresponding to $t \simeq 20$ s, and a temperature of $T \simeq 0.25$ MeV.

The next stage of the process is the formation of deuterium. The rate for the process $np \to D\gamma$ exceeds the expansion rate of the universe until temperatures of order 10^{-3} MeV, so that deuterium will be present in this epoch with its equilibrium abundance. Using the non-relativistic number densities, analogous to (4.19), gives the Saha equation for the deuterium abundance n_D

$$\frac{n_D}{n_n n_p} = \frac{N_D}{N_n N_p} \left(\frac{2\pi m_D}{m_n m_p T}\right)^{3/2} e^{\Delta_D/T} \quad (1.154)$$

where $N_D = 3$ and $N_p = N_n = 2$ are the statistical factors, defined in section 2.2, for the deuteron and nucleons, and

$$\Delta_D = m_p + m_n - m_D \simeq 2.23 \text{ MeV} \quad (1.155)$$

is the binding energy of the deuteron. Then the corresponding mass fractions

$$X_i \equiv \frac{n_i A_i}{n_N} \quad (1.156)$$

where $A_n = A_p = 1$ and $A_D = 2$ are, respectively, the mass numbers of the nucleons and deuteron, satisfy

$$\frac{X_D}{X_n X_p} = \frac{24\zeta(3)}{\sqrt{\pi}} \left(\frac{T}{m_p}\right)^{3/2} \eta e^{\Delta_D/T} \quad (1.157)$$

where η is defined in (1.129). A rough estimate of the temperature T_{ns} at which nucleosynthesis starts may be made by determining when $X_D/X_n X_p$ becomes of order one. Taking logarithms of (1.157) gives

$$\frac{\Delta_D}{T_{ns}} = -\ln \eta + 6.27 + \frac{3}{2} \ln \frac{\Delta_D}{T_{ns}} \tag{1.158}$$

which may be solved iteratively. With the (inferred) value given in (1.129), we get

$$T_{ns} \simeq \frac{\Delta_D}{33} \simeq 0.068 \text{ MeV.} \tag{1.159}$$

The temperature at which nucleosynthesis starts is so much less than the deuteron binding energy because η is so small. Since there are of order 10^{10} photons per nucleon, there are enough high-energy photons in the Wien tail of the Planck distribution to dissociate the deuterons until the temperature drops to much less than the binding energy. A more careful estimate can be made [6] using the rate equation for the deuterium abundance, with the onset of nucleosynthesis being defined by $\dot{X}_D = 0$. This gives $T_{ns} \simeq 0.086$ MeV. Using (1.121), the temperature–time relation (1.111) gives

$$t \simeq 1.32 \left(\frac{T}{\text{MeV}} \right)^{-2} \text{ s} \tag{1.160}$$

so that nucleosynthesis begins when

$$t_{ns} \simeq 178 \text{ s} \tag{1.161}$$

as immortalized by Weinberg [5]. The neutrons that survived when the weak interactions decoupled have been depleted by beta-decay during the intervening period. Thus, the relative abundance of neutrons surviving until the onset of nucleosynthesis is

$$X_n(t_{ns}) \simeq X_n(y \to \infty) e^{-t_{ns}/\tau_n} \simeq 0.12. \tag{1.162}$$

As explained earlier, nearly all of these neutrons wind up in ^{4}He, because of its large binding energy. If we assume that all of the neutrons are captured in ^{4}He, the mass fraction of primordial ^{4}He, denoted $Y_p(^4\text{He})$, is simply given by

$$Y_p(^4\text{He}) \simeq 2X_n(t_{ns}) \simeq 0.24 \tag{1.163}$$

in excellent agreement with the data [8]

$$0.214 < Y_p(^4\text{He}) < 0.242. \tag{1.164}$$

The foregoing calculation shows how the primordial ^{4}He abundance is determined by the baryon asymmetry η. In principle, the same calculation determines η once the helium abundance has been measured. However, since $Y_p(^4\text{He})$ depends only logarithmically on η, as is apparent from (1.158), this does not yield a very precise value for η. In contrast, when similar techniques are applied to the (much smaller) primordial abundances of the other light nuclei, specifically D, ^3He, and ^7Li, the dependence upon η becomes a power. It is a testament to the success of the standard cosmological model that a single value of η simultaneously fits the data on all primordial abundances and allows a much more precise determination of the parameter. The value (1.129) of the parameter that we use in chapter 4 is obtained from a simultaneous fit of the cosmological parameters to all of these and other data, including the recent WMAP microwave anisotropy data [1].

1.12 Exercises

1. Show that the radial path from coordinate (r, θ, ϕ) at $t = 0$ to coordinate $(0, \theta, \phi)$ at time t is a geodesic.
2. Verify that the Ricci tensor for the Robertson–Walker metric (1.1) has the non-zero components given in (1.30).
3. Verify that the $\mu = 0$ component of (1.43) gives (1.45). What information is provided by the $\mu = i$ components?
4. For a flat FRW universe with pressureless matter and a negative cosmological constant Λ, show that the universe collapses to a point singularity in a time

$$T = \frac{2\pi}{\sqrt{3|\Lambda|}}.$$

5. Show that the solution of the balance equation (1.141) for the relative neutron abundance has the form (1.148) with the integrating factor as given in (1.150). Using the approximation (1.147), verify that the function $K(y)$ is given by (1.151).

1.13 General references

The books and review articles that we have found most useful in preparing this chapter are:

- Kolb E W and Turner M S 1990 *The Early Universe* (Reading, MA: Addison-Wesley)
- Weinberg S 1972 *Gravitation and Cosmology: Principles and Applications of the General Theory of Relativity* (New York: Wiley)
- Weinberg S 1989 *Rev. Mod. Phys.* **61** 1
- Sarkar S 1996 *Rep. Prog. Phys.* **59** 1493, arXiv:hep-ph/9602260

Bibliography

[1] Spergel D N *et al* 2003 *Astrophys. J. Suppl.* **148** 175, arXiv:astro-ph/0302209
[2] Knop R A *et al* The supernova cosmology project *Astrophys. J.* to be published, arXiv:astro-ph/0309368
[3] See, for example, Bailin D and Love A 1994 *Supersymmetric Gauge Field Theory and String Theory* (Bristol: IOP) ch 1
[4] Weinberg S 1987 *Phys. Rev. Lett.* **59** 2607
[5] Weinberg S 1977 *The First Three Minutes* (London: André Deutsch)
[6] Bernstein J, Brown L M and Feinberg G 1989 *Rev. Mod. Phys.* **61** 25
[7] Sarkar S 1996 *Rep. Prog. Phys.* **59** 1493, arXiv:hep-ph/9602260
[8] Pagel B E J *et al* 1992 *Mon. Not. R. Astron. Soc.* **255** 325

Chapter 2

Phase transitions in the early universe

2.1 Introduction

Elementary particle theory possesses gauge symmetries that are spontaneously broken by scalar fields belonging to non-trivial representations of the gauge group when these fields develop non-zero expectation values at the minimum of the effective potential. In particular, the $SU(2)_L \times U(1)_Y$ gauge group of electroweak theory is spontaneously broken to the $U(1)_{em}$ of electromagnetism by the electroweak Higgs scalar expectation value. If grand unification to a group larger than the $SU(3)_c \times SU(2)_L \times U(1)_Y$ of the standard model, e.g. to $SU(5)$, occurs at some energy scale, then the grand unified gauge group breaks spontaneously to the standard model gauge group before this gauge group in turn breaks to the $U(1)_{em}$ gauge group. Things may be more complicated than this, with a sequence of spontaneous symmetry breakings to subgroups of the original grand unified group.

As we shall see later in this chapter, finite temperature effects may result in some other minimum of the effective potential being deeper than the absolute minimum of the zero-temperature theory. Then, as the universe cools, it may undergo a series of first- or second-order phase transitions between different minima of the effective potential. Such transitions will occur at temperatures corresponding to the scales of energy associated with the various spontaneous symmetry breakings. In the case of the electroweak phase transition to the phase with only $U(1)_{em}$ unbroken, the scale of temperature for the phase transition will be of order 10^2–10^3 GeV and, in the case of a grand unified phase transition to a phase with $SU(3)_c \times SU(2)_L \times U(1)_Y$ unbroken, the phase transition will occur at a temperature closer to the Planck scale, perhaps at about 10^{16} GeV. If the grand unified theory breaks to $SU(3)_c \times SU(2)_L \times U(1)_Y$ in stages through a sequence of phase transitions, the additional phase transitions will occur at intermediate scales.

In later chapters we shall see that these phase transitions can have a profound effect on the history of the universe through a number of different processes. For example, topologically stable objects such as domain walls, cosmic strings and magnetic monopoles can be formed when the 'alignment' of the spontaneous symmetry breaking expectation value is different in adjacent causal domains. These can make substantial contributions to the energy density of the universe. Moreover, if supercooling occurs before the phase transition is completed, the reheating that takes place when the phase transition occurs can greatly modify pre-existing particle densities. In addition, if the universe spends some time with positive vacuum energy (cosmological constant) before relaxing to a minimum with zero vacuum energy, then rapid expansion can occur. Such an 'inflationary' stage in the history of the universe, to be discussed in later chapters, may explain the extreme isotropy, homogeneity and flatness of the present day observed universe. For all of these reasons it is important to understand any phase transitions that may have occurred as the universe cooled.

In this chapter we shall begin by developing the partition function and the effective potential for the gauge field theories at finite temperature [1–5] before applying these methods to the Higgs model, as a warm-up, and then to electroweak theory and grand unified theory. In each case, the nature of the phase transitions that occur as the temperature of the universe drops will be studied. We shall then extend the discussion to gauge theories with global supersymmetry and local supersymmetry (supergravity). Finally, the nucleation of (stable) 'true' vacuum from (metastable) 'false' vacuum in first-order phase transitions will be considered. This nucleation rate will control the extent of any supercooling that occurs before the phase transition is complete.

2.2   Partition functions

One of the fundamental objects in the statistical thermodynamics of a finite temperature system is the partition function Z defined by

$$Z = \mathrm{Tr}\, e^{-\beta \hat{H}} \tag{2.1}$$

where $\hat{H}$ is the Hamiltonian operator and

$$\beta = (k_B T)^{-1} = T^{-1} \tag{2.2}$$

in units where the Boltzmann constant k_B is set equal to 1. The trace in (2.1) means that we are to sum over the (diagonal) matrix elements of $e^{-\beta \hat{H}}$ for all independent states of the system. Once the partition function has been evaluated, the (Helmholtz) free energy F is given by

$$Z = e^{-\beta F} \tag{2.3}$$

where, as usual in thermodynamics, F is related to the internal energy E and the entropy S by

$$F = E - TS. \tag{2.4}$$

The pressure P and entropy are obtained from the free energy as

$$P = -\frac{\partial F}{\partial V}\bigg|_T \tag{2.5}$$

and

$$S = -\frac{\partial F}{\partial T}\bigg|_V. \tag{2.6}$$

As follows immediately from (2.4), the energy density ρ is given by

$$\rho = \mathcal{F} + Ts \tag{2.7}$$

where $\mathcal{F}$ and s are the free energy and entropy densities, with

$$E = \int d^3x\, \rho \tag{2.8}$$

and so forth. Thus, in particular, a calculation of the partition function will provide us with a determination of the energy density.

The partition function in a gauge field theory is most efficiently calculated using path integral methods. It is not the business of the present book to develop such methods which can be found developed at length elsewhere [1–5]. It will suffice for our purposes here to to summarize the outcome for the various contributions to the partition function.

The simplest contribution comes from the free (neutral) real scalar fields. The Lagrangian density for such a field ϕ having mass m is given by

$$\mathcal{L}(\phi, \partial_\mu \phi) = \frac{1}{2}\left(\frac{\partial \phi}{\partial t}\right)^2 - \frac{1}{2}(\nabla\phi)^2 - \frac{1}{2}m^2\phi^2. \tag{2.9}$$

In field theory at finite temperature, scalar fields $\phi(t, x)$ are replaced by fields $\phi(\tau, x)$ periodic in τ with period β,

$$\phi(\tau = 0, x) = \phi(\tau = \beta, x) \tag{2.10}$$

where

$$\tau = it. \tag{2.11}$$

We shall use the usual convention of referring to non-zero temperature as 'finite temperature'. The partition function is formulated in terms of these periodic fields as

$$Z = \tilde{N}(\beta) \int_{\text{periodic}} \mathcal{D}\phi \exp\left[\int_0^\beta d\tau \int d^3x\, \mathcal{L}(\phi, \bar{\partial}_\mu \phi)\right] \tag{2.12}$$

where

$$\bar{\partial}_\mu \phi \equiv \left(i\frac{\partial \phi}{\partial \tau}, \nabla\phi\right) \tag{2.13}$$

and $\bar{N}(\beta)$ is a temperature-dependent normalization. The integral $\int \mathcal{D}\phi$ is a path integral. Such integrals may be thought of as the generalization of an integration $\int_{-\infty}^{\infty} dy_1 \int_{-\infty}^{\infty} dy_2 \dots \int_{-\infty}^{\infty} dy_n$ over the finite number of components of an n-component column vector y to an integration over the continuous infinity of components of a function $\phi(\tau, x)$. Evaluation of the path integral gives for the contribution of a real scalar field to the free energy

$$-\beta F = \ln Z$$
$$= -\int d^3 x \int \frac{d^3 p}{(2\pi)^3} \left(\frac{\beta}{2} \sqrt{p^2 + m^2} + \ln \left[1 - \exp\left(-\beta \sqrt{p^2 + m^2} \right) \right] \right).$$
$$(2.14)$$

When the mass of the scalar field is negligible compared with the temperature (an ideal ultra relativistic gas of scalar particles), the free energy density simplifies to

$$\mathcal{F} = -\frac{\pi^2 T^4}{90} \qquad \text{when } T \gg m. \qquad (2.15)$$

For gauge vector bosons, there are some subtleties because, for a typical choice of gauge, the Lagrangian involves all four degrees of freedom of the gauge field $A^\mu(x)$ and also involves the Fadeev–Popov ghost fields which occur in the construction of a consistent renormalizable theory but are not physical particles. However, a massless vector field has only two degrees of freedom and the extra degrees of freedom are not physical and cannot be in equilibrium with a heat bath nor, of course, can the Fadeev–Popov ghosts. Fortunately there exist gauges in which each gauge field has only two degrees of freedom and in which there are no Fadeev–Popov ghosts and the partition function can be related to the Lagrangian density in such a gauge. In any other gauge, it can be shown that one may continue to use this expression for Z but with the form of the Lagrangian appropriate for that gauge. In an arbitrary gauge the contribution of the two non-physical degrees of freedom of the gauge field cancels the contribution from the Fadeev–Popov ghosts. Then, the contribution to the free energy density from a massless vector gauge field is found to be

$$\mathcal{F} = -\frac{2\pi^2 T^4}{90}. \qquad (2.16)$$

In the case of Dirac fields ψ, the corresponding development at finite temperature involves fields $\psi(\tau, x)$ that are *anti*-periodic in τ in the interval $(0, \beta)$,

$$\psi(\tau = 0, x) = -\psi(\tau = \beta, x) \qquad (2.17)$$

and the contribution to the free energy density is

$$\mathcal{F} = -\frac{7\pi^2 T^4}{180} \qquad \text{when } T \gg m. \qquad (2.18)$$

For massless fermions (with only one helicity state of the particle), the calculation (using Weyl spinors) gives half of this answer.

Putting all of this together, the free energy density of an ideal ultra relativistic gas $(T \gg m)$ is given by

$$\mathcal{F} = -\left(N_B + \frac{7}{8}N_F\right)\frac{\pi^2 T^4}{90} \tag{2.19}$$

where N_B and N_F are, respectively, the numbers of bosonic and fermionic degrees of freedom. ($N_B = 1$ for a real scalar field, $N_B = 2$ for a real gauge field, $N_F = 4$ for a Dirac particle where there are two helicity states for the particle and two for the antiparticle and $N_F = 2$ for a Weyl field.) The pressure, entropy density and energy density follow from (2.5), (2.6) and (2.4).

$$P = \left(N_B + \frac{7}{8}N_F\right)\frac{\pi^2 T^4}{90} \tag{2.20}$$

$$s = \left(N_B + \frac{7}{8}N_F\right)\frac{2\pi^2 T^3}{45} \tag{2.21}$$

and

$$\rho = \left(N_B + \frac{7}{8}N_F\right)\frac{\pi^2 T^4}{30}. \tag{2.22}$$

2.3 The effective potential at finite temperature

In quantum field theory at zero temperature, the expectation value ϕ_c of a scalar field ϕ (also referred to as the classical field) is determined by minimizing the effective potential $V(\phi_c)$. The effective potential contains a tree-level potential term, which can be read off from the Hamiltonian density, and quantum corrections from various loop orders. The one-loop quantum correction is calculated by shifting the fields ϕ by their expectation values ϕ_c and isolating the terms $\mathcal{L}_{\text{quad}}(\phi_c, \tilde{\phi})$ in the Lagrangian density which are quadratic in the shifted fields $\tilde{\phi}$. If we write

$$V(\phi_c) = V_0(\phi_c) + V_1(\phi_c) \tag{2.23}$$

where V_0 is the tree-level contribution and V_1 is the one-loop quantum correction then, for a single scalar field,

$$\exp\left(-\mathrm{i}\int \mathrm{d}^4x\, V_1(\phi_c)\right) = \int \mathcal{D}\tilde{\phi}\exp\left(\mathrm{i}\int \mathrm{d}^4x\, \mathcal{L}_{\text{quad}}(\phi_c, \tilde{\phi})\right) \tag{2.24}$$

where, as in section 2.2, $\int \mathcal{D}\tilde{\phi}$, denotes a path integral. (The derivation and evaluation of (2.24) can be found elsewhere [1].)

At finite temperature, as discussed in section 2.2, scalar fields $\phi(t, x)$ are replaced by fields $\phi(\tau, x)$ periodic in τ with period β, where β is given by (2.2). We now write the finite-temperature effective potential $\bar{V}(\phi_c)$ as

$$\bar{V}(\phi_c) = \bar{V}_0(\phi_c) + \bar{V}_1(\phi_c) \tag{2.25}$$

where $\bar{V}_0$ and $\bar{V}_1$ are the tree-level and one-loop terms and the expectation value ϕ_c is now a thermal average. Then (2.24) is modified to

$$\exp\left(-\int_0^\beta d\tau \int d^3x \, \bar{V}_1(\phi_c)\right) = \int_{\text{periodic}} \mathcal{D}\tilde{\phi} \exp\left(\int_0^\beta d\tau \int d^3x \, \mathcal{L}_{\text{quad}}(\phi_c, \tilde{\phi})\right). \tag{2.26}$$

If gauge fields and fermion fields are included (but with only scalar fields being given expectation values to avoid breaking Lorentz invariance), then (2.26) also contains path integrals over the gauge fields and their associated Fadeev–Popov ghosts, and over antiperiodic fermion fields.

It is convenient to separate the one-loop terms into the temperature-independent part $\bar{V}_1^0$ (which is identical in form to V_1) and the temperature-dependent part $\bar{V}_1^T$ and write

$$\bar{V}_1 = \bar{V}_1^0 + \bar{V}_1^T \tag{2.27}$$

In general, for a theory involving scalar fields ϕ_i, gauge fields A_a^μ and Dirac fermions ψ_r, after shifting the scalar fields by their expectation values, the terms in the Lagrangian of quadratic order in the fields are of the form

$$\begin{aligned}
\mathcal{L}_{\text{quad}}(\phi_c, \tilde{\phi}) = &-\tfrac{1}{2}[\hat{M}_S^2(\phi_c)]_{ij}\tilde{\phi}_i\tilde{\phi}_j + \tfrac{1}{2}[\hat{M}_V^2(\phi_c)]_{ab}A_a^\mu A_{b\mu} \\
&- [\hat{M}_F(\phi_c)]_{rs}\bar{\psi}_r\psi_s + \tfrac{1}{2}\bar{\partial}_\mu\tilde{\phi}_i\partial^\mu\tilde{\phi}_i \\
&- \tfrac{1}{4}(\bar{\partial}^\mu A_a^\nu - \bar{\partial}^\nu A_a^\mu)(\bar{\partial}_\mu A_{a\nu} - \bar{\partial}_\nu A^{a\mu}) \\
&- \frac{1}{2\xi}(\bar{\partial}_\mu A_a^\mu)^2 + \bar{\partial}_\mu\eta_a^*\bar{\partial}^\mu\eta_a.
\end{aligned} \tag{2.28}$$

In (2.28), ϕ_c denotes the complete set of expectation values of scalar fields, $\tilde{\phi}_i$ denotes the shifted scalar fields and $\bar{\partial}_\mu$ is as in (2.13). Also, η_a are the Fadeev–Popov ghost fields which have to be introduced in the construction of a consistent renormalizable theory of gauge fields but do not correspond to physical particles and ξ is the gauge-fixing parameter. For convenience, we have adopted the Landau gauge $\xi \to 0$ in which couplings of scalar fields to Fadeev–Popov ghosts are avoided.

If the eigenvalues of the mass-squared matrices $\hat{M}_S^2$, $\hat{M}_V^2$ and $\hat{M}_F^2$ are $(M_S^2)_i, (M_V^2)_a$ and $(M_F^2)_r$ then the temperature-dependent one-loop term in the effective potential $\bar{V}_1^T$ takes the form

$$\bar{V}_1^T(\phi_c) = \frac{T^4}{2\pi^2} \int_0^\infty dy \, y^2 \left\{ \sum_i \ln\left[1 - \exp\left(-\sqrt{y^2 + T^{-2}(M_S^2)_i}\right)\right] \right.$$

$$+ \sum_a \left(3 \ln \left[1 - \exp \left(-\sqrt{y^2 + T^{-2}(M_V^2)_a} \right) \right] - \ln(1 - e^{-y}) \right)$$

$$- 4 \sum_r \ln \left[1 + \exp \left(-\sqrt{y^2 + T^{-2}(M_F)_r^2} \right) \right] \Bigg\}. \tag{2.29}$$

There are two limits in which $\overline{V}_1^T$ is particularly simple. First, in the limit when all mass-squared eigenvalues are very much greater than T^2 all terms in $\overline{V}_1^T$ approach zero exponentially and $\overline{V}_1^T$ becomes negligible. (It is not obvious that this is true of the $\ln(1 - e^{-y})$ term in (2.29). However, in a general gauge, this term is replaced by $\ln[1 - \exp(-\sqrt{y^2 + \xi T^{-2}(M_V^2)_a})]$, where ξ is the gauge parameter. If the limit $T^{-2}(M_V^2)_a \to \infty$ is taken before the limit $\xi \to 0$, to recover the Landau gauge, this term vanishes.)

Second, in the high-temperature limit where T^2 is very much greater than the mass-squared eigenvalues, we may use

$$\frac{T^4}{2\pi^2} \int_0^\infty dy\, y^2 \ln \left[1 - \exp\left(-\sqrt{y^2 + RT^{-2}} \right) \right]$$

$$= -\frac{\pi^2 T^4}{90} + \frac{RT^2}{24} - \frac{R^{3/2}T}{12\pi} - \frac{R^2}{64\pi^2} \ln\left(\frac{R}{a_b T^2} \right)$$

$$- \frac{R^2}{16\pi^{5/2}} \sum_{\ell=1}^\infty (-1)^\ell \frac{\varsigma(2\ell+1)}{(l+1)!} \left(\frac{R}{4\pi^2 T^2} \right)^\ell$$

where

$$a_b = 16\pi^2 \ln(\tfrac{3}{2} - 2\gamma_E) \qquad \ln a_b = 5.4076 \tag{2.30}$$

and

$$\frac{T^4}{2\pi^2} \int_0^\infty dy\, y^2 \ln \left[1 + \exp\left(-\sqrt{y^2 + RT^{-2}} \right) \right]$$

$$= \frac{7\pi^2 T^4}{720} - \frac{RT^2}{48} - \frac{R^2}{64\pi^2} \ln\left(\frac{R}{a_f T^2} \right)$$

$$- \frac{R^2}{16\pi^{5/2}} \sum_{\ell=1}^\infty (-1)^\ell \frac{\varsigma(2\ell+1)}{(l+1)!} (1 - 2^{-2\ell-1}) \Gamma\left(\ell + \frac{1}{2} \right) \left(\frac{R}{4\pi^2 T^2} \right)^\ell$$

where

$$a_f = \pi^2 \ln\left(\frac{3}{2} - 2\gamma_E \right) = \frac{a_b}{16} \qquad \ln a_f = 2.6351. \tag{2.31}$$

Thus, in the high-temperature limit,

$$\overline{V}_1^T(\phi_c) \simeq -\frac{\pi^2 T^4}{90} \left(N_B + \frac{7}{8} N_F \right)$$

$$+ \frac{T^2}{24}\left[\sum_i (M_S^2)_i + 3\sum_a (M_V^2)_a + 2\sum_r (M_F)_r^2 \right]$$

$$- \frac{T}{12\pi}\left[\sum_i (M_S^3)_i + 3\sum_a (M_V^3)_a \right] + \cdots$$

$$= - \frac{\pi^2 T^4}{90}\left(N_B + \frac{7}{8}N_F \right)$$

$$+ \frac{T^2}{24}[\mathrm{tr}\,\hat{M}_S^2(\phi_c) + 3\,\mathrm{tr}\,\hat{M}_V^2(\phi_c) + 2\,\mathrm{tr}\,\hat{M}_F^2(\phi_c)]$$

$$- \frac{T}{12\pi}[\mathrm{tr}\{\hat{M}_S^2(\phi_c)\}^{3/2} + 3\,\mathrm{tr}\{\hat{M}_V^2(\phi_c)\}^{3/2}] + \cdots. \qquad (2.32)$$

where $\hat{M}_S^2(\phi_c)$, $\hat{M}_V^2(\phi_c)$ and $\hat{M}_F^2(\phi_c)$ are the scalar, vector and Dirac fermion mass matrices of (2.28). (For fermions described by Weyl spinor fields there should be no factor of 2 in front of the $\hat{M}_F^2$ term in (2.32).) The T^4 term in (2.32) is just the free energy density for an ideal ultra relativistic gas (in agreement with (2.19) with N_B and N_F respectively the number of bosonic and fermionic degrees of freedom, in the sense described following (2.19). If some fields are heavy and some are light on the scale of the temperature T, then N_B and N_F should be interpreted as the degrees of freedom of light fields, and the traces over the mass matrices should be evaluated only for light fields, since heavy fields do not contribute, as discussed earlier.

2.4   Phase transitions in the Higgs model

Before studying phase transitions in electroweak theory and grand unified theory, we warm up on the simpler case of the Higgs model. The Higgs model is the theory of a complex scalar field coupled to a $U(1)$ gauge field, which may be taken to be the electromagnetic field, with the $U(1)$ gauge symmetry spontaneously broken. In other words, it is scalar electrodynamics with spontaneously broken electromagnetic gauge symmetry. The finite-temperature Lagrangian density is

$$\mathcal{L} = \bar{D}_\mu \phi \bar{D}^\mu \phi^* - m^2 \phi^* \phi - \frac{\lambda}{4}(\phi^*\phi)^2 - \frac{1}{4}\bar{F}_{\mu\nu}\bar{F}^{\mu\nu}$$

$$- \frac{1}{2\xi}(\bar{\partial}_\mu A^\mu)^2 + \bar{\partial}_\mu \eta^* \bar{\partial}^\mu \eta \qquad (2.33)$$

where

$$\bar{F}_{\mu\nu} \equiv \bar{\partial}_\mu A_\nu - \bar{\partial}_\nu A_\mu \qquad (2.34)$$

$$\bar{D}_\mu \phi \equiv (\bar{\partial}_\mu + ieA_\mu)\phi \qquad (2.35)$$

and

$$\bar{D}_\mu \phi^* \equiv (\bar{\partial}_\mu - ieA_\mu)\phi^* \qquad (2.36)$$

with $\bar{\partial}_\mu$ as in (2.13). The Fadeev–Popov ghosts η cancel contributions to the free energy from the two non-physical degrees of freedom of the gauge field A^μ, as discussed in section 2.2, and ξ is the gauge-fixing parameter. For spontaneous symmetry-breaking to occur (without requiring radiative corrections to drive it), m^2 must be negative.

To derive the finite-temperature effective potential using the methods of section 2.3 it is necessary to shift the scalar field by its expectation value. We write

$$\langle \phi \rangle = \frac{\phi_c}{\sqrt{2}} \tag{2.37}$$

where the factor of $1/\sqrt{2}$ has no significance but has simply been introduced for convenience, and ϕ_c may be taken to be real because of gauge invariance. Then real fields ϕ_1 and ϕ_2 are introduced through

$$\phi = \frac{1}{\sqrt{2}}(\phi_c + \phi_1 + i\phi_2). \tag{2.38}$$

The quadratic terms in the shifted Lagrangian density are

$$\begin{aligned}
\mathcal{L}_{\text{quad}} = &-\frac{1}{2}\left(m^2 + \frac{3\lambda}{4}\phi_c^2\right)\phi_1^2 - \frac{1}{2}\left(m^2 + \frac{\lambda}{4}\phi_c^2\right)\phi_2^2 \\
&+ \frac{e^2}{2}\phi_c^2 A_\mu A^\mu + \frac{1}{2}(\bar{\partial}_\mu\phi_1)^2 + \frac{1}{2}(\bar{\partial}_\mu\phi_2)^2 \\
&- \frac{1}{2\xi}(\bar{\partial}_\mu A^\mu)^2 + \bar{\partial}_\mu\eta^*\bar{\partial}^\mu\eta
\end{aligned} \tag{2.39}$$

where we have adopted the Landau gauge $\xi \to 0$ which removes an $A^\mu\bar{\partial}_\mu\phi_2$ cross term. Then, in the notation of (2.28),

$$\hat{M}_S^2(\phi_c) = \text{diag}\left(m^2 + \frac{3\lambda}{4}\phi_c^2, m^2 + \frac{\lambda}{4}\phi_c^2\right) \tag{2.40}$$

and

$$\hat{M}_V^2(\phi_c) = e^2\phi_c^2. \tag{2.41}$$

The nature of the phase transition depends on the relative sizes of e^4 and λ.

2.4.1 $e^4 \ll \lambda$

The tree-level contribution $\bar{V}_0(\phi_c)$ to the effective potential may be read from (2.33) by replacing ϕ by its expectation value $\frac{1}{\sqrt{2}}\phi_c$ and is

$$\bar{V}_0(\phi_c) = \frac{m^2}{2}\phi_c^2 + \frac{\lambda}{16}\phi_c^4. \tag{2.42}$$

The zero-temperature one-loop correction $\overline{V}_1^0$ of (2.27) and (2.23) has contributions proportional to λ^2 and e^4 from different loop diagrams. Provided that λ is small and that, in addition, $e^4 \ll \lambda$, $\overline{V}_1^0$ may be neglected compared to the tree term (2.42). In the high-temperature limit,

$$T^2 \gg \lambda\phi_c^2 \qquad e^2\phi_c^2 \qquad -m^2 \tag{2.43}$$

the one-loop temperature-dependent contribution to the effective potential $\overline{V}_1^T$ obtained from (2.32) is given by

$$\overline{V}_1^T(\phi_c) = -\frac{4\pi^2 T^4}{90} + \frac{(\lambda + 3e^2)T^2}{24}\phi_c^2 - \frac{CT}{3}\phi_c^3 + \cdots \tag{2.44}$$

where

$$4\pi C = \text{tr}[\hat{M}_S^2(\phi_c)]^{3/2} + 3\,\text{tr}[\hat{M}_V^2(\phi_c)]^{3/2}$$

$$= \left(m^2\phi_c^{-2} + \frac{3\lambda}{4}\right)^{3/2} + \left(m^2\phi_c^{-2} + \frac{\lambda}{4}\right)^{3/2} + 3e^3 \tag{2.45}$$

$$\simeq \left(\frac{3\lambda}{4}\right)^{3/2} + \left(\frac{\lambda}{4}\right)^{3/2} + 3e^3 \tag{2.46}$$

when $\lambda\phi_c^2 \gg m^2$. Thus, the complete effective potential to one-loop order is given by

$$\overline{V}(\phi_c) = -\frac{4\pi^2 T^4}{90} + \frac{1}{2}m^2(T)\phi_c^2 - \frac{CT}{3}\phi_c^3 + \frac{\lambda}{16}\phi_c^4 \tag{2.47}$$

where a temperature-dependent mass $m^2(T)$ has been defined by

$$m^2(T) = m^2 + \frac{(\lambda + 3e^2)T^2}{12}. \tag{2.48}$$

The expectation value ϕ_c of the scalar field is obtained by minimizing the effective potential. For sufficiently high temperatures, there is only one solution of

$$\frac{\partial \overline{V}}{\partial \phi_c} = 0 \tag{2.49}$$

namely

$$\phi_c = 0 \tag{2.50}$$

and this is a minimum so long as $m^2(T)$ is positive. From (2.48), we see that this is the case provided the temperature T exceeds T_0 where

$$T_0^2 \equiv \frac{-12m^2}{\lambda + 3e^2}. \tag{2.51}$$

(See figure 2.1, curve A.) We may write

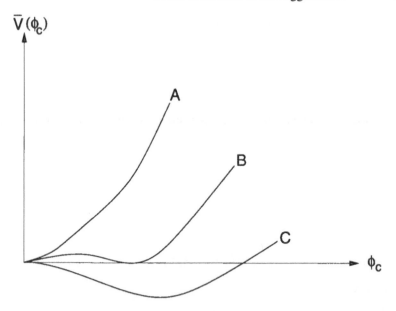

Figure 2.1. The finite-temperature effective potential for the Higgs model when $e^4 \ll \lambda$. Curves A,B,C are for $T > T_1$, $T = T_c$ and $T < T_0$ respectively.

$$m^2(T) = m^2 \left(1 - \frac{T^2}{T_0^2} \right). \tag{2.52}$$

A maximum and a second (local) minimum of $\overline{V}$ develop at non-zero values of ϕ_c when

$$m^2(T) \le \frac{C^2 T^2}{\lambda} \tag{2.53}$$

and this happens when the temperature drops below T_1 where

$$T_1^2 \equiv \frac{T_0^2}{1 + C^2 T_0^2 / \lambda m^2} = \frac{-12\lambda m^2}{\lambda(\lambda + 3e^2) - 12C^2} > T_0^2. \tag{2.54}$$

The second minimum is at

$$\phi_c = v(T) \equiv v \left[\frac{CT}{\sqrt{\lambda}|m|} + \left(1 - \frac{T^2}{T_1^2} \right)^{1/2} \right] \tag{2.55}$$

where

$$v \equiv \frac{2|m|}{\sqrt{\lambda}} \tag{2.56}$$

and the mass $m_H(T)$ of the Higgs particle associated with the fluctuations around this minimum is given by

$$m_H^2(T) \equiv \left. \frac{\partial^2 \overline{V}}{\partial \phi_c^2} \right|_{\phi_c = v(T)} = CTv(T) - 2m^2(T). \qquad (2.57)$$

In the same way, we may define a temperature-dependent vector boson mass by

$$m_V(T) \equiv ev(T). \qquad (2.58)$$

When the second local minimum first arises, the global minimum of $\overline{V}$ is still at $\phi_c = 0$. However, as the temperature falls, there is a critical temperature T_c at which the second minimum becomes degenerate with the global minimum. (See figure 2.1, curve B.) This occurs when

$$\tfrac{8}{9} C^2 T^2 = \lambda m^2(T) \qquad (2.59)$$

which gives

$$T_c^2 \equiv \frac{T_0^2}{1 + 8C^2 T_0^2 / 9\lambda m^2} = \frac{9T_0^2}{1 + 8T_0^2 / T_1^2} = \frac{-12\lambda m^2}{\lambda(\lambda + 3e^2) - 32C^2/3} \qquad (2.60)$$

so that $T_1 > T_c > T_0$. At temperatures below the critical temperature, the minimum at non-zero ϕ_c is the global minimum of $\overline{V}$ and the system is in a phase with spontaneous symmetry breaking, referred to as the asymmetric phase. The value of ϕ_c at the global minimum changes *discontinuously* from $\phi_c = 0$ to

$$\phi_c = v(T_c) = \frac{8CT_c}{3\lambda} \qquad (2.61)$$

as the temperature passes through $T = T_c$, so that there is a first-order phase transition. As the temperature falls below $T = T_0$, $m^2(T)$ becomes negative, the local minimum at $\phi_c = 0$ turns into a local maximum and the only minimum is at the non-zero value of $\phi_c = v(T)$. (See figure 2.1, curve C.) All of this occurs because $C \neq 0$. For future reference, we note that if C is zero or so small that $v(T_c) \ll v$, then $T_1 \simeq T_c \simeq T_0$, and there is effectively a (continuous) second-order phase transition at temperature $T = T_c$.

2.4.2 $e^4 \gg \lambda$

When the gauge coupling constant is larger relative to the ϕ^4 coupling constant, there are two differences in the treatment required. First, the zero-temperature correction to the effective potential may no longer be negligible and, second, it may not be correct to make the high-temperature approximation that T is very much larger than the masses of all (shifted) fields.

If we now include the zero-temperature radiative correction, the complete finite-temperature effective potential becomes

$$\bar{V}(\phi_c) = \frac{m^2}{2}\phi_c^2 + \frac{\lambda}{16}\phi_c^4 + B\phi_c^4\left[\ln\left(\frac{\phi_c^2}{M^2}\right) - \frac{25}{6}\right] + \bar{V}_1^T(\phi_c) \qquad (2.62)$$

where M is a renormalization scale which may, if we wish, be eliminated in favour of the value of ϕ_c at the zero-temperature minimum of $\bar{V}$ and

$$B = \frac{1}{64\pi^2}\left(\frac{5}{8}\lambda^2 + 3e^4\right). \qquad (2.63)$$

Renormalization has been carried out according to

$$\left.\frac{d^2\bar{V}}{d\phi_c^2}\right|_{\phi_c=0} = m^2 \qquad \left.\frac{d^4\bar{V}}{d\phi_c^4}\right|_{\phi_c=M} = \frac{3}{2}\lambda. \qquad (2.64)$$

(Details of the derivation of the zero-temperature radiative correction may be found elsewhere [1].) If λ is small and $\lambda \lesssim e^4$, then λ^4 is negligible compared to e^4 and B simplifies to

$$B \simeq \frac{3e^4}{64\pi^2}. \qquad (2.65)$$

With the mass-squared matrices of (2.40) and (2.41) for the (shifted) scalar and vector fields, and *not* making the high-temperature approximation,

$$\bar{V}_1^T(\phi_c) = \frac{T^4}{2\pi^2}\int_0^\infty dy\, y^2\left\{\ln\left[1 - \exp\left(-\sqrt{y^2 + T^{-2}(m^2 + 3\lambda\phi_c^2/4)}\right)\right]\right.$$
$$+ \ln\left[1 - \exp\left(-\sqrt{y^2 + T^{-2}(m^2 + \lambda\phi_c^2/4)}\right)\right]$$
$$\left. + 3\ln\left[1 - \exp\left(-\sqrt{y^2 + T^{-2}e^2\phi_c^2}\right)\right] - \ln(1 - e^{-y})\right\}. \qquad (2.66)$$

We now ask whether we should use the high-temperature approximation to study the phase transitions when $e^4 \gg \lambda$. If we *do* use the high-temperature approximation (and neglect the zero-temperature radiative correction for the moment) then, as before, the critical temperature is given by (2.60). If $e^2 < 1$ (as in scalar electrodynamics), then, when $e^4 \gg \lambda$, we certainly have $e^2 \gg \lambda$ and $e^3 \gg \lambda^{3/2}$, so

$$C \simeq \frac{3e^3}{4\pi}. \qquad (2.67)$$

and

$$T_c^2 \simeq \frac{-4m^2}{e^2} \qquad (2.68)$$

At the zero-temperature asymmetric minimum, still neglecting the radiative corrections,

$$\phi_c^2 = v^2 = \frac{-4m^2}{\lambda} \tag{2.69}$$

so that

$$T_c^2 \gg m^2 + \frac{\lambda}{4}\phi_c^2 = 0 \tag{2.70}$$

and

$$T_c^2 \gg m^2 + \frac{3\lambda}{4}\phi_c^2 = -2m^2. \tag{2.71}$$

However, when $e^4 \gg \lambda$,

$$T_c^2 \ll m_V^2 = e^2v^2 = \frac{-4m^2}{e^2}\frac{e^4}{\lambda}. \tag{2.72}$$

Therefore, it is not correct to use the high-temperature approximation for the vector boson terms when $e^4 \gg \lambda$.

Taking account of this observation, we now compute the values of the effective potential at the symmetric and asymmetric minima to decide which is the absolute minimum when both exist. At the symmetric minimum, ϕ_c is zero and the high-temperature approximation is valid provided only that $T^2 \gg -m^2$. Thus,

$$\overline{V}(\phi_c = 0) = \overline{V}_1^T(\phi_c = 0) \simeq -\frac{4\pi^2 T^4}{90}. \tag{2.73}$$

At the asymmetric minimum, $\phi_c = v$, the contribution to $\overline{V}_1^T(\phi_c)$ involving the gauge field mass $e\phi_c$ is exponentially suppressed but the high-temperature expansion may still be used for the scalar field terms. Thus,

$$\overline{V}_1^T(\phi_c = v) \simeq -\frac{2\pi^2 T^4}{90} + \frac{T^2}{24}(-2m^2) - \frac{T}{12\pi}(-2m^2)^{3/2}. \tag{2.74}$$

Dropping the zero-temperature radiative correction for the moment,

$$\overline{V}(\phi_c = v) = -\frac{m^4}{\lambda} - \frac{\pi^2 T^4}{45} - \frac{m^2 T^2}{12} - \frac{T}{12\pi}(-2m^2)^{3/2} \tag{2.75}$$

where we have assumed that the value of the effective potential at the asymmetric minimum is the same as at zero temperature apart from the terms proportional to T^4, T^2 and T. This can be shown to be correct apart from corrections of higher order in e^2. Neglecting the $m^2 T^2$ term and the $(-2m^2)^{3/2}$ term compared with the T^4 term, it can now be seen that the symmetric minimum is at a lower value of $\overline{V}$ than the asymmetric minimum when

$$T > \left(\frac{45}{\pi^2\lambda}\right)^{1/4}|m| \equiv T_{c1} \tag{2.76}$$

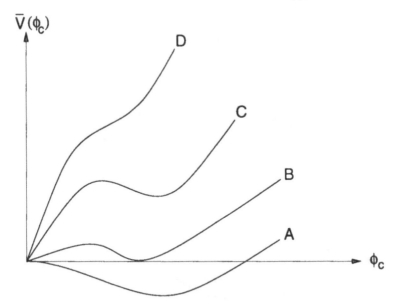

Figure 2.2. Development of asymmetric minimum with temperature in Higgs models for $4\pi^2\lambda/11 \gg e^4 \gg \lambda$. Curve A is at zero temperature, curve B is at T_c, the critical temperature for the first order phase transition, and C and D correspond to higher temperatures.

This is illustrated in figure 2.2.

The phase transition now occurs, in principle, at $T = T_{c1}$ and, because the value of the expectation value undergoes a discontinuous change from 0 to v, the phase transition is now first order. As the system cools the phase transition to the asymmetric phase may not occur in practice until T is much less than T_{c1} because of the need to tunnel through the potential barrier between the symmetric minimum and the asymmetric minimum.

When the zero-temperature radiative corrections are taken into account the situation can change. The effective potential of (2.62) can be recast (exercise 1) in terms of the value v of ϕ_c at the zero-temperature (asymmetric) minimum of the effective potential, including radiative corrections, as

$$\bar{V}(\phi_c) = B\left(\frac{\alpha}{2}v^2\phi_c^2 - \frac{\alpha+2}{4}\phi_c^4 + \phi_c^4 \ln\frac{\phi_c^2}{v^2}\right) + \bar{V}_1^T(\phi_c) \qquad (2.77)$$

where

$$\alpha = 2B^{-1}\left(\frac{22}{3}B - \frac{\lambda}{8}\right). \qquad (2.78)$$

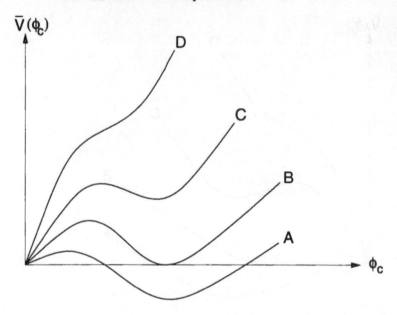

Figure 2.3. Development of asymmetric minimum with temperature in Higgs models for $e^4 \gg 4\pi^2\lambda/11$. Curve A is at zero temperature, curve B is at T_c, the critical temperature for the first-order phase transition and C and D correspond to higher temperatures.

The physical Higgs scalar mass m_H is given by

$$m_H^2 = \frac{d^2\bar{V}}{d\phi_c^2}\bigg|_{\phi_c=v} = 2Bv^2(4-\alpha). \qquad (2.79)$$

The case of symmetry breaking at zero temperature driven by radiative corrections, discussed by Coleman and Weinberg [6], corresponds to $\alpha = 0$, i.e. to

$$m_H^2 = m_{CW}^2 = 8Bv^2. \qquad (2.80)$$

When $m_H^2 < m_{CW}^2$ ($\alpha > 0$), the situation differs qualitatively from that just described because there is a local minimum of the effective potential at $\phi_c = 0$ due to radiative corrections already present at $T = 0$. Then, we must take account of the zero-temperature radiative corrections in deriving the development of the effective potential with temperature. This is illustrated in figure 2.3.

The case $\alpha > 0$, where this is necessary, corresponds to

$$e^4 > \frac{4}{11}\pi^2\lambda. \qquad (2.81)$$

2.5   Phase transitions in electroweak theory

In electroweak theory, the phase transition to be studied is often from a phase where the $SU(2)_L \times U(1)_Y$ gauge group is unbroken to one in which only $U(1)_{em}$ is unbroken, i.e. it is from a phase in which the gauge fields mediating the weak interactions are massless as well as the photon, so that the weak interactions are long range like the electromagnetic interaction, to a phase in which only the electromagnetic interaction is long range. The Lagrangian density for electroweak theory has a pure gauge field part

$$\mathcal{L}_{\text{gauge}} = -\tfrac{1}{4} W^a_{\mu\nu} W^{a\mu\nu} - \tfrac{1}{4} B_{\mu\nu} B^{\mu\nu} \tag{2.82}$$

where the gauge-fixing term and the Fadeev–Popov ghosts have been omitted,

$$W^a_{\mu\nu} \equiv \partial_\mu W^a_\nu - \partial_\nu W^a_\mu - g\epsilon^{abc} W^b_\mu W^c_\nu \tag{2.83}$$

and

$$B_{\mu\nu} \equiv \partial_\mu B_\nu - \partial_\nu B_\mu. \tag{2.84}$$

In (2.82) and (2.83), $W^w_\mu \,(a = 1, 2, 3)$ are the three gauge fields associated with the generators of $SU(2)_L$ and B_μ is the gauge field associated with $U(1)_Y$. The electromagnetic field is the superposition

$$A_\mu = \cos\theta_W B_\mu + \sin\theta_W W^3_\mu. \tag{2.85}$$

After spontaneous symmetry breaking to $U(1)_{em}$, A_μ remains massless but the othogonal combination

$$Z_\mu = -\sin\theta_W B_\mu + \cos\theta_W W^3_\mu \tag{2.86}$$

together with

$$W^\pm_\mu = W^1_\mu \pm i W^2_\mu \tag{2.87}$$

becomes massive. When A_μ is correctly coupled to the electromagnetic current with strength e, the Weinberg angle θ_W obeys

$$g \sin\theta_W = e = g' \cos\theta_W \tag{2.88}$$

where g and g' are the gauge coupling constants for $SU(2)_L$ and $U(1)_Y$.

The $SU(2)_L \times U(1)_Y$ gauge symmetry is spontaneously broken by the Higgs doublet (under $SU(2)_L$) introduced by writing

$$H = \begin{pmatrix} H^+ \\ H^0 \end{pmatrix} \tag{2.89}$$

with weak hypercharge $Y = 1/2$. The Higgs boson part of the Lagrangian density (including the couplings to the gauge fields) is given by

$$\mathcal{L}_{\text{Higgs}} = (D_\mu H)^\dagger (D^\mu H) - m^2 H^\dagger H - \lambda (H^\dagger H)^2 \tag{2.90}$$

where the covariant derivative of H is

$$D_\mu H = (\partial_\mu + ig\tfrac{1}{2}\tau_a W_\mu^a + ig'\tfrac{1}{2}B_\mu)H \tag{2.91}$$

and τ_a, $a = 1, 2, 3$, are the Pauli matrices. If the expectation value for the Higgs doublet is introduced by writing

$$\langle H \rangle = \begin{pmatrix} 0 \\ \phi_c/\sqrt{2} \end{pmatrix} \tag{2.92}$$

the zero-temperature tree-level term in the effective potential is

$$\overline{V}_0(\phi_c) = \frac{m^2}{2}\phi_c^2 + \frac{\lambda}{4}\phi_c^4. \tag{2.93}$$

Including the zero-temperature and one-loop contributions, and assuming that T^2 is large compared with all shifted masses, we have, for the complete effective potential,

$$\overline{V}(\phi_c) = \frac{1}{2}m^2(T)\phi_c^2 - \frac{1}{3}CT\phi_c^3 + \frac{\lambda}{4}\phi_c^4 - \left(N_B + \frac{7}{8}N_F\right)\frac{\pi^2 T^4}{90}$$

$$+ B\phi_c^4\left(\ln\frac{\phi_c^2}{M^2} - \frac{25}{6}\right) \tag{2.94}$$

with

$$m^2(T) = m^2 + \left(\frac{\lambda}{2} + \frac{e^2(1 + 2\cos^2\theta_W)}{4\sin^2 2\theta_W} + \sum_f \frac{h_f^2}{12}\right)T^2 \tag{2.95}$$

where h_f are the Yukawa couplings of the fermions, with the largest contribution coming from the top quark. Retaining only the gauge boson contributions to the ϕ_c^3 term, which is expected to be a reasonable approximation provided that the mass of the Higgs particle is not too much larger than the Z boson, we have

$$C = \frac{3e^3(1 + 2\cos^3\theta_W)}{4\pi\sin^3 2\theta_W}. \tag{2.96}$$

Dropping the λ^2 contributions to the radiative corrections, which are always perturbatively negligible compared with the tree terms,

$$B = \frac{3}{4}\left(\frac{e^2}{4\pi}\right)^2\frac{1 + 2\cos^4\theta_W}{\sin^4 2\theta_W} - \frac{1}{64\pi^2}\sum_f h_f^2. \tag{2.97}$$

To study the nature of the phase transition, it is important to determine the sign of B. At zero temperature, (2.94) is of the same form as (2.62) with λ replaced by

4λ and can be cast in the form (2.77) with the same replacement in (2.78). A little calculation (exercise 2) shows that, when $B > 0$ and $\alpha > 0$, the zero-temperature effective potential always has a minimum at $\phi_c = 0$, in contrast to the situation when radiative corrections are neglected where there is a maximum at $\phi_c = 0$ and a minimum at $\phi_c^2 = -m^2/\lambda$.

However, when $B < 0$, there can only be a minimum at $\phi_c = 0$ if $\alpha < 0$. However, $\alpha < 0$ is ruled out by the requirement that the mass-squared m_H^2 of the physical neutral Higgs scalar particle given in (2.79) should be positive. Thus, there is never a minimum at $\phi_c = 0$ when $B < 0$ and the situation is then qualitatively the same as in the absence of the zero-temperature radiative corrections: the phase transition is first order.

The sign of B may be determined as follows. We shall retain only the largest Yukawa coupling which is the top quark coupling h_t. (Retaining more Yukawas only strengthens the conclusion.) The corresponding Lagrangian is

$$\mathcal{L}_{\text{top}} = -h_t Q_t^T i\tau^2 H t_R + \text{h.c.} \tag{2.98}$$

where Q_t is the $SU(2)_L$ doublet

$$Q_t = \begin{pmatrix} t \\ b \end{pmatrix}_L. \tag{2.99}$$

Taking account of the three colours of top quark contributing to the loop, $\sum_f G_f^4$ is replaced by $3h_t^4$ in (2.97). The top-quark mass term deriving from (2.98) is $h_t t_L t_R \phi_c/\sqrt{2} + \text{h.c.}$, so the top quark mass is

$$m_t = \frac{1}{\sqrt{2}} h_t \phi_c. \tag{2.100}$$

Also, the W and Z masses are given at tree-level by

$$m_W^2 = \tfrac{1}{8} g^2 \phi_c^2 \qquad m_Z^2 = \tfrac{1}{4} \sec^2 \theta_W g^2 \phi_c^2. \tag{2.101}$$

With a measured top quark mass of about 175 GeV, and an empirical value of $\sin^2 \theta_W$ determined by the measured W and Z masses of 0.243, so that $\phi_c \simeq 263$ GeV, we find that

$$h_t \simeq 0.94. \tag{2.102}$$

With $e^2/4\pi = 1/137$, we find from (2.97) that B is negative.

We also have to decide whether it is correct to assume that T^2 is large compared with all (shifted) masses. Using the high-temperature expansion and neglecting the zero-temperature radiative corrections, we see from (2.95) that, as in (2.60),

$$\frac{1}{T_c^2} = -\frac{1}{m^2}\left[\frac{\lambda}{2} + \frac{e^2(1 + 2\cos^2\theta_W)}{4\sin^2 2\theta_W} + \frac{h_t^2}{4} - \frac{e^6}{8\pi^2\lambda}\left(\frac{1 + 2\cos^3\theta_W}{\sin^3 2\theta_W}\right)^2\right] \tag{2.103}$$

where, again, we have retained only the top-quark Yukawa coupling. Provided $e^2 \gg \lambda \gg e^6$, we estimate

$$T_c^2 \simeq -2m^2. \tag{2.104}$$

As in the Higgs model, the shifted masses for the Higgs scalars are of order $-m^2$ and the high-temperature approximation is not too unreasonable. The greatest danger for the high-temperature approximation in the Higgs model with $e^4 \gg \lambda$ arose from the gauge field mass. In the present case, dropping $\lambda/2$ in the numerator of (2.103) and using the value of ϕ_c^2 at the zero-temperature asymmetric minimum $\phi_c^2 = -m^2/\lambda$, we can estimate that

$$\frac{T_c^2}{m_W^2} \simeq 2\frac{\lambda}{e^2} \qquad \frac{T_c^2}{m_Z^2} \simeq 1.5\frac{\lambda}{e^2}. \tag{2.105}$$

Thus, if $\lambda \gtrsim e^2 \simeq 0.09$, the high-temperature approximation may again be not too unreasonable.

This discussion suggests that the electroweak phase transition is effectively second order, because C, defined in (2.45), is small in the sense discussed at the end of section 2.4.1. For $T > T_c$, the system is in the symmetric phase in which $\phi_c = 0$ and all gauge bosons are massless. For $T < T_c$, the system is in the asymmetric phase for which $\phi_c \neq 0$, the $W^\pm$ and Z gauge bosons acquire a mass and the symmetry is broken from $SU(2)_L \times U(1)_Y$ to $U(1)_{em}$. The critical temperature T_c given by (2.103) is of the same order of magnitude as the zero-temperature value of ϕ_c at the asymmetric minimum of the effective potential provided that $\lambda \gtrsim e^2$. It was estimated before (2.102) that $\phi_c \simeq 263$ GeV and so T_c should be of this order of magnitude.

2.6   Phase transitions in grand unified theories

Electroweak theory combines the weak and electromagnetic interactions in a single model with $SU(2)_L \times U(1)_Y$ gauge group but achieves no unification of these interactions with the strong interaction. It is possible that the weak, electromagnetic and strong interactions are unified in a theory involving a larger gauge group (a grand unified theory or GUT), perhaps with a single gauge coupling constant. Once such a unification has been assumed, the coupling constants g, g' and g_s (the QCD coupling constant) are related by group theory factors to a single GUT coupling constant g_G for the grand unified group. The values of the renormalized coupling constants depend on the renormalization scale M and, if the coupling constants $g(M)$, $g'(M)$ and $g_s(M)$ obey the gauge theoretic relationships of the grand unified group at one such scale $M = M_G$, they cannot obey these relationships at lower energy scales. This is because at energy scales below M_G the extra gauge fields associated with the enlargement of the gauge group to the grand unified group may be ignored. (They acquire masses on the scale of M_G.) Then $g(M)$, $g'(M)$ and $g_s(M)$ run differently with M when the renormalization group is deployed to derive their dependence on M. However, for energy scales greater than M_G, all gauge fields are on the same footing and there is a single gauge coupling constant g_G developing in accordance with the renormalization group equation of the GUT.

The simplest example of a grand unified group that is large enough to contain $SU(3)_c \times SU(2)_L \times U(1)_Y$ of the standard model is $SU(5)$ and we shall use this example to illustrate phase transitions in GUTs. If the renormalization group equations are used to run the low-energy values of the gauge coupling constants to the scale $M = M_G$ at which the $SU(5)$ relationships

$$g_s(M_G) = g(M_G) = \sqrt{\tfrac{5}{3}} g'(M_G) = g_G(M_G) \qquad (2.106)$$

hold, then upon inputting the values of the strong and electromagnetic coupling constants at $M = m_Z$, the unification scale is found to be of order 10^{15} GeV. In addition, there is a prediction for $\sin^2 \theta_W(m_Z)$ at $M = m_Z$ of around 0.21, which differs significantly from the observed value of around 0.23. Nevertheless, we shall use the $SU(5)$ GUT as a simple illustration of the way in which phase transitions work in a GUT. In section 2.7, we shall consider supersymmetric GUTs in which the prediction for $\sin^2(m_Z)$ can be brought into line with experiment to a high degree of accuracy.

In the $SU(5)$ GUT, the grand unified phase transition is from the $SU(5)$ symmetric phase to the $SU(3)_c \times SU(2)_L \times U(1)_Y$ symmetric phase and is followed at a lower temperature by the electroweak phase transition described in the previous section. We expect the critical temperature for the grand unified phase transition to be of order 10^{15} GeV (the energy at which the spontaneous symmetry breaking occurs) and, at such high temperatures, the expectation values of the electroweak Higgs scalars (of order 200 GeV) are negligible. Thus, to describe the grand unified phase transition we need only retain the Higgs scalars responsible for breaking the $SU(5)$ gauge group, whose expectation values are on the 10^{15} GeV scale.

The grand unified Higgs scalars Φ belong to the 24-dimensional adjoint representation of $SU(5)$:

$$\Phi = \sum_{a=1}^{24} \phi_a t_a \qquad (2.107)$$

where t_a are the $SU(5)$ generators in the fundamental five-dimensional representation. Suppressing the gauge-fixing term and the Fadeev–Popov ghost term, the finite-temperature Lagrangian density (apart from a possible tr Φ^3 term) is

$$\mathcal{L} = -m_1^2 \operatorname{tr} \Phi^2 - \lambda_1 (\operatorname{tr} \Phi^2)^2 - \lambda_2 \operatorname{tr} \Phi^4 + \operatorname{tr}(\bar{D}_\mu \Phi)^2 - \tfrac{1}{2} \operatorname{tr}(\bar{F}_{\mu\nu} \bar{F}^{\mu\nu}). \quad (2.108)$$

In (2.108), the covariant derivative $D_\mu \Phi$ is given by

$$D_\mu \Phi = \partial_\mu \Phi + i g_G [A_\mu, \Phi] \qquad (2.109)$$

where

$$A_\mu \equiv \sum_{a=1}^{24} A_{a\mu} t_a \tag{2.110}$$

with $A_{a\mu}$ the gauge fields in the adjoint representation of $SU(5)$, and the transition to a finite-temperature theory is made by the replacement of ∂_μ by $\bar\partial_\mu$, as in (2.13). The gauge field strength $F_{\mu\nu}$ is given by

$$F_{\mu\nu} \equiv \partial_\mu A_\nu - \partial_\nu A_\mu + ig_G[A_\mu, A_\nu] \tag{2.111}$$

with the transition to the finite temperature being made in the same way. The fermionic terms have been dropped in (2.108) because quark and lepton masses are negligible on the grand unified scale.

For the breaking of $SU(5)$ to $SU(3)_c \times SU(2)_L \times U(1)_Y$, we take the expectation value of the field Φ to be of the form

$$\langle \Phi \rangle = \frac{\phi_c}{\sqrt{15}} \, \text{diag}\left(1, 1, 1, -\frac{3}{2}, -\frac{3}{2}\right). \tag{2.112}$$

($\langle \Phi \rangle$ must be traceless because the matrices t_a are.) This can be shown to be the lowest energy state at zero temperature for

$$\lambda_1 > -\frac{7}{30}\lambda_2 \qquad \lambda_2 > 0. \tag{2.113}$$

The finite-temperature effective potential can then be written, for temperatures large compared to all masses, as

$$\bar V(\phi_c) = \frac{1}{2}m_1^2(T)\phi_c^2 + \frac{1}{4}\left(\lambda_1 + \frac{7}{30}\lambda_2\right)\phi_c^4 - \left(N_B + \frac{7}{8}N_F\right)\frac{\pi^2 T^4}{90}$$
$$+ B\phi_c^4\left[\ln\left(\frac{\phi_c^2}{M^2}\right) - \frac{25}{6}\right] \tag{2.114}$$

where

$$m_1^2(T) = m_1^2 + \frac{1}{60}(130\lambda_1 + 47\lambda_2 + 75g_G^2)T^2 \tag{2.115}$$

$$B = \frac{25}{256\pi^2}g_G^4 \tag{2.116}$$

and the zero-temperature radiative correction has been renormalized at mass M as in section 2.4. In (2.114) the λ_1^2 and λ_2^2 contributions are always small compared with the tree terms in (2.114) and have been dropped.

If $g_G^4 \ll \lambda_1, \lambda_2$ we may neglect the zero-temperature radiative correction. There is then a second-order phase transition with critical temperature T_c given by

$$T_c^2 = \frac{-60m_1^2}{130\lambda_1 + 47\lambda_2 + 75g_G^2}. \tag{2.117}$$

For $T > T_c$, the system is in the $SU(5)$ symmetric phase, for which $\phi_c = 0$, and all gauge bosons are massless. For $T < T_c$, ϕ_c is non-zero, the system is in the $SU(3)_c \times SU(2)_L \times U(1)_Y$ symmetric phase, and only the electroweak gauge bosons are massless. By the same sort of argument as in the previous section, T_c should be of order 10^{15} GeV.

However, if $g_G^4 \gg \lambda_1, \lambda_2$, then a discussion similar to that given for the Higgs model in section 2.4 shows that a first-order phase transition takes place. In the case of a GUT, this conclusion is not negated by Yukawa couplings of fermions giving additional contributions to the coefficient B, because quarks and leptons do not couple to the grand unified Higgses. This is an important difference compared with the electroweak phase transition.

2.7 Phase transitions in supersymmetric GUTs

If elementary particle theories possess supersymmetry, then each (complex scalar) spin-0 particle is paired with one chirality of a spin-$\frac{1}{2}$ particle in the same so-called 'chiral supermultiplet' and each spin-1 vector particle is paired with a spin-$\frac{1}{2}$ particle of a single chirality in the same so-called 'vector supermultiplet'. The quarks and leptons have supersymmetric partners referred to as 'squarks' and 'sleptons', the Higgs scalars have supersymmetric partners referred to as 'Higgsinos' and the gauge bosons are paired with 'gauginos'. In the absence of supersymmetry breaking, particles in the same supermultiplet have the same mass. Of course, since at the time of writing we have not observed supersymmetric partners of the known particles ('sparticles'), there must be some (spontaneous) supersymmetry breaking to produce substantial mass splittings within supermultiplets.

The presence of these extra sparticles can be very important for the discussion of phase transitions at temperatures large compared to the sparticle masses. In addition, the supersymmetry transformations transforming particles of different spin within a supermultiplet into each other strongly constrain the form of the Lagrangian and the tree-level effective potential, with further implications for phase transitions. These supersymmetry transformations may be local or global depending respectively on whether the parameters of the transformation do or do not depend on the point in spacetime. In this section the case of globally supersymmetric GUTs will be discussed [7–12] and the locally supersymmetric (supergravity) case will be discussed in the next section. (For a systematic development of globally and locally supersymmetric theories see [14].)

In general, in a supersymmetric theory, the Lagrangian and the tree-level effective potential are determined once the superpotential W is given. For example, for a theory with a single scalar field ϕ together with its supersymmetric partner, the superpotential for a renormalizable theory takes the form

$$W = \tfrac{1}{2}m\phi^2 + \tfrac{1}{3}\lambda\phi^3 \qquad (2.118)$$

the bosonic part of the Lagrangian density is

$$\mathcal{L}_{\text{bosonic}} = \partial_\mu \phi^\dagger \partial^\mu \phi - \left| \frac{\partial W}{\partial \phi} \right|^2 \tag{2.119}$$

and the tree-level effective potential is

$$V = \left| \frac{\partial W}{\partial \phi} \right|^2 = |m\phi + \lambda\phi^2|^2. \tag{2.120}$$

In a supersymmetric $SU(5)$ GUT, the generalization of this renormalizable superpotential is

$$W = \tfrac{1}{2}m \operatorname{tr} \Phi^2 + \tfrac{1}{3}\lambda \operatorname{tr} \Phi^3 \tag{2.121}$$

with Φ defined in (2.107), and the tree-level effective potential is

$$V = \tfrac{1}{2} \operatorname{tr} |m\Phi + \lambda(\Phi^2 - \tfrac{1}{3}\operatorname{tr}\Phi^2 I)|^2 + g_G^2 \operatorname{tr}([\Phi, \Phi^\dagger]^2). \tag{2.122}$$

The last term in (2.122) is the so-called 'D-term' that arises in a supersymmetric gauge theory and the first term, which is independent of the gauge group, is referred to as the 'F-term'. At $T = 0$ (and in the absence of supersymmetry breaking), the effective potential (2.122) has degenerate minima (exercise 3) with $V = 0$, namely

$$\langle \Phi \rangle = 0 \tag{2.123}$$

$$\langle \Phi \rangle = \frac{m}{3\lambda} \operatorname{diag}(1, 1, 1, 1, -4) \tag{2.124}$$

and

$$\langle \Phi \rangle = \frac{m}{\lambda} \operatorname{diag}(2, 2, 2, -3, -3). \tag{2.125}$$

The minima (2.123), (2.124) and (2.125) correspond respectively to $SU(5)$ symmetric, $SU(4) \times U(1)$ symmetric and $SU(3) \times SU(2) \times U(1)$ symmetric phases.

At finite temperature, the degeneracy of these supersymmetric minima is lifted by the T^4 terms and by $T^2 \operatorname{tr}(\Phi^\dagger \Phi) = \sum_a \tfrac{1}{2}|\phi_a|^2 T^2$ terms. As in (2.32), the coefficient of the T^4 term depends on the value of $N_B + \tfrac{7}{8}N_F$ for states light on the scale of the temperature T. For the $SU(5)$ symmetric phase, the 24 gauge fields together with their gauginos contribute 90 towards $N_B + \tfrac{7}{8}N_F$. Each fermion generation has three doublets of left-chiral quarks, one for each of the three colours, six right-chiral quarks, a left-chiral lepton doublet, and a right-chiral (charged) lepton. These give $n_G = 3$ copies of the $\bar{5} + 10$ representation of $SU(5)$. These three generations of quarks and leptons, together with their associated squarks and sleptons, contribute $\frac{675}{4}$ to $N_B + \tfrac{7}{8}N_F$. In total, this gives the coefficient of the T^4 term in the temperature-dependent effective potential

$$-\frac{\pi^2}{90}\left(N_B + \frac{7}{8}N_F\right) = -\frac{23}{8}\pi^2 \qquad SU(5) \text{ symmetric phase.} \tag{2.126}$$

For the $SU(3) \times SU(2) \times U(1)$ symmetric phase, the 12 gauge fields together with their associated gauginos contribute 45 towards $N_B + \frac{7}{8}N_F$ and, for the $SU(4) \times U(1)$ symmetric phase, the 16 gauge fields together with their gauginos contribute 60 towards $N_B + \frac{7}{8}N_F$. In each of these two cases, the matter field content is the same as for the $SU(5)$ symmetric phase and we find that

$$\frac{\pi^2}{90}\left(N_B + \frac{7}{8}N_F\right) = \frac{-19}{8}\pi^2 \qquad SU(3) \times SU(2) \times U(1) \text{ symmetric phase}$$

$$(2.127)$$

$$\frac{\pi^2}{90}\left(N_B + \frac{7}{8}N_F\right) = \frac{-61}{24}\pi^2 \qquad SU(4) \times U(1) \text{ symmetric phase.} \quad (2.128)$$

If a copy of $5 + \bar{5}$ is included to provide the two electroweak Higgs doublets (plus Higgsinos) needed to give masses to both up-like and down-like quarks in a supersymmetric theory, then there is an additional contribution $-5\pi^2/12$ to (2.126) and $-\pi^2/6$ to (2.127), in the latter case from the two $SU(2)_L$ doublets that are all that survive from the $5 + \bar{5}$ after spontaneous symmetry breaking of $SU(5)$ by the expectation values of the adjoint Higgs scalars. (The surviving adjoint Higgs scalar states are too heavy to contribute to the temperature-dependent corrections to the effective potential for temperatures below the grand unification scale.) In the $SU(4) \times U(1)$ phase, the complete $5+\bar{5}$ becomes massive and fails to contribute to the temperature-dependent corrections. The values of $\frac{\pi^2}{90}(N_B + \frac{7}{8}N_F)$ in (2.126),(2.127) and (2.128) are then modified to $\frac{79}{24}\pi^2$, $\frac{61}{24}\pi^2$ and $\frac{61}{24}\pi^2$, respectively. Thus, the T^4 term favours the $SU(5)$ symmetric phase over the $SU(3) \times SU(2) \times U(1)$ and $SU(4) \times U(1)$ symmetric phases, which remain on the same footing. This conclusion is only strengthened by the inclusion of the adjoint Higgs supermutiplet which provides extra light states in the $SU(5)$ symmetric phase. If the theory contains more than one pair of Higgs multiplets coming from $5 + \bar{5}$, then the $SU(5)$ symmetric phase continues to be favoured over the other two phases but the $SU(3) \times SU(2) \times U(1)$ symmetric phase is favoured over the $SU(4) \times U(1)$ symmetric phase, which is the assumption we shall make in what follows.

Clearly, the $SU(5)$ symmetric phase, for which the scalar expectation value is zero, minimizes the $T^2\phi_c^2$ term as well as the T^4 term. Thus, the theory appears to favour the $SU(5)$ symmetric phase at all temperatures. However, at temperatures below 100 GeV–1 TeV the (non-perturbative) supersymmetry breaking mechanism will lift the degeneracy of the three phases more than the temperature-dependent terms and may favour the $SU(3) \times SU(2) \times U(1)$ symmetric phase. At higher temperatures the temperature-dependent terms dominate. This suggests that the universe is in an $SU(5)$ symmetric phase down to temperatures of 100 GeV–1 TeV.

This conclusion is modified by the running of the gauge coupling constant g_5 for $SU(5)$ with temperature. This may result in g_5 becoming strong at temperatures of order 10^9–10^{10} GeV. Then, confinement may result in fewer massless states. (The $SU(5)$ coupling becomes strong at a higher temperature than the $SU(4)$ and $SU(3)$ couplings.) Then one of the other two phases may become the absolute minimum and, eventually, tunnelling may occur to one of the other phases. In general, running coupling constants $g_a(\mu)$ and $g_a(M)$ at energy scales μ and M are related by

$$16\pi^2 g_a^{-2}(\mu) = 16\pi^2 g_a^{-2}(M) + b_a \ln\left(\frac{M^2}{\mu^2}\right) \tag{2.129}$$

where the renormalization group coefficient b_a is given by

$$b_a = -\tfrac{11}{3}c_1(G_a) + \tfrac{2}{3}\sum_{R_a} c_2(R_a) + \tfrac{1}{3}\sum_{S_a} c_2(S_a). \tag{2.130}$$

In (2.130), the group theory factor $c_1(G_a)$ for the group G_a is related to the structure constants $f_{\alpha\beta\gamma}$ by

$$c_1\delta_{\alpha\beta} = f_{\alpha\gamma\delta}f_{\beta\gamma\delta} \tag{2.131}$$

the group theory factor $c_2(R_a)$ for the representation R_a of the group is given in terms of the matrices T_α representing the generators of G_a in the representation R_a by

$$c_2\delta_{\alpha\beta} = \mathrm{tr}(T_\alpha T_\beta) \tag{2.132}$$

and the summations are over chiral fermion representations R_a of G_a and scalars in representations S_a of G_a. For a supersymmetric theory, each gauge field is accompanied by a gaugino in the adjoint representation, so that

$$-\tfrac{11}{3}c_1(G_a) \rightarrow -\tfrac{11}{3}c_1(G_a) + \tfrac{2}{3}c_1(G_a) = -3c_1(G_a). \tag{2.133}$$

Also, each chiral fermion is accompanied by a complex scalar so that

$$\tfrac{2}{3}c_2(R_a) \rightarrow \tfrac{2}{3}c_2(R_a) + \tfrac{1}{3}c_2(R_a) = c_2(R_a). \tag{2.134}$$

Thus, in a supersymmetric theory

$$b_a = -3c_1(G_a) + \sum_{R_a} c_2(R_a) \tag{2.135}$$

where the sum is over all chiral supermultiplets. Identifying the energy scale μ in (2.129) with T, and recasting the renormalization group equations in terms of $\alpha_a \equiv g_a^2/4\pi$, gives

$$\alpha_a^{-1}(T) = \alpha_a^{-1}(M) + \frac{b_a}{2\pi}\ln\left(\frac{M}{T}\right). \tag{2.136}$$

For $SU(5)$,

$$c_1(SU(5)) = 5 \qquad c_2(5) = c_2(\bar{5}) = \tfrac{1}{2} \qquad c_2(10) = \tfrac{3}{2}. \tag{2.137}$$

With $n_G = 3$ generations in $\bar{5} + 10$ and N_H sets of Higgs scalars in 5, we have

$$b_5 = -9 + \tfrac{1}{2}N_H \tag{2.138}$$

and, for $N_H = 2$,

$$b_5 = -8. \tag{2.139}$$

It is convenient to choose

$$M = M_X = 2 \times 10^{16} \text{ GeV} \tag{2.140}$$

which is the energy scale at which the low-energy (supersymmetric) $SU(3) \times SU(2) \times U(1)$ coupling constants reach a common value so that grand unification may occur. At this scale,

$$\alpha_5(M_X) \simeq \tfrac{1}{25}. \tag{2.141}$$

If we take the criterion for the $SU(5)$ coupling constant to become strong to be $\alpha_5(T) \simeq 1$, then the corresponding temperature is

$$T \simeq 6.5 \times 10^{-9}M \simeq 10^8 \text{ GeV}. \tag{2.142}$$

It is at this temperature that we expect that either the $SU(3) \times SU(2) \times U(1)$ or the $SU(4) \times U(1)$ symmetric phase becomes the absolute minimum. Eventually tunnelling will occur to whichever of these phases is the absolute minimum. If it is the $SU(3) \times SU(2) \times U(1)$ symmetric phase, then the universe will continue in this phase until the coupling constant g_4 becomes strong at some lower temperature.

In these globally supersymmetric theories, because the zero-temperature effective potential is zero when supersymmetry is unbroken, the cosmological constant is zero in each of the $SU(5)$, $SU(4) \times U(1)$ and $SU(3) \times SU(2) \times U(1)$ phases until supersymmetry breaking becomes non-negligible (with respect to T) for temperatures below 10^2–10^3 GeV.

2.8 Phase transitions in supergravity theories

Up to now we have been discussing theories with global supersymmetry. A theory with local supersymmetry is necessarily a theory which contains gravity (supergravity). The reason is that the supersymmetry algebra contains the generator P_μ of translations and when we allow supersymmetry transformations that depend on the point in spacetime (local supersymmetry), we have to consider, among other things, translations that vary from point to point in spacetime. Thus, local supersymmetry contains general coordinate transformations of spacetime and so is a theory of gravity.

In phenomenologically acceptable theories, the supersymmetry breaking scale M_s is large (typically 10^{10}–10^{11} GeV) where M_s^2 is the expectation value of the auxiliary field of the scalar responsible for supersymmetry breaking. For example, in theories with F-term supersymmetry breaking, at tree level, a fermion has a supersymmetric partner of lower mass than itself as well as one of higher mass than itself. Since this does not occur in the real world, it is necessary for there to be significant quantum corrections to avoid this problem, though not so big that the hierarchy problem is no longer solved. This is the origin of the large supersymmetry breaking scale. In these circumstances, the effects of quantum gravity can no longer be neglected. In particular, in the presence of supersymmetry breaking, scalar particles acquire masses of order M_s^2/m_P (where $m_P \simeq 1.2 \times 10^{19}$ GeV is the Planck mass) which are of order 10^2 GeV.

Once gravitational effects are important, we should allow not only that the superpotential may contain non-renormalizable terms but also that there may be non-renormalizable kinetic terms. Thus, for example, the scalar field kinetic terms take the form

$$\frac{\partial^2 K}{\partial \phi_i \partial \phi_j^*} \partial_\mu \phi_i \partial^\mu \phi_j^* \tag{2.143}$$

where $K(\phi_i, \phi_i^*)$ is referred to as the Kähler potential. It turns out [13] that the complete supergravity Lagrangian can be expressed in terms of

$$G = K + \ln |W|^2 \tag{2.144}$$

apart from couplings to gauge fields, which involve the gauge kinetic function. It will often be convenient to work in units where the reduced Planck mass $M_P = 1$, where M_P is defined by

$$M_P^{-2} \equiv 8\pi G_N \tag{2.145}$$

where G_N is Newton's constant, so that

$$M_P \simeq 2.44 \times 10^{18} \text{ GeV}. \tag{2.146}$$

In these units, the zero-temperature effective potential takes the form

$$V = e^G (G_i (G^{-1})_j^i G^j - 3) \tag{2.147}$$

where the scalar fields have been written as ϕ_i, their adjoints as ϕ^{i*} and derivatives of G as

$$G^i \equiv \frac{\partial G}{\partial \phi_i} \qquad G_i \equiv \frac{\partial G}{\partial \phi^{i*}} \tag{2.148}$$

and

$$G_j^i \equiv \frac{\partial^2 G}{\partial \phi_i \partial \phi_j^*}. \tag{2.149}$$

The inverse $(G^{-1})_j^i$ obeys

$$(G^{-1})_j^i G_k^j = \delta_k^i. \tag{2.150}$$

In particular, in the case of a single gauge-singlet chiral superfield Φ with minimal kinetic terms arising from

$$G = \phi^* \phi + \ln |W|^2 \tag{2.151}$$

we have

$$V = e^{\phi^* \phi} \left(\left| \frac{\partial W}{\partial \phi} + \phi^* W \right|^2 - 3|W|^2 \right) \qquad (2.152)$$

instead of $V = |\partial W/\partial \phi|^2$ for the globally supersymmetric case, as in (2.120). Consideration of the supersymmetry transformation laws of scalar fields and their fermionic superpartners shows that the criterion for supersymmetry breaking is that $\partial W/\partial \phi + \phi^* W$ should be non-zero [14]. Thus, whereas the globally supersymmetric theory vacua with unbroken supersymmetry had $V = 0$, in the locally supersymmetric case (with minimal kinetic terms) supersymmetric vacua have

$$V = -3e^{\phi^* \phi}|W|^2 \qquad (2.153)$$

More generally, we may consider gauge non-singlet chiral superfields Φ_i. In that case, the supergravity Lagrangian [13] also involves the gauge kinetic function f_{ab}. For the minimal choice of gauge kinetic function

$$f_{ab} = \delta_{ab} \qquad (2.154)$$

the gauge kinetic term

$$-\tfrac{1}{4} \operatorname{Re} f_{ab} F_{a\mu\nu} F_b^{\mu\nu} \qquad (2.155)$$

simplifies to $-\tfrac{1}{4} F_{a\mu\nu} F_a^{\mu\nu}$. Then, with minimal kinetic terms, the zero-temperature effective potential takes the form

$$V = e^{\phi_j^* \phi_j} \left(\left| \frac{\partial W}{\partial \phi_i} + \phi^{i*} W \right|^2 - 3|W|^2 \right) + \tfrac{1}{2} g^2 G^i (T_a)_{ij} \phi_j G^k (T_a)_{kl} \phi_l \quad (2.156)$$

where we have assumed a simple gauge group with gauge coupling constant g, and

$$G^i = \phi^{i*} + \frac{1}{W} \frac{\partial W}{\partial \phi_i}. \qquad (2.157)$$

If supergravity is unbroken, study of the supersymmetry transformation laws shows that we must have

$$G^i (T_a)_{ij} \phi_j = 0 \qquad \text{and} \qquad \frac{\partial W}{\partial \phi_i} + \phi^{i*} W = 0 \qquad (2.158)$$

and supersymmetric vacua have

$$V = -3e^{\phi_j^* \phi_j}|W|^2. \qquad (2.159)$$

There is no longer any requirement that supersymmetric minima should be degenerate in energy at $T = 0$ nor that they should have lower energy than all other vacua.

In the high-temperature limit, (2.32) still applies to the one-loop temperature-dependent correction to the effective potential provided that, in the case that the kinetic terms are non-minimal, we first construct fields with canonical kinetic terms by field redefinition. This means that, for scalar fields ϕ_i (and their fermionic superpartners), we have to write

$$\phi_i = (G^{-1/2})_i^j (\phi_j)_N \tag{2.160}$$

and for the gauginos

$$\lambda_a = (\text{Re } f_{ab})^{-1/2}(\lambda_b)_N \tag{2.161}$$

where $(\phi_j)_N$ and $(\lambda_b)_N$ are the normalized fields. The relevant mass matrices are obtained from the standard supergravity Lagrangian. The outcome [15–18] is particularly simple in the case of minimal kinetic terms. Then

$$\begin{aligned}
\overline{V}_1^T = &-\frac{\pi^2 T^4}{90}\left(N_B + \frac{7}{8}N_F\right) \\
&+ \frac{T^2}{12}e^G\left[\frac{3}{2}(A+B) + (C+N)(C-2) + \frac{1}{2}C^2 + C - 1\right]
\end{aligned} \tag{2.162}$$

where

$$A = G^i G_{ij} G^j + G_i G^{ij} G_j \tag{2.163}$$

$$B = G_{ij} G^{ij} \tag{2.164}$$

$$C = G_i G^i \tag{2.165}$$

and N is the number of chiral superfields. In practice, N is often large. For example, if the matter field content is that of the minimal $SU(5)$ GUT, there are $n_G = 3$ generations in the $\bar{5} + 10$ representation contributing 45 to N, two copies of 5 or $\bar{5}$ for electroweak Higgs contributing 10 to N and one copy of 24 for the grand unified Higgs scalars contributing 24 to N, leading to a total value of $N = 79$.

Provided that all of the couplings in the superpotential are of the same order of magnitude, in units where the reduced Planck mass is 1, we can then take the large-N limit to obtain

$$\overline{V}_1^T = -\frac{\pi^2 T^4}{90}\left(N_B + \frac{7}{8}N_F\right) + N\frac{T^2}{12}e^G(G_i G^i - 2). \tag{2.166}$$

Provided that the changes in $\overline{V}_1^T$ generated by non-minimal kinetic terms do not introduce extra factors of N (which is true for most choices of G), this is still the large-N limit of $\overline{V}_1^T$ in that case. For any particular choice of Kähler potential and superpotential and, therefore, of G, the discussion of phase transitions proceeds as before with the modified finite-temperature corrections to the effective potential of (2.162).

2.9   Nucleation of true vacuum

In sections 2.4 and 2.6, we have found that first-order phase transitions may occur as the universe cools. If the phase transition is first order, it will be necessary for the universe to tunnel out of the metastable minimum [19–21] (false vacuum) to reach the absolute minimum (true vacuum). If the tunnelling rate is small, this may occur at temperatures very much lower than the temperature T_{c1} of (2.76) at which the energy of the zero- (and low-) temperature vacuum drops below that of the high-temperature vacuum. In what follows, we shall approximate the tunnelling rate by its $T = 0$ value and study, for simplicity, the case of a single scalar field ϕ.

In the semi-classical limit (small $\hbar$) the probability per unit time per unit volume for formation of a bubble of true vacuum Γ is given by [19]

$$\Gamma = Ae^{-B/\hbar} \tag{2.167}$$

where

$$B = S_E \tag{2.168}$$

with S_E the Euclidean action for a solution of the Euclidean Euler–Lagrange equations which satisfies the boundary conditions that ϕ approaches the false vacuum (metastable minimum) as the Euclidean time $t_E \to \pm\infty$, and with zero Euclidean time derivative at $t_E = 0$, where

$$t_E \equiv it. \tag{2.169}$$

This is referred to as the 'bounce' solution (because it turns around and bounces back to the false vacuum.) The tunnelling is dominated by the solution for ϕ that gives the smallest value of S_E. The derivation of this result is by studying the imaginary part of the effective potential in the false vacuum. The coefficient A is, in general, more difficult to calculate [20]. However, since it does not appear in an exponent, an estimate on dimensional grounds is sufficient. At $T = 0$, we may expect A to be of order M^4 where M is an appropriate mass scale, such as the height of the potential barrier to be tunnelled through or (see [21]) the value of $(d^2V/d\phi^2)^{1/2}$ at the metastable minimum, which will usually be of the same order of magnitude.

For a single real scalar field, the Euclidean action takes the form

$$S_E = \int d^4x \left(\tfrac{1}{2}\partial_\mu\phi\partial^\mu\phi + V(\phi)\right) \tag{2.170}$$

with the metric the positive-definite metric of four-dimensional Euclidean space, and the Euclidean Euler–Lagrange equation is

$$\partial_\mu\partial^\mu\phi = \frac{\partial^2\phi}{\partial t_E^2} + \nabla^2\phi = V'(\phi). \tag{2.171}$$

It can be shown that the bounce which minimizes S_E is $O(4)$ symmetric, i.e. ϕ is a function of the four-dimensional radial variable ρ alone, where

$$\rho^2 = t_E^2 + \mathbf{x}^2. \tag{2.172}$$

Then

$$S_E = 2\pi^2 \int_0^\infty \mathrm{d}\rho \, \rho^3 \left[\frac{1}{2} \left(\frac{\mathrm{d}\phi}{\mathrm{d}\rho} \right)^2 + V(\phi) \right] \tag{2.173}$$

and the equation of motion is (exercise 4)

$$\frac{\mathrm{d}^2\phi}{\mathrm{d}\rho^2} + \frac{3}{\rho} \frac{\mathrm{d}\phi}{\mathrm{d}\rho} = V'(\phi). \tag{2.174}$$

In terms of this variable, the boundary conditions for a bounce solution are $\phi \to \phi_+$ as $\rho \to \infty$, and $\mathrm{d}\phi \, \mathrm{d}\rho = 0$ when $\rho = 0$, where ϕ_+ is the value of ϕ at the metastable minimum.

An explicit bounce solution is most easily obtained in the so-called 'thin-wall' approximation [19] which treats the energy difference ϵ between the two vacua as small compared with the height of the potential barrier between them. Then, we write

$$V(\phi) = V_0(\phi) + O(\epsilon) \tag{2.175}$$

where $V_0(\phi)$ is the effective potential in the limit that we neglect the energy difference between the two vacua, and

$$\epsilon = V(\phi_+) - V(\phi_-) \tag{2.176}$$

where ϕ_+ and ϕ_- are, respectively, the values of ϕ at the metastable and absolute minima. The Euclidean equation of motion is approximated first by replacing $V'(\phi)$ by $V_0'(\phi)$ in (2.174). We shall see later that, in this approximation, it is also correct to neglect $\frac{1}{\rho} \frac{\mathrm{d}\phi}{\mathrm{d}\rho}$, in which case the equation of motion to be solved for the bounce solution becomes

$$\frac{\mathrm{d}^2\phi}{\mathrm{d}\rho^2} = V_0'(\phi). \tag{2.177}$$

It is not difficult to show that the solution is

$$\rho = \int \frac{\mathrm{d}\phi}{\sqrt{2V_0(\phi)}}. \tag{2.178}$$

A simple example is obtained by taking

$$V_0(\phi) = \frac{\lambda}{8} \left(\phi^2 - \frac{\mu^2}{\lambda} \right)^2 \tag{2.179}$$

which has degenerate minima at $\phi = \pm\mu/\sqrt{\lambda}$ with $V_0 = 0$. Then (2.178) leads (exercise 5) to

$$\phi - \phi_0 = \frac{\mu}{\sqrt{\lambda}} \tanh\left[\frac{\mu}{2}(\rho - \rho_0)\right] \tag{2.180}$$

where ϕ_0 is the value of ϕ at some reference value ρ_0 of ρ. Choosing ρ_0 to be the value at which ϕ takes the average of its values in the true and false vacua, namely $\phi_0 = 0$, then

$$\phi = \frac{\mu}{\sqrt{\lambda}} \tanh\left[\frac{\mu}{2}(\rho - \rho_0)\right]. \tag{2.181}$$

Assuming that $\rho_0 \gg \mu^{-1}$, the length scale on which ϕ varies, then

$$\phi \to -\frac{\mu}{\sqrt{\lambda}} \qquad \text{as } \rho \to 0 \tag{2.182}$$

and, in any case,

$$\phi \to \frac{\mu}{\sqrt{\lambda}} \qquad \text{as } \rho \to \infty. \tag{2.183}$$

It will be seen later that this is correct for the bounce solution that minimizes S_E. If (after lifting the degeneracy of the two vacua using the $O(\epsilon)$ term in V) $\phi = \phi_- = -\mu/\sqrt{\lambda}$ is the true vacuum and $\phi = \phi_+ = \mu/\sqrt{\lambda}$ is the false vacuum, then the bounce solution describes a bubble of true vacuum embedded in the false vacuum with wall thickness of order μ^{-1}, where the rapid variation of ϕ occurs, separating the two regions. Under the assumption that $\rho_0 \gg \mu^{-1}$, the radius of the bubble is large compared with the thickness of the wall, which explains the 'thin-wall' approximation terminology.

The next steps are to calculate B and to justify the various assumptions made. In the thin-wall approximation,

$$\phi(\rho) = -\frac{\mu}{\sqrt{\lambda}} \qquad \text{for } \rho \ll \rho_0 \tag{2.184}$$

$$= \frac{\mu}{\sqrt{\lambda}} \tanh\left[\frac{\mu}{2}(\rho - \rho_0)\right] \qquad \text{for } \rho \simeq \rho_0 \tag{2.185}$$

$$= \frac{\mu}{\sqrt{\lambda}} \qquad \text{for } \rho \gg \rho_0. \tag{2.186}$$

The contribution to S_E from outside the wall ($\rho \gg \rho_0$) is zero because here $\frac{1}{2}(d\phi/d\rho)^2 + V(\phi) \simeq 0$. The contribution from inside the wall ($\rho \ll \rho_0$) is obtained by first noting that here

$$\frac{1}{2}\left(\frac{d\phi}{d\rho}\right)^2 + V(\phi) \simeq -\epsilon. \tag{2.187}$$

As a consequence, the contribution to S_E from inside the wall is

$$S_E \simeq -\frac{\pi^2}{2}\epsilon\rho_0^4. \tag{2.188}$$

The remaining contribution to S_E comes from the region within the wall. Noting that (2.178) implies that

$$\frac{d\phi}{d\rho} = \sqrt{2V_0} \tag{2.189}$$

this contribution to S_E may be written as

$$S_E \simeq 2\pi^2 \rho_0^3 \int d\rho \, 2V_0(\phi) = 2\pi^2 \rho_0^3 \int_{\phi_-}^{\phi_+} d\phi \, \sqrt{2V_0(\phi)}. \tag{2.190}$$

Thus, the total value of S_E in the thin-wall approximation is

$$S_E = -\frac{\pi^2}{2}\epsilon\rho_0^4 + 2\pi^2\rho_0^3 I \tag{2.191}$$

where

$$I \equiv \int_{\phi_-}^{\phi_+} d\phi \, \sqrt{2V_0(\phi)}. \tag{2.192}$$

Minimizing S_E with respect to ρ_0, to find the bounce solution that dominates the tunnelling, gives

$$\rho_0 \simeq \frac{3I}{\epsilon}. \tag{2.193}$$

Thus, when ϵ is small, ρ_0 is large compared with μ^{-1}, which justifies an earlier assumption. It also follows that the small energy difference between the two phases does indeed correspond to a bubble of true vacuum with a thin wall. Moreover, the neglect of $\frac{1}{\rho}\frac{d\phi}{d\rho}$ is also justified because outside or inside the bubble $\frac{d\phi}{d\rho}$ is negligible because ϕ is slowly varying, and within the bubble wall $\frac{1}{\rho}\frac{d\phi}{d\rho} \simeq \frac{1}{\rho_0}\frac{d\phi}{d\rho}$ is negligible because ρ_0 is large. With ρ_0 given by (2.193), the minimum value of S_E deriving from (2.191) is

$$S_E = \frac{27\pi^2 I^4}{2\epsilon^3} \tag{2.194}$$

which provides the value of B for the bounce solution that dominates the tunnelling rate (2.167). With $\phi_\pm = \pm\mu/\sqrt{\lambda}$, and V_0 given by (2.179), it is straightforward to evaluate I to obtain

$$I = \frac{2\mu^3}{3\lambda}. \tag{2.195}$$

Then the tunnelling rate is given by (2.167) with

$$B = \frac{8\pi^2\mu^{12}}{\epsilon^3\lambda^4}. \tag{2.196}$$

Once the bubble of the true vacuum has materialized, it can be shown [22] that (in the thin-wall approximation) it materialises with radius $\rho = \rho_0$ and that the development in time of the bubble can be obtained by continuing in time from Euclidean time to real time. Thus, whereas in Euclidean time the surface of the bubble was at

$$\rho = \sqrt{t_E^2 + x^2} = \rho_0 \tag{2.197}$$

in real time the surface of the bubble is at

$$\sqrt{x^2 - c^2 t^2} = \rho_0 \tag{2.198}$$

(restoring the explicit speed of light c which we have been setting to 1). The quantity ρ_0 is, in general, on a sub-microscopic scale and so negligible. Thus, to a good approximation, the surface of the bubble is at $x^2 = c^2 t^2$. Consequently, the radius of the bubble grows with the speed of light.

2.10 Exercises

1. Recast the effective potential of (2.62) in the form (2.77).
2. Show that the zero-temperature effective potential deriving from (2.94) always has a minimum at $\phi_c = 0$ when $B > 0$ and $\alpha > 0$. Also, show that when $B < 0$ there can only be a minimum at the origin when $\alpha < 0$.
3. Check that the $SU(5)$ symmetric, $SU(4) \times U(1)$ symmetric and $SU(3) \times SU(2) \times U(1)$ symmetric minima of (2.123), (2.124) and (2.125) all have $V = 0$.
4. Derive the Euclidean action (2.173) and the equation of motion (2.174) when ϕ is a function of the four-dimensional radial variable ρ alone.
5. Check that the bounce solution satisfies (2.178) and that, for V_0 given by (2.179), this leads to the explicit solution (2.180).

2.11 General references

The books and review articles that we have found most useful in preparing this chapter are:

- Kolb E W and Turner M S 1990 *The Early Universe* (Reading, MA: Addison-Wesley)
- Olive K A 1990 *Phys. Rep.* **190** 307
- Linde A D 1979 *Rep. Prog. Phys.* **42** 389
- Bailin D and Love A 1993 *Introduction to Gauge Field Theory* (Bristol: IOP)
- Bailin D and Love A 1994 *Supersymmetric Gauge Field Theory and String Theory* (Bristol: IOP)

Bibliography

[1] Bailin D and Love A 1993 *Introduction to Gauge Field Theory* (Bristol: IOP) ch 17

[2] Kirzhnits D A and Linde A D 1972 *Phys. Lett.* B **42** 477
[3] Weinberg S 1974 *Phys. Rev.* D **9** 3320
[4] Dolan L and Jackiw R 1974 *Phys. Rev.* D **9** 3357
[5] Bernard C W 1974 *Phys. Rev.* D **9** 3312
[6] Coleman S and Weinberg E 1973 *Phys. Rev.* D **7** 1888
[7] Nanopoulos D V and Tamvakis K 1982 *Phys. Lett.* B **110** 449
[8] Srednicki M 1982 *Nucl. Phys.* B **202** 327
[9] Ellis J, Llewellwyn Smith C H and Ross G G 1982 *Phys. Lett.* B **114** 227
[10] Nanopoulos D V, Olive K A and Tamvakis K 1982 **115** 15
[11] Srednicki M 1982 *Nucl. Phys.* B **206** 132
[12] Bailin D and Love A 1984 *Nucl. Phys.* B **239** 277
[13] Cremmer E, Ferrara S, Girardello L and van Proeyen A 1983 *Nucl. Phys.* B **212** 413
[14] Bailin D and Love A 1994 *Supersymmetric Gauge Field Theory and String Theory* (Bristol: IOP)
[15] Gelmini G B, Nanopoulos D V and Olive K A 1983 *Phys. Lett.* B **131** 52
[16] Olive K A and Srednicki M A 1984 *Phys. Lett.* B **148** 437
[17] Enqvist K *et al* 1985 *Phys. Lett.* B **152** 181
[18] Binetruy P and Gaillard M K 1985 *Phys. Rev.* D **32** 931
[19] Coleman S 1977 *Phys. Rev.* D **15** 2929
[20] Callan C and Coleman S 1977 *Phys. Rev.* D **16** 1762
[21] Linde A D 1983 *Nucl. Phys.* B **216** 421
[22] Coleman S and de Luccia F 1980 *Phys. Rev.* D **21** 3305

Chapter 3

Topological defects

3.1 Introduction

When a phase transition occurs in the early universe, the alignment of the spontaneous symmetry breaking expectation value may be different in adjacent causal domains. In that case, topologically stable objects such as domain walls, cosmic strings and magnetic monopoles, referred to as 'topological defects', can be formed [1]. Similar topological defects are familiar in condensed state physics. For example, a ferromagnet is, in general, divided into domains where the spontaneous magnetization is aligned in a definite direction. At the boundary between two such domains, the direction of the magnetization evolves continuously between its direction in one domain and its direction in the adjacent domain to form a so-called 'domain wall'. Another example of a topological defect is provided by a magnetic flux line in a type II superconductor. In this case, the phase of a complex scalar field (associated with the Cooper-paired electron condensate) changes by $2\pi n$, where n is an integer, in going around a closed loop surrounding the flux line and the flux line carries n units of a quantum of magnetic flux. In addition to these two-dimensional and one-dimensional topological defects, there are also point defects. These are most familiar in a particle physics context as magnetic monopole solutions though analogous objects occur in superfluid ^{3}He. In the context of particle physics and early cosmology, the corresponding topological defects are associated with the vacuum expectation values (VEVs) in electroweak or grand unified theories or with the various moduli that occur in superstring theories.

Once such topological defects have been formed at a phase transition in the early universe, they can manifest themselves in various ways. The simplest manifestations are as a potentially substantial contribution to the energy density of the universe or, in the case of magnetic monopoles, as a relic density of particles carrying magnetic charge. More subtle manifestations are also possible. For example, relic cosmic strings may act as gravitational lenses or produce temperature fluctuations in the cosmic microwave background radiation.

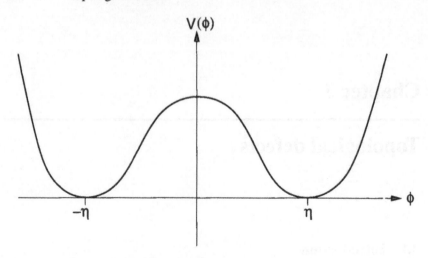

Figure 3.1. *Effective potential for ϕ^4 theory.*

In the following sections, we shall develop the theory of domain walls, cosmic strings and magnetic monopoles in turn.

3.2 Domain walls

When the effective potential in a field theory has two degenerate minima with $V = 0$, two-dimensional solutions of the field equations with finite energy per unit area can occur. The simplest examples of such domain-wall solutions [2] are obtained from the Lagrangian density

$$\mathcal{L} = \tfrac{1}{2}\partial_\mu\phi\partial^\mu\phi - V(\phi) \tag{3.1}$$

with

$$V(\phi) = \frac{\lambda}{4}(\phi^2 - \eta^2)^2 \tag{3.2}$$

where ϕ is a real scalar field and λ and η are real constants. The potential V has minima with $V = 0$ at $\phi = \pm\eta$ and a maximum at $\phi = 0$ (see figure 3.1). The idea is to construct a static solution for which ϕ evolves from one minimum for $z \rightarrow -\infty$ to the other minimum for $z \rightarrow +\infty$. In that case, the domain wall is in the x–y plane. Clearly, we can construct such solutions with the domain wall in any chosen plane.

In general, for a static solution where ϕ depends only on z, the field equation is

$$\frac{\mathrm{d}^2\phi}{\mathrm{d}z^2} = V'(\phi). \tag{3.3}$$

As for the (approximate) bounce in (2.177), the solution satisfies

$$\frac{1}{2}\left(\frac{d\phi}{dz}\right)^2 = V + c \tag{3.4}$$

where c is a constant. The energy (per unit area) of the domain wall is given by

$$E = \int_{-\infty}^{\infty} \left[\frac{1}{2}\left(\frac{d\phi}{dz}\right)^2 + V(\phi)\right] dz \tag{3.5}$$

and so to obtain a solution with finite energy density, it is necessary to require that $d\phi/dz \to 0$ as $z \to \pm\infty$. ($V(\phi)$ will already approach zero as $z \to \pm\infty$ if we succeed in constructing a solution of the type we are looking for.) Thus, we must take $c = 0$. Then integrating (3.4) gives

$$z = \pm \int \frac{d\phi}{\sqrt{2V}} \tag{3.6}$$

(analogously to (2.177)). For the choice (3.2), this gives

$$z - z_0 = \mp \frac{\sqrt{2}}{\eta\sqrt{\lambda}} \text{arctanh}\left(\frac{\phi}{\eta}\right) \tag{3.7}$$

where z_0 is an integration constant. Different choices of this constant amount to moving the centre of the domain wall along the z-axis. Inverting (3.7) gives

$$\phi = \phi_{\mp} \equiv \mp\eta \tanh\left[\frac{\eta\sqrt{\lambda}}{\sqrt{2}}(z - z_0)\right]. \tag{3.8}$$

As $z \to \infty$, $\phi_{\mp} \to \mp\eta$ and, as $z \to -\infty$, $\phi_{\mp} \to \pm\eta$. The two solutions $\phi = \phi_+$ and $\phi = \phi_-$ are referred to, respectively, as the 'kink' and the 'antikink'. The kink evolves from the minimum at $\phi = -\eta$ for $z \to -\infty$ to the minimum at $\phi = +\eta$ for $z \to +\infty$ (see figure 3.2) and the antikink evolves conversely (see figure 3.3). Both domain walls have their centre at $z = z_0$ in the sense that $\phi = 0$ when $z = z_0$. The 'thickness' of each domain wall is of order $\sqrt{(2/\lambda)}\eta^{-1}$. This is a balance between the desire of the potential energy to make the wall as thin as possible and the desire of the gradient energy to make the wall as thick as possible. The total energy per unit cross-sectional area of a kink or antikink is finite because in (3.5) $V(\phi)$ and $d\phi/dz$ go to zero sufficiently fast as $z \to \pm\infty$. Substituting the explicit solutions (3.8) into (3.5) gives the finite energy per unit area of a kink or antikink (exercise 1):

$$E = \frac{2}{3}\sqrt{(2)}\lambda\eta^3. \tag{3.9}$$

The stability of a domain wall is associated with a topological principle. The Lagrangian (3.1) possesses a discrete Z_2 symmetry

$$Z_2 : \phi \to -\phi \tag{3.10}$$

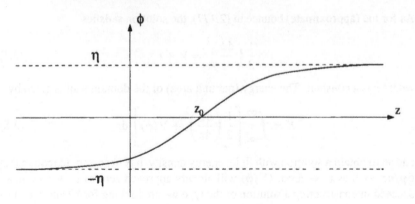

Figure 3.2. Kink soliton solution.

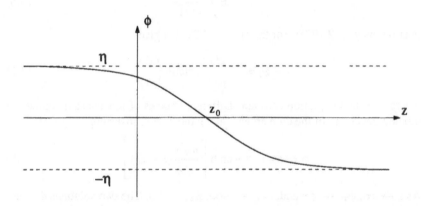

Figure 3.3. Antikink soliton solution.

which is broken by the vacuum expectation value of ϕ. The domain walls connect the two distinct minima which constitute the complete space of minima. There is no way to continuously deform a domain wall to a new object that does not have its ends at distinct minima, while keeping the energy density finite.

The formation of such topological defects can be understood in terms of the Kibble mechanism [1], which we now describe. As discussed in chapter 2, symmetries are expected to be restored at high temperatures. As the universe cools, it passes through a phase transition and different regions of the universe undergo phase transitions to different minima of the zero-temperature effective potential. There will be some correlation length ξ such that the VEVs for points of space separated by more than about ξ are uncorrelated. The length ξ cannot be larger than the particle horizon since no influence can have propagated over a distance greater than this from the big bang to the time of the phase transition.

In the present case, at high temperatures we expect $\langle\phi\rangle$ to be zero and, after the phase transition, some regions will have $\langle\phi\rangle = \eta$ and some will have $\langle\phi\rangle = -\eta$. In this way topological defects can form. Although a domain wall can change its shape, topology prevents it from disappearing once formed because its ends are at discrete minima of V. In general, there will be curved domain walls as well as flat domain walls, with the curved walls enclosing a region of space inside which the VEV of ϕ differs from the VEV of ϕ outside.

Many domain walls will form constituting a random network whose evolution with time may be studied. The result for non-relativistic domain walls is that the energy density (i.e. the energy per unit volume) of the domain walls scales as R^{-1}, where $R(t)$ is the scale factor of the universe. This should be compared with the energy density due to radiation which scales as R^{-4} and that due to matter which scales as R^{-3} (as in section 1.3). Consequently, as R increases with time, the energy density of the universe comes to be dominated by domain walls. The total energy associated with a plane domain wall with area H_0^{-2}, where H_0 is the present day Hubble constant, is far larger than the estimated total energy due to matter within the Hubble radius. For example, for λ not too much different from 1 and $\eta \sim 100$ GeV, the former is larger by 12 orders of magnitude. A larger expectation value for the scalar field makes things worse (exercise 2). Thus, domain walls appear undesirable. This suggests that a theory is needed which does not have disconnected vacuum states, such as $\langle\phi\rangle = \pm\eta$ in the present model, to avoid the existence of domain walls. Alternatively, a period of inflation (see chapter 7) is needed to dilute the domain wall density.

3.3   Global cosmic strings

One-dimensional topological defects (cosmic strings) can also be produced by phase transitions in the early universe. The simplest example of a cosmic string [3] may be derived from the Lagrangian density for a complex scalar field ϕ:

$$\mathcal{L} = \partial_\mu\phi^*\partial^\mu\phi - V(\phi) \tag{3.11}$$

with

$$V(\phi) = \frac{\lambda}{2}(\phi^*\phi - \eta^2)^2 \tag{3.12}$$

and λ and η are real constants. This Lagrangian possesses a global $U(1)$ symmetry under

$$\phi \to e^{i\alpha}\phi \tag{3.13}$$

where α is an arbitrary constant real number. The potential V of (3.12) has a maximum at $\phi = 0$ and minima with $V = 0$ when

$$\phi = \eta e^{i\beta} \tag{3.14}$$

where β is an arbitrary real number. The vacuum VEV (3.14) breaks the global $U(1)$ symmetry because it is not invariant under the transformation (3.13).

It is possible to construct other extended static solutions as follows. Take cylindrical polar coordinates (ρ, θ, z). The solutions in question have the form

$$\phi = \eta e^{in\theta} f(\rho) \qquad (3.15)$$

for some integer n and the function $f(\rho)$ is to be determined from the field equations. In cylindrical polar coordinates, these are (exercise 3)

$$\frac{d^2 f}{d\xi^2} + \frac{1}{\xi}\frac{df}{d\xi} - \frac{n^2}{\xi^2}f = f(f^2 - 1) \qquad (3.16)$$

where

$$\xi \equiv \lambda^{1/2}\eta\rho. \qquad (3.17)$$

The phase of ϕ will become undefined at $\rho = 0$ unless $|\phi| \to 0$ as $\rho \to 0$. Thus, the boundary condition

$$f(\rho) \to 0 \qquad \text{as } \rho \to 0 \qquad (3.18)$$

is required to ensure a single-valued field ϕ. Also, there is the boundary condition

$$f(\rho) \to 1 \qquad \text{as } \rho \to \infty \qquad (3.19)$$

so that ϕ approaches one of its continuum of minima (3.14) in order to minimize the energy. Equation (3.16) may be solved numerically with these boundary conditions. The scale of the distance is set by $\lambda^{1/2}\eta$ so the vortex line or cosmic string has a core of radius of order $\lambda^{-1/2}\eta^{-1}$ outside of which ϕ approaches its minima as $\rho \to \infty$ and inside of which $\phi \to 0$ as $\rho \to 0$.

The energy E of the vortex line or cosmic string is given by

$$E = \int d^3x \left[\nabla\phi^* \cdot \nabla\phi + V(\phi)\right]. \qquad (3.20)$$

Taking cylindrical polar coordinates, the energy per unit length (along the z-direction) of a cosmic string of length l is

$$\frac{E}{l} = \int_0^\infty \rho d\rho \int_0^{2\pi} d\theta \left(\frac{\partial\phi^*}{\partial\rho}\frac{\partial\phi}{\partial\rho} + \frac{1}{\rho^2}\frac{\partial\phi^*}{\partial\theta}\frac{\partial\phi}{\partial\theta}\right) \qquad (3.21)$$

because (3.14) is independent of z. The last term in (3.21) gives a contribution to E/l proportional to $\int_0^\infty f^2\rho^{-1} d\rho$ and, because $f \to 1$ as $\rho \to \infty$, this contribution is logarithmically divergent. Considering the energy inside a cylinder of radius R and recalling that $\lambda^{-1/2}\eta^{-1}$ sets the length scale, we must get

$$\frac{E}{l} \sim \ln(\lambda^{1/2}\eta R). \qquad (3.22)$$

This global cosmic string resembles the vortex line in superfluid ^{4}He where ϕ is the condensate wavefunction.

Like a domain wall, a (global) cosmic string is stabilized by topological considerations. A cosmic string with asymptotic behaviour $e^{in\theta}$ as $\rho \to \infty$ is said to have winding number n. The space of vacuum states (minima of V) is characterized by $e^{i\beta}$ (as in (3.14)) and so is just a circle S^1. If (for fixed z) we draw a circular path of large radius in real space encircling the core of the cosmic string (where $\phi = 0$), then, as we go once around this path in real space, the field ϕ goes n times around the circle S^1 which is the space of vacuum states. Provided that the cosmic string is either of infinite length or forms a closed loop, this is a property of a cosmic string that cannot be changed by continuous deformations. It is a topological quantum number which, at least at the classical level, guarantees the continued existence of a vortex line once formed, unless it encounters other vortex lines or divides into more vortex lines in such a way that n is conserved (e.g. into n vortex lines with unit winding number.)

If we denote the space of true vacua (minima of V) by $\mathcal{M}$, then the topological entity involved is the homotopy group $\pi_1(\mathcal{M})$. In the present case, the relevant homotopy group is $\pi_1(S^1)$ which is known to be Z, i.e. isomorphic to the integers. The winding number $n \in Z$ expresses this fact.

3.4   Local cosmic strings

If a complex scalar field is coupled to a gauge field, e.g. the electromagnetic field, then the Lagrangian possesses a local symmetry, rather than a global symmetry as in section 3.3, and a so-called 'local cosmic string' [4] or gauge string can occur as a solution of the field equations. The simplest example is provided by the Higgs model, which is the theory of a complex scalar field coupled to a $U(1)$ gauge field which we may take to be the electromagnetic field. The Lagrangian is that of section 3.3 amended to incorporate the gauge coupling. Thus,

$$\mathcal{L} = (D_\mu \phi)^*(D^\mu \phi) - \tfrac{1}{4}F_{\mu\nu}F^{\mu\nu} - V(\phi) \tag{3.23}$$

with $V(\phi)$ given by (3.12) and

$$D_\mu \phi \equiv (\partial_\mu + ieA_\mu)\phi \tag{3.24}$$

$$F_{\mu\nu} = \partial_\mu A_\nu - \partial_\nu A_\mu \tag{3.25}$$

where A_μ is the electromagnetic four-potential and e is the charge of the scalar field. This is the same model as that studied in section 2.4, after adding a constant to V and with a different definition of λ.

As in the discussion of global cosmic strings, V has a minimum at $\phi = \eta e^{i\beta}$. However, the Lagrangian now possesses a local $U(1)$ symmetry under

$$\phi \to e^{i\Lambda(x)}\phi \tag{3.26}$$

$$A_\mu \to A_\mu - \frac{1}{e}\partial_\mu \Lambda(x) \tag{3.27}$$

which is broken by the VEV of ϕ. Using cylindrical polar coordinates, we look for solutions for ϕ of the form (3.15), as before, but we must now also determine A_μ. For large ρ, with the boundary condition (3.19) so that ϕ approaches one of its minima,

$$\phi \sim \eta e^{in\theta} \quad \text{as } \rho \to \infty. \tag{3.28}$$

If we also arrange that A_μ has the boundary condition

$$A_\mu \sim -ie^{-1}\partial_\mu \ln(\phi/\eta) \qquad \text{as } \rho \to \infty \tag{3.29}$$

then $D_\mu\phi$ and $F_{\mu\nu}$ both approach zero for large values of ρ and the energy density vanishes for large ρ. Indeed, when the complete solution with these boundary conditions is constructed numerically, in a way that we shall discuss shortly, the energy density approaches zero fast enough as $\rho \to \infty$ that this cosmic string has finite energy density per unit length.

The local cosmic string or gauge string carries magnetic flux. The amount of flux may be determined by integrating over the area of a circle of large radius R in the (ρ, θ) plane with the asymptotic form (3.29). Then (exercise 5)

$$\oint B \cdot dS = \oint A \cdot dl = 2\pi n e^{-1}. \tag{3.30}$$

Thus, the local cosmic string characterized by winding number n carries n units of magnetic flux $2\pi e^{-1}$. Local cosmic strings are, therefore, quantized tubes of magnetic flux analogous to flux lines in a superconductor.

To construct the required (static) solution for ϕ and A_μ, we take

$$\phi = \eta e^{in\theta} f(\rho) \tag{3.31}$$

as for the global cosmic string. Then, since

$$\nabla\phi = \frac{\partial\phi}{\partial\rho}\hat{\rho} + \frac{1}{\rho}\frac{\partial\phi}{\partial\theta}\hat{\theta} + \frac{\partial\phi}{\partial z}\hat{k} \tag{3.32}$$

the boundary condition (3.29) suggests that we should take A_μ to have non-zero components only in the $\hat{\rho}$ and $\hat{\theta}$ directions. Working in a gauge in which the component $A_\rho = 0$, we take

$$A = \frac{n}{e\rho}a(\rho)\hat{\theta}. \tag{3.33}$$

The functions $a(\rho)$ and $f(\rho)$ are then determined numerically by solving the field equations

$$D_\mu D^\mu\phi + \lambda(\phi^*\phi - \eta^2)\phi = 0 \tag{3.34}$$

$$\partial_\nu F^{\mu\nu} + ie(\phi^* D^\mu\phi - \phi(D^\mu\phi)^*) = 0 \tag{3.35}$$

subject to the boundary conditions (3.28) and (3.29).

There are now two length scales in the problem instead of one as in the case of the global string. There is the mass m_ϕ of the scalar field after spontaneous spontaneous symmetry breaking which is obtained by substituting $\langle \phi \rangle$ in $V(\phi)$. This gives

$$m_\phi^2 = \eta^2 \lambda. \tag{3.36}$$

It is m_ϕ that controls the rate of variation of ϕ for large ρ and, therefore, the variation of $f(\rho)$. There is also the mass m_A of the gauge field after spontaneous symmetry breaking which is obtained by substituting $\langle \phi \rangle$ in the $(D_\mu \phi)^*(D^\mu \phi)$ term. This gives

$$m_A^2 = e^2 \eta^2. \tag{3.37}$$

It is m_A that controls the rate of variation of A for large ρ and, therefore, variation of $a(\rho)$. The approximate solution is found to be of the form

$$f \sim 1 - f_1 \xi^{-1/2} \exp\left(-\frac{m_\phi \xi}{m_A}\right) \qquad \text{as } \xi \to \infty \tag{3.38}$$

$$a \sim 1 - a_1 \xi^{1/2} \exp(-\xi) \qquad \text{as } \xi \to \infty \tag{3.39}$$

where f_1 and a_1 are constants and

$$\xi \equiv m_A \rho. \tag{3.40}$$

It can be seen from (3.38) and (3.39) that ϕ is localized on a scale m_ϕ^{-1} and A is localized on a scale m_A^{-1}. As a consequence, the energy density is localized without introducing a cut-off.

The energy per unit length of a local cosmic string may be estimated as follows. The cosmic string has an inner core where ϕ is approximately zero (i.e. a core of false vacuum) with radius

$$R_\phi \simeq m_\phi^{-1} = \lambda^{-1/2} \eta^{-1} \tag{3.41}$$

and a tube of magnetic flux of radius

$$R_A \simeq m_A^{-1} = e^{-1} \eta^{-1}. \tag{3.42}$$

The energy density obtained by putting $\phi = 0$ in $V(\phi)$ is $\frac{1}{2} \lambda \eta^4$. Thus, there is an energy density per unit length from the inner core of the cosmic string of order $\frac{1}{2} \lambda \eta^4 \pi R_\phi^2 \sim \eta^2$. There is also an energy per unit length from the tube of magnetic flux of order $B^2 R_A^2$ where B is the magnitude of the magnetic field strength. From (3.30), with one unit of flux, we estimate

$$B \sim R_A^{-2} e^{-1}. \tag{3.43}$$

Thus, using (3.42), we find that $B^2 R_A^2 \sim \eta^2$ and both the magnetic and inner-core contributions to the energy per unit length μ are of the same order. The total has order of magnitude

$$\mu \sim \eta^2. \tag{3.44}$$

3.5   Gravitational fields of local cosmic strings

To calculate the gravitational field due to a cosmic string, the energy–momentum tensor produced by the string is required. (See section 5 of the review by Vilenkin in the general references at the end of this chapter.) For cosmological purposes, we are interested in cosmic strings of length much greater than the radius of the inner core or the flux tube. We therefore average the energy–momentum tensor over the core of the cosmic string and treat the string as having zero radius. Thus, for a long straight line with axis along the z-direction, we replace the energy-momentum tensor $T_{\mu\nu}$ by $\tilde{T}_{\mu\nu}$ where

$$\tilde{T}_{\mu\nu} = \delta(x)\delta(y) \int_{\text{core}} T_{\mu\nu} \, dx \, dy. \tag{3.45}$$

Invariance under Lorentz boosts along the z-direction shows that $\tilde{T}_{00} = \tilde{T}_{33}$ and there are no off-diagonal components. The conservation law for the energy-momentum tensor

$$D^\nu T_{\mu\nu} = 0 \tag{3.46}$$

must also be imposed; D^ν is the gravitational covariant derivative. Then, by considering $\int D^j T_{ij} x^k \, dx \, dy$ and integrating by parts, we conclude that

$$\tilde{T}_{ik} = 0 \qquad \text{for } i, k = 1, 2. \tag{3.47}$$

Also, since the total energy per unit length is μ, we can now write

$$\tilde{T}_{\mu\nu} = \mu \, \text{diag}(1, 0, 0, 1)\delta(x)\delta(y) \tag{3.48}$$

so that $\int T_{00} \, dx \, dy = \mu$.

With the energy-momentum tensor (3.48) for the local cosmic string, Einstein's field equations can be solved in the limit $G_N \mu \ll 1$ for the metric in the region outside an infinitely long straight string [6]. In cylindrical polar coordinates (ρ, θ, z), the result for the proper-time element $d\tau$ is

$$d\tau^2 = dt^2 - dz^2 - d\rho^2 - (1 - 4G_N\mu)^2 \rho^2 \, d\theta^2. \tag{3.49}$$

This can be recast as the metric of flat Minkowski space by the transformation

$$\hat{\theta} = (1 - 4G_N\mu)\theta. \tag{3.50}$$

However, for $0 \le \theta < 2\pi$, we have

$$0 \le \hat{\theta} < 2\pi(1 - 4G_N\mu) \tag{3.51}$$

which limits the range of $\hat{\theta}$. This is called a 'conical singularity'. Space with a conical singularity is the same as flat space $0 \le \hat{\theta} < 2\pi$ with the angular region

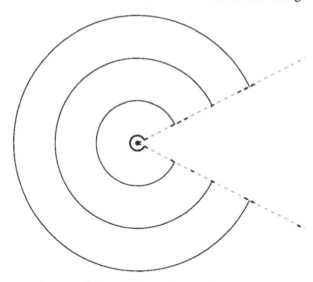

Figure 3.4. Conical singularity on the space outside a local cosmic string. ⊙ is the axis of the string.

between $2\pi(1-4G_N\mu)$ and 2π removed. In addition, because $\theta = 0$ and $\theta = 2\pi$ are to be identified, we must identify $\hat{\theta} = 0$ and $\hat{\theta} = 2\pi - \Delta\theta$ where

$$\Delta\theta \equiv 8\pi G_N\mu. \qquad (3.52)$$

Then, looking along the z-axis, space looks as in figure 3.4 with the hatched area removed and points at the same value of ρ on the dotted lines identified. The conical singularity in the space outside a long straight local cosmic string has several striking consequences:

3.5.1 Double images [5]

A galaxy located behind a local cosmic string, from the perspective of the observer, will acquire a double image. Consider, for simplicity, an infinitely long straight cosmic string normal to the plane of the page in figure 3.5 with the observer ω and the galaxy g being observed in the plane of the page. Because points A and B are identified, as discussed earlier, two images of the galaxy are seen, emanating from A and B. The angle between the two images $\Delta\alpha$ is given by

$$d_1\Delta\alpha = d_2\Delta\theta \qquad (3.53)$$

where d_1 and d_2 are, respectively, the distances from the observer to the galaxy and from the cosmic string to the galaxy. As a consequence of (3.52),

$$\Delta\alpha = \frac{d_2}{d_1}8\pi G_N\mu. \qquad (3.54)$$

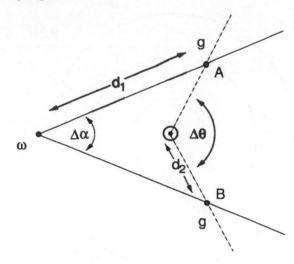

Figure 3.5. Double image of a galaxy behind a cosmic string. ⊙ is the axis of the string, ω is the observer and g is the galaxy.

3.5.2 Temperature discontinuities [6]

Consider a local cosmic string moving perpendicularly to the line of sight of an observer observing the cosmic microwave background radiation coming from far off ($d_1 \simeq d_2$ in the previous discussion). There are two images P_1 and P_2 of the same point separated (see figure 3.6) by an angle $\Delta\alpha \simeq 8\pi G_N\mu$. If the relative velocity of the cosmic string and observer is v, then P_1 and P_2 have a component of velocity of order $\Delta\alpha$ antiparallel or parallel to v respectively. As a consequence, there is a Doppler shift in the temperature of the radiation between the two points. This results in a discontinuity $\delta T/T$ from one side of the cosmic string to the other of order $8\pi G_N\mu|v|$.

3.5.3 Cosmic string wakes [7]

A long straight cosmic string moving with velocity v across the universe will deflect particles of matter. A wedge of matter with opening angle $8\pi G_N\mu$ and radius vt forms as a wake in time t. This may be relevant to structure formation.

3.6 Dynamics of local cosmic strings

Once a network of cosmic strings has formed, following a phase transition in the early universe, its evolution depends on the emission of gravitational radiation by string loops. In principle, the classical field equations (3.34) and (3.35) provide the equations of motion for a local cosmic string. In practice, when we are

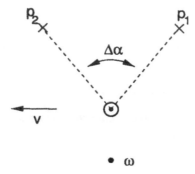

Figure 3.6. Temperature discontinuity due to a cosmic string. ⊙ is the axis of the string, ω is the observer and P_1 and P_2 are two images of the same point.

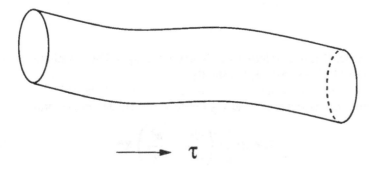

Figure 3.7. World sheet for a cosmic string.

neglecting the radius of the string core, it is much simpler to use the equations of motion for a relativistic string of zero radius which derive from the Nambu–Goto action and are identical to the equations of motion for the fundamental bosonic string.

We denote by X^μ the position in spacetime of points on the axis of the cosmic string (where the Higgs field ϕ is zero.) Whereas a point particle may be described by degrees of freedom $X^\mu(\tau)$ depending only on a timelike coordinate τ, to describe a string we need, in addition, a spacelike coordinate σ which we may take, for convenience, to be in the range $0 \leq \sigma \leq \pi$. Then, the string degrees of freedom $X^\mu(\tau, \sigma)$ trace out a curve (see figure 3.7) as σ varies at fixed τ. The action S for a relativistic string propagating in Minkowski spacetime is of the form

$$S = -\frac{T}{2} \int_{\tau_i}^{\tau_f} d\tau \int_0^\pi d\sigma \, (-\det h)^{1/2} h^{\alpha\beta} \eta_{\mu\nu} \partial_\alpha X^\mu \partial_\beta X^\nu \tag{3.55}$$

where T is the string tension,

$$\eta_{\mu\nu} = \text{diag}(1, -1, \ldots, -1) \tag{3.56}$$

and $h_{\alpha\beta}(\tau, \sigma)$ is a world-sheet metric of signature $(+, -)$ where $\alpha = 0$ and 1 refer to τ and σ respectively. This action displays two-dimensional world-sheet reparametrization invariance and also possesses conformal invariance under a local rescaling of the world-sheet metric

$$\delta h_{\alpha\beta} = \Lambda(\tau, \sigma) h_{\alpha\beta} \qquad \delta X^{\mu} = 0. \tag{3.57}$$

With the aid of world-sheet reparametrization invariance, the world-sheet metric may be reduced to the form

$$h_{\alpha\beta}(\tau, \sigma) = e^{\gamma(\tau, \sigma)} \eta_{\alpha\beta} \tag{3.58}$$

where

$$\eta_{\alpha\beta} = \mathrm{diag}(1, -1). \tag{3.59}$$

With the aid of conformal invariance, it may be further reduced to

$$h_{\alpha\beta} = \eta_{\alpha\beta}. \tag{3.60}$$

The gauges (3.60) are referred to as 'covariant' gauges. There is still further gauge freedom which we shall exploit shortly.

In a covariant gauge, the equations of motion of the string, obtained by varying with respect to X^{μ} and $h_{\alpha\beta}$ (exercise 6), take the simple form

$$\partial_{\alpha}\partial^{\alpha} X^{\mu} = \left(\frac{\partial^2}{\partial\tau^2} - \frac{\partial^2}{\partial\sigma^2}\right) X^{\mu} = 0 \tag{3.61}$$

with the constraints

$$\frac{\partial X^{\mu}}{\partial\tau}\frac{\partial X_{\mu}}{\partial\tau} + \frac{\partial X^{\mu}}{\partial\sigma}\frac{\partial X_{\mu}}{\partial\sigma} = 0 \tag{3.62}$$

$$\frac{\partial X^{\mu}}{\partial\tau}\frac{\partial X_{\mu}}{\partial\sigma} = 0. \tag{3.63}$$

For a closed string loop, there is the boundary condition

$$X^{\mu}(\tau, \sigma + \pi) = X^{\mu}(\tau, \sigma). \tag{3.64}$$

The remaining gauge degrees of freedom may be used to choose the 'temporal' gauge in which τ is identified with $X^0 \equiv t$ (Minkowski time.) Then, the equations of motion and constraints become

$$\left(\frac{\partial^2}{\partial t^2} - \frac{\partial^2}{\partial\sigma^2}\right) X = 0 \tag{3.65}$$

$$\frac{\partial X}{\partial t} \cdot \frac{\partial X}{\partial\sigma} = 0 \tag{3.66}$$

$$\left(\frac{\partial X}{\partial t}\right)^2 + \left(\frac{\partial X}{\partial\sigma}\right)^2 = 1. \tag{3.67}$$

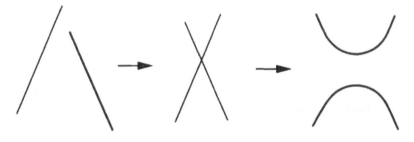

Figure 3.8. Intercommuting cosmic strings.

Figure 3.9. Cosmic string loop intercommuting with itself.

These equations have oscillatory solutions and these will allow string loops to radiate. Because the string loops are relativistic, the quadrupole formula for gravitational radiation cannot be used. A relativistic calculation shows that the power P emitted in gravitational radiation by a string loop is given by

$$P = \gamma G_N \mu^2 \qquad (3.68)$$

where G_N is Newton's constant and γ is a number of order 100 which depends on the particular loop [8]. As in section 3.5, μ is the energy per unit length of the cosmic string.

This gravitational radiation is important for the development in time of the network of cosmic strings that formed at a phase transition. Also important is the process of intercommuting [1,9] as in figure 3.8. In particular, a string loop may intercommute with itself as in figure 3.9 to produce two smaller loops. When the evolution of a string network is studied [10], allowing for these two effects, it is found that strings will not dominate the present day energy density of the universe. However, apart from individual relic strings producing the observable effects discussed earlier, the evolution of the string network will leave a relic gravitational wave background as a result of gravitational radiation emission by oscillating string loops. Since the gravitational emission is controlled by $G_N\mu^2$, this will set a limit on μ if this gravitational background is not to undo the predictions of the standard model for nucleosynthesis. This is found to require [11] $G_N\mu \lesssim 10^{-5}$. There is, however, a tighter bound [12] set by the magnitude of the cosmic microwave background fluctuation of $G_N\mu \lesssim 10^{-6}$. It is possible that particle production rather than gravitational wave emission dominates the energy loss from oscillating cosmic string loops. In that case [13], there is an even tighter bound $G_N\mu \lesssim 10^{-9}$.

3.7 Magnetic monopoles

It is also possible for point topological defects, magnetic monopoles [14], to form at phase transitions in the early universe. The simplest model exhibiting this is an $SO(3)$ gauge field theory with $SO(3)$ spontaneously broken to $U(1)$ by the expectation value of a scalar field ϕ in the three-dimensional representation of $SO(3)$. The Lagrangian density for this model is

$$\mathcal{L} = D_\mu \phi \cdot D^\mu \phi - \frac{1}{4} F^a_{\mu\nu} F^{\mu\nu}_a - \frac{\lambda}{8} (\phi \cdot \phi - \eta^2)^2 \qquad (3.69)$$

where the gauge field strength is

$$F^a_{\mu\nu} = \partial_\mu A^a_\nu - \partial_\nu A^a_\mu - g \epsilon_{abc} A^b_\mu A^c_\nu. \qquad (3.70)$$

The covariant derivative of the scalar field is

$$D_\mu \phi_a = \partial_\mu \phi_a - g \epsilon_{abc} A^b_\mu \phi_c \qquad (3.71)$$

and a, b, c take the values 1, 2, 3. Minimization of the effective potential

$$V = \frac{\lambda}{8} (\phi \cdot \phi - \eta^2)^2 \qquad (3.72)$$

fixes

$$|\phi| = \eta. \qquad (3.73)$$

However, because of the $SO(3)$ symmetry, the direction of ϕ is not fixed.

The magnetic monopole solution [16] is a spherically symmetric solution for ϕ of the form

$$\phi = \eta f(r) \hat{r} \qquad (3.74)$$

which is a mapping from ordinary space to the $SO(3)$ space, with the asymptotic behaviour

$$\phi \sim \eta \hat{r} \qquad \text{as } r \to \infty \qquad (3.75)$$

The spatial variation of ϕ will be determined by the covariant derivative $D_\mu \phi$ and so $g\eta$ must enter the r-dependence. On dimensional grounds, we can write, without loss of generality,

$$\phi = \frac{H(\xi)}{\xi} \eta \hat{r} \qquad (3.76)$$

where

$$\xi \equiv g \eta r. \qquad (3.77)$$

The required behaviour as $r \to \infty$ is obtained if

$$\frac{H(\xi)}{\xi} \to 1 \qquad \text{as } \xi \to \infty. \qquad (3.78)$$

In the absence of gauge fields, the contribution of the scalar kinetic term to the energy is given by

$$E = \frac{1}{2} \int \partial_i \phi \cdot \partial_i \phi \, d^3 x \tag{3.79}$$

where $i = 1, 2, 3$ is a (summed) spatial index. In spherical polar coordinates (r, θ, ϕ),

$$\nabla \phi_a = \frac{\partial \phi_a}{\partial r} \hat{r} + \frac{1}{r} \frac{\partial \phi_a}{\partial \theta} \hat{\theta} + \frac{1}{r \sin \theta} \frac{\partial \phi_a}{\partial \phi} \hat{\phi}. \tag{3.80}$$

Since $\hat{r}$ is a function of θ and ϕ, but not of r, the large r behaviour of the integral (3.79) for the energy of the monopole is controlled by

$$E \sim \int dr \left[\left(\frac{\partial \hat{r}}{\partial \theta} \right)^2 + \frac{1}{\sin^2 \theta} \left(\frac{\partial \hat{r}}{\partial \phi} \right)^2 \right]. \tag{3.81}$$

Thus, in the absence of a gauge field contribution to E, the energy of the monopole solution would be infinite.

To find a finite-energy solution, we need the gauge field contribution to the covariant derivative to produce a cancellation to 'improve' the behaviour of $\nabla \phi_a$ for large r. This possibility may be studied by making the following ansatz for the gauge field expectation value:

$$A_i^a = \frac{\epsilon_{ail} r^l}{gr^2} (K(\xi) - 1). \tag{3.82}$$

Then (exercise 7)

$$D_i \phi_a = \frac{K(\xi) H(\xi)}{gr^4} (r^2 \delta_{ai} - r_a r_i) + (\xi H'(\xi) - H(\xi)) \frac{r_a r_i}{gr^4}. \tag{3.83}$$

A dangerous term of the type $H(\xi) r^2 \delta_{ai} / gr^4$ has cancelled between the $\partial_i \phi_a$ and $-g \epsilon_{abc} A_i^b \phi_c$ contributions to $D_i \phi_a$. For $\frac{1}{\xi} H(\xi) \to 1$ as $\xi \to \infty$, this term would have had the unwelcome asymptotic behaviour r^{-1} as $r \to \infty$. The surviving terms are of order r^{-2} as $r \to \infty$, provided $K(\xi) H(\xi)$ and $\xi H'(\xi) - H(\xi)$ are finite for $\xi \to \infty$, and a divergent contribution to the energy of the monopole is avoided.

With the ansatz (3.82) for the expectation value of the gauge field, the field strength is given by

$$g F_{ij}^a = \frac{K^2 - 1}{r^2} \epsilon_{iaj} + \left(\frac{K'}{r^2} - \frac{K^2 - 1}{r^4} \right) (\epsilon_{iap} r^p r_j - \epsilon_{jap} r^p r_i). \tag{3.84}$$

The energy of the magnetic monopole solution may now be written in terms of H and K as

$$E = \frac{4\pi \eta}{g} \int_0^\infty d\xi \, \xi^{-2} \left[\frac{1}{2} (\xi H' - H)^2 + H^2 K^2 + (\xi K')^2 \right. \tag{3.85}$$

$$\left. + \frac{1}{2} (K^2 - 1)^2 + \frac{\lambda}{8g} (H^2 - \eta^2)^2 \right]. \tag{3.86}$$

Minimizing with respect to variation of H and K gives (exercise 8)

$$\xi^2 K'' = K H^2 + K(K^2 - 1) \tag{3.87}$$

$$\xi^2 H'' = 2K^2 H + \frac{\lambda}{2g^2} H(H^2 - \xi^2). \tag{3.88}$$

There is an analytic solution [15] in the limit $\lambda/g^2 \to 0$,

$$H(\xi) = \xi \coth \xi - 1 \tag{3.89}$$

$$K(\xi) = \xi \operatorname{cosech} \xi. \tag{3.90}$$

Note that $K(\xi)H(\xi)$ and $\xi H'(\xi) - H(\xi)$ are finite as $\xi \to \infty$, as required earlier to avoid a divergent contribution to the energy of the monopole solution. In fact,

$$K(\xi)H(\xi) \to 0 \qquad \text{as } \xi \to \infty \tag{3.91}$$

in this limit. The corresponding energy, which is (at least at the classical level) the mass m_M of the monopole, is given by

$$m_M = \frac{4\pi\eta}{g}. \tag{3.92}$$

More generally, it has the form

$$m_M = \frac{4\pi\eta}{g} h\left(\frac{\lambda}{g^2}\right) \tag{3.93}$$

where h turns out to be a slowly varying function.

The monopole solution carries a magnetic charge, which we now evaluate in the limit $\lambda/g^2 \to 0$. The solution (3.90) implies that

$$K(\xi) \to 0 \qquad \text{as } \xi \to \infty. \tag{3.94}$$

Thus, as $\xi \to \infty$, (3.82) reduces to

$$A_i^a = -\frac{\epsilon_{ail} r_l}{gr^2} \qquad \text{as } \xi \to \infty. \tag{3.95}$$

We shall refer to

$$B_i^a = (\nabla \times A^a)_i = \tfrac{1}{2}\epsilon_{ijk} F_{jk}^a \tag{3.96}$$

as the 'magnetic' field, though we are not dealing with electroweak theory here because the gauge group is $SO(3)$ rather than $SU(2) \times U(1)$. With the gauge field expectation value given by (3.95), the magnetic field is (exercise 9)

$$B_i^a = \frac{\hat{r}_i \hat{r}_a}{gr^2}. \tag{3.97}$$

The surviving $U(1)$ gauge group after spontaneous symmetry breaking by the scalar field expectation value (3.74) is the group of rotations about the direction $\hat{r}$. The magnetic field should, therefore, be identified with the component B of B^a in this direction:

$$B = \frac{\hat{r}}{gr^2} \tag{3.98}$$

Thus, the magnetic field is the field of a magnetic monopole with magnetic charge $4\pi/g$.

3.8 Monopole topological quantum number

The asymptotic form $\phi = \eta\hat{r}$ of the magnetic monopole cannot be deformed continuously to the trivial configuration $\phi = \eta\hat{z}$ and, for this reason, the magnetic monopole is topologically conserved once formed. This is reflected in the existence of a topological quantum number defined as follows. The sphere $\mathcal{M}$ of all solutions for the expectation value of ϕ which minimizes the tree-level effective potential is given by

$$|\phi| = \eta. \tag{3.99}$$

This defines the surface of a sphere and $\phi(r)$ can be thought of as a mapping from the surface Σ of a sphere in coordinate space to the surface $\mathcal{M}$ of the sphere in the space of solutions. If we parametrize the surface Σ by parameters u and v, then the element of surface on the space $\mathcal{M}$ is

$$dS = dSn = \left(\frac{\partial\phi}{\partial u} \times \frac{\partial\phi}{\partial v}\right) du\, dv. \tag{3.100}$$

The unit normal n to the sphere is $\hat{r}$ and so

$$\phi \cdot \left(\frac{\partial\phi}{\partial u} \times \frac{\partial\phi}{\partial v}\right) du\, dv = \eta\, dS. \tag{3.101}$$

As the parameters u and v are varied to allow $\hat{r}$ to sweep out the surface Σ of a sphere in coordinate space, ϕ sweeps out the surface $\mathcal{M}$ of a sphere in the space of solutions. For the single monopole solution that we are discussing, as the first sphere is swept out once so is the second sphere. More generally, for a multi-monopole solution, it is possible for the second sphere to be swept out N times, where N is an integer, as the first sphere is swept out once. Thus,

$$\int_\Sigma \phi \cdot \left(\frac{\partial\phi}{\partial u} \times \frac{\partial\phi}{\partial v}\right) du\, dv = 4\pi\eta^3 N. \tag{3.102}$$

If we continuously deform ϕ, N cannot change because it is an integer, one in the case of the single-monopole solution. It is a topological quantum number, which reflects the fact that a magnetic monopole configuration once formed cannot be continuously deformed to the trivial configuration. Thus, magnetic monopole configurations are stabilized in a topological way.

The topological quantum number N is related to the magnetic charge. This can be demonstrated as follows. Denote the elements of surface and the unit normal on the sphere in coordinate space by $d\tilde{S}$ and $\tilde{n}$. Then

$$d\tilde{S}\,\tilde{n} = \left(\frac{\partial r}{\partial u} \times \frac{\partial r}{\partial v} \right) du\, dv \qquad (3.103)$$

from which it follows that

$$\epsilon_{ijk}\, d\tilde{S}\,\tilde{n}_i = \frac{\partial(r_j, r_k)}{\partial(u, v)}\, du\, dv. \qquad (3.104)$$

Noting that

$$\frac{\partial \phi}{\partial u} \times \frac{\partial \phi}{\partial v} = \frac{1}{2} \left(\frac{\partial \phi}{\partial r_j} \times \frac{\partial \phi}{\partial r_k} \right) \frac{\partial(r_j, r_k)}{\partial(u, v)} \qquad (3.105)$$

and using (3.103), the topological quantum number N may be recast in terms of coordinate-space derivatives of ϕ as

$$4\pi \eta^3 N = \frac{1}{2} \int_{\Sigma} \epsilon_{ijk}\tilde{n}_i \phi \cdot \left(\frac{\partial \phi}{\partial r_j} \times \frac{\partial \phi}{\partial r_k} \right) d\tilde{S}. \qquad (3.106)$$

The integral (3.106) may be formulated as a surface integral of $\phi \cdot F_{jk}$ on a sphere Σ of large radius, where the F_{jk}^a are the spatial components of the gauge field strength defined in (3.70). For this purpose, we need a solution for F_{jk} in terms of ϕ valid for large r. As discussed in section 3.7, for a finite-energy solution there is a cancellation between the two terms in the covariant derivative

$$D_i \phi = \partial_i \phi - g(A_i \times \phi) \qquad (3.107)$$

such that the covariant derivative is of order r^{-2} for $r \to \infty$, whereas, separately, the two terms are of order r^{-1}. Thus, for large r,

$$\partial_i \phi \simeq g(A_i \times \phi) \qquad (3.108)$$

from which it follows that

$$A_i = \frac{1}{g\eta^2}(\phi \times \partial_i \phi) + \frac{1}{\eta^2}\alpha_i \phi \qquad (3.109)$$

where

$$\alpha_i \equiv \phi \cdot A_i. \qquad (3.110)$$

The corresponding expression for $\phi \cdot F_{jk}$ obtained from (3.70) is

$$\phi \cdot F_{jk} = \frac{1}{g\eta^2}\phi \cdot (\partial_j \phi \times \partial_k \phi) + (\partial_j \alpha_k - \partial_k \alpha_j). \qquad (3.111)$$

Combining (3.106) with (3.111), the topological quantum number N may be written in terms of the field strength as

$$4\pi \eta^3 N = \frac{g\eta^2}{2} \int_\Sigma \epsilon_{ijk} \tilde{n}_i \boldsymbol{\phi} \cdot F_{jk} \, \mathrm{d}\tilde{S}. \tag{3.112}$$

Finally, N may be recast as a surface integral of the magnetic field on a sphere Σ of large radius. As discussed in section 3.7, the magnetic field in the model with $SO(3)$ gauge group should be identified with the component B_i^a in (3.96) along the direction about which the surviving $U(1)$ gauge symmetry is the group of rotations. In general, this is the direction $|\boldsymbol{\phi}|^{-1}\boldsymbol{\phi}$. Consequently, the magnetic field B is given by

$$B_i = \frac{1}{2\eta} \epsilon_{ijk} \boldsymbol{\phi} \cdot F_{jk}. \tag{3.113}$$

It follows from (3.112) that

$$N = \frac{g}{4\pi} \int_\Sigma B \cdot \tilde{n} \, \mathrm{d}\tilde{S}. \tag{3.114}$$

Thus, the topological quantum number N measures the magnetic charge in units of $4\pi/g$.

3.9 Magnetic monopoles in grand unified theories

In general, if we start with a grand unified group G and the symmetry is spontaneously broken to H (at a phase transition), then the action of any element of H leaves a vacuum state invariant. Consequently, distinct vacuum states correspond to the coset manifold G/H. The logic behind this is that G invariance of the effective potential means that starting from any vacuum state, we can generate further vacuum states (degenerate in energy) by acting with elements of G. However, when the element of G in question is an element of the subgroup H, it does not produce a new vacuum state. The topological entity underlying the existence of stable magnetic monopoles is the second homotopy group for G/H denoted by $\pi_2(G/H)$, whose elements are inequivalent mappings from the surface of a two-sphere S^2 to G/H, i.e. mappings which cannot be continuously deformed into each other. There is a theorem that $\pi_2(G/H)$ can be identified with $\pi_1(H)/\pi_1(G)$. Here, $\pi_1(G)$ is the first homotopy group of G whose elements are inequivalent mappings from a circle S^1 to G and similarly for H.

In the example just considered, $G = SO(3)$ and $H = U(1)$. Also $\pi_1(U(1)) = Z$, the integers, with the value of the integer being the winding number, i.e. the number of times we wind around the circle defined by $U(1)$ as we wind once around the circle in coordinate space. Less obviously, $\pi_1(SO(3)) = Z_2$, the integers modulo 2. In this case, therefore, $\pi_2(G/H) = Z/Z_2$ or the even integers. This is why we found magnetic charges in multiples of $4\pi/g$ which is twice the Dirac magnetic monopole charge.

When H is $SU(3) \times SU(2) \times U(1)$, which will be the case for a phase transition in which the grand unified theory breaks spontaneously to the standard model, then $\pi_1(H)$ is just $\pi_1(U(1)) = Z$, because $\pi_1(SU(3))$ and $\pi_1(SU(2))$ are both trivial. Thus, for such a spontaneous symmetry breaking, the resulting second homotopy group is $Z/\pi_1(G)$. In particular, if $\pi_1(G)$ is trivial, as is the case for the $SU(5)$ grand unified group discussed in section 2.6, then $\pi_2(G/H) = Z$ and we have magnetic monopole solutions.

We now ask what masses are possessed by the magnetic monopoles in grand unified theories. By analogy with (3.92), the magnetic monopole mass will be of order $4\pi \eta/g_G$, where η is the expectation value of the Higgs scalar responsible for breaking the grand unified symmetry and g_G is the value of the gauge coupling constant for the grand unified group at the unification scale. In the case of the $SU(5)$ grand unified theory of section 2.6, η, which is identified with ϕ_c, is of order 10^{15} GeV and g_G is of order 1. Thus, we expect the magnetic monopole mass m_M to be of order 10^{16} GeV. In the case of the supersymmetric $SU(5)$ grand unified theory of section 2.7, with a unification scale of 2×10^{16} GeV, which is 1.5 orders of magnitude greater than in the non-supersymmetric case, a magnetic monopole mass of order 10^{17}–10^{18} GeV is to be expected.

3.10   Abundance of magnetic monopoles

Magnetic monopoles form as the phase transition from the $SU(5)$ symmetric phase to the standard model $SU(3) \times SU(2) \times U(1)$ phase occurs. This is the result of the expectation values of the Higgs field only being correlated over some finite distance. The expectation values of the Higgs field at different points in space will not be aligned to produce a uniform Higgs field over distances greater than this. Thus, we can expect topologically non-trivial configurations to be produced, in particular, magnetic monopoles. The number of magnetic monopoles formed [16, 17] should be determined as to order of magnitude by the distance over which the Higgs expectation values are correlated [1,9].

There are two effects which can limit the range over which this correlation occurs. The first is the statistical-mechanical thermal average over the product of the two Higgs fields. For a second-order phase transition, this correlation length is of order T_c^{-1} but can be larger for a first-order phase transition, which proceeds through the formation of bubbles of the low-temperature phase which then coalesce. The second effect is the general-relativistic particle horizon d_H. Correlations cannot occur over distances greater than the distance d_H that light has been able to travel since the big bang. For a Friedman–Robertson–Walker (FRW) universe, as discussed in section 1.2, the proper distance at time t from any point to the particle horizon is

$$d_H(t) = R(t) \int_0^t \frac{dt'}{R(t')}. \tag{3.115}$$

Barring cosmological inflation (to be discussed in later chapters), the growth of $R(t)$ with time is according to the power law

$$R(t) \sim t^n \tag{3.116}$$

(with $n = \frac{1}{2}$ for a radiation-dominated universe). Then

$$d_H(t) = \frac{t}{1-n} \tag{3.117}$$

provided $n \neq 0$. Thus, $d_H(t)$ is of order t. We require the particle horizon at time t_c that the phase transition is completed. A slightly different discussion is required for second-order (or weakly first-order) and first-order phase transitions.

For a second-order phase transition, the phase transition is completed at the critical temperature T_c. For the radiation-dominated era of the FRW universe, there is the connection (see section 1.3) between the time t since the big bang and temperature T:

$$t \simeq 0.3 N_*^{-1/2} \frac{m_P}{T^2} \tag{3.118}$$

where m_P is the Planck mass ($\sim 10^{19}$ GeV) and N_* is the effective number of degrees of freedom at temperature T:

$$N_* = N_B + \tfrac{7}{8} N_F. \tag{3.119}$$

N_B and N_F are, respectively, the numbers of bosonic and fermionic degrees of freedom for particles with mass small compared to T, in the sense described after (2.19). For approximately one monopole per horizon volume, the number density n_M of monopoles should be

$$n_M(T_c) \sim (d_H(t_c))^{-3} \sim t_c^{-3} \sim N_*^{3/2} T_c^6 m_P^{-3} (0.6)^{-3} \tag{3.120}$$

where we have taken

$$d_H(t_c) = 2t_c \tag{3.121}$$

for the radiation-dominated era. If we compare this with the entropy density of (2.21),

$$s = \frac{2\pi^2}{45} N_* T^3 \tag{3.122}$$

then

$$\frac{n_M(T_c)}{s(T_c)} \sim 10.6 N_*^{1/2} T_c^3 m_P^{-3}. \tag{3.123}$$

At temperatures below T_c but above the electroweak phase transition, the appropriate value of N_* is that for the $SU(3) \times SU(2) \times U(1)$ standard model:

$$N_* = 106.75. \tag{3.124}$$

Thus,

$$\frac{n_M(T_c)}{s(T_c)} \sim 10^2 \left(\frac{T_c}{m_P}\right)^3. \tag{3.125}$$

Assuming that the expansion of the universe for $T < T_c$ is adiabatic, then $s \propto R^{-3}$ and the ratio $n_M(T)/s(T)$ does not change. As a consequence, the monopole contribution $\Omega_M h^2$ to $\Omega_m h^2$ today is predicted to be many orders of magnitude greater than the observational bound of about 0.15. For a non-supersymmetric GUT theory with T_c of order 10^{15} GeV and a magnetic monopole mass m_M of order 10^{16} GeV, $\Omega_M h^2$ is 14 orders of magnitude greater than this upper bound. For a supersymmetric GUT theory with T_c of order 10^{16} GeV and a magnetic monopole mass m_M of order 10^{17}–10^{18} GeV, the situation is even worse with $\Omega_M h^2$ some 18–19 orders of magnitude greater than the upper bound (exercise 10).

In the case of a first-order phase transition, the transition does not proceed until some temperature below T_c at which the bubble nucleation rate for bubbles of the low-temperature phase is of the same order as the expansion rate H for the universe. We expect the Higgs expectation values to be correlated within a bubble but uncorrelated between any two bubbles. Thus, the number density of monopoles (or antimonopoles) produced should be of the order of $(\frac{4}{3}\pi r_b^3)^{-1}$, where r_b is the average radius of a bubble at a time when the bubbles have expanded to just fill the whole of space. The universe supercools at the first-order phase transition but reheats when the bubbles coalesce, so that the entropy density after reheating is $2\pi^2 N_* T_c^3/45$, as in the second-order case. Thus, for a first-order phase transition,

$$\frac{n_M}{s} \sim \frac{45}{2\pi^2} N_*^{-1} T_c^{-3} \left(\frac{4}{3}\pi r_b^3\right)^{-1}. \tag{3.126}$$

The value of r_b has been estimated [18] leading to a value of $\Omega_M h^2$ even larger than in the second-order phase transition case.

In either case, if magnetic monopoles form at a grand unified phase transition, some mechanism is required to dilute the monopole density by many orders of magnitude. The most obvious mechanism would be annihilation of monopoles and antimonopoles. However, this has been estimated [16] and there is no significant effect for $n_M/s \lesssim 10^{-10}$ and, for larger values of n_M/s, the annihilation process cannot reduce n_M/s much below 10^{-10}. For the non-supersymmetric case, this mechanism is ineffective and, for the supersymmetric case, it can do no more than reduce the monopole abundance closer to that for the non-supersymmetric case. A possible mechanism that can do the trick is cosmological inflation, which will be discussed in later chapters.

After some mechanism has reduced the monopole abundance to a value compatible with the bound on $\Omega_m h^2$, any residual monopole density can have important astrophysical consequences [19]. For instance, because the expectation value of the grand unified Higgs field approaches zero as the centre of the monopole is approached, the $SU(5)$ grand unified symmetry is essentially unbroken in the core of the monopole. Consequently, when a nucleon encounters a magnetic monopole, decay of the nucleon can be induced by the baryon-number non-conserving interactions of the lepto-quark fields of the $SU(5)$ grand unified theory. In this way, the magnetic monopoles collected by stars in the course of time will cause emission of radiation from neutron stars. This puts severe limits on the flux of magnetic monopoles. More details of this and other astrophysical effects may be found elsewhere [19].

3.11 Exercises

1. Calculate the energy of a kink or antikink soliton solution.
2. Compare the total energy associated with a plane domain wall with $\eta = 100 \, \text{GeV}$ and area H_0^{-2} with the known mass of the universe within a Hubble volume H_0^{-3}.
3. Derive equation (3.16) for the dependence of a global cosmic string on the cylindrical polar coordinate ρ.
4. Check that the Lagrangian (3.23) possesses the local $U(1)$ gauge symmetry (3.26) and (3.27).
5. Show that the magnetic flux carried by a local cosmic string is given by (3.30).
6. Derive the string equation of motion (3.61) and the constraints (3.62) and (3.63) by varying the action (3.55) with respect to X^μ and $h_{\alpha\beta}$.
7. Derive the covariant derivative (3.83) for a monopole solution.
8. Derive equations (3.87) and (3.88) for the form of a monopole solution.
9. Derive equation (3.96) for the magnetic field due to a magnetic monopole.
10. Estimate the monopole contribution to Ωh^2 in non-supersymmetric and supersymmetric grand unified theories.

3.12 General references

The books and review articles that we have found most useful in preparing this chapter are:

- Hindmarsh M B and Kibble T W B 1995 *Rep. Prog. Phys.* **58** 477
- Vilenkin A 1985 *Phys. Rep.* **121** 263
- Kolb E W and Turner M S 1990 *The Early Universe* (Reading, MA: Addison-Wesley)

Bibliography

[1] Kibble T W B 1976 *J. Phys. A: Math. Gen.* **9** 1387
[2] Zeldovich Ya B, Kobzarev I B and Okun L B 1975 *Sov. Phys.–JETP* **40** 1

[3]	Vilenkin A and Everett A E 1982 *Phys. Rev. Lett.* **48** 1867

[4]	Nielsen H B and Olesen P 1973 *Nucl. Phys.* B **61** 45

[5]	Vilenkin A 1981 *Phys. Rev.* D **23** 852

[6]	Kaiser N and Stebbins A 1984 *Nature* **310** 391

[7]	Silk J and Vilenkin A 1984 *Phys. Rev. Lett.* **53** 1700

[8]	Vachaspati T and Vilenkin A 1985 *Phys. Rev.* D **31** 3052

[9]	Kibble T W K 1980 *Phys. Rep.* **67** 183

[10]	Albrecht A and Turok N 1985 *Phys. Rev. Lett.* **54** 1868
	Bennett D B and Bouchet F R 1988 *Phys. Rev. Lett.* **60** 257

[11]	Davis R L 1985 *Phys. Lett.* B **161** 285
	Bennett D P 1986 *Phys. Rev.* D **34** 3592

[12]	Albrecht A, Battye R and Robinson J 1997 *Phys. Rev. Lett.* **79** 4736
	Pen U L, Seljak U and Turok N 1997 *Phys. Rev. Lett.* **79** 1611
	Alen B *et al* 1997 *Phys. Rev. Lett.* **79** 2624

[13]	Vincent G R, Antunes N and Hindmarsh M B 1998 *Phys. Rev. Lett.* **80** 2277
	Hindmarsh M B 1997 Talk given at *COSMO97*, Ambleside, September 1997

[14]	't Hooft G 1974 *Nucl. Phys.* B **79** 276
	Polyakov A M 1974 *JETP Lett.* **20** 194

[15]	Prasad M and Sommerfield C M 1975 *Phys. Rev. Lett.* **35** 760

[16]	Preskill J P 1979 *Phys. Rev. Lett.* **43** 1365

[17]	Guth A H and Tye S-H H 1980 *Phys. Rev. Lett.* **44** 631
	Langacker P and Pi S-Y 1980 *Phys. Rev. Lett.* **45** 1

[18]	Guth A H and Weinberg E 1983 *Nucl. Phys.* B **212** 321

[19]	Kolb E W and Turner M S 1990 *The Early Universe* (Reading, MA: Addison-Wesley)
	ch 7, section 7.6

Chapter 4

Baryogenesis

4.1 Introduction

The success of the standard model in describing the fundamental interactions has the consequence, among many others, of verifying the TCP invariance of nature. This requires that, for each particle X having mass m_X, decay width Γ_X and quantum numbers Q_X etc, there is an antiparticle $\bar{X}$ with the same mass and width, $m_{\bar{X}} = m_X$, $\Gamma_{\bar{X}} = \Gamma_X$ but with opposite quantum numbers $Q_{\bar{X}} = -Q_X$ etc. One might, therefore, suppose that the world we inhabit would share this symmetry and contain equal numbers N_X of particles and antiparticles $N_{\bar{X}} = N_X$. This is clearly not the case. We know that the solar system is made of matter (protons, neutrons, electrons) and not antimatter, and the experimental bound on antihelium is [1]

$$\frac{n_{\overline{^4\text{He}}}}{n_{^4\text{He}}} < 3.1 \times 10^{-6} \qquad \text{at 95\% CL.} \tag{4.1}$$

Any region of antimatter must be well separated from regions of matter, since, in any region where protons and antiprotons coexisted, their annihilation into pions with the subsequent $\pi^0 \rightarrow 2\gamma$ decays would significantly distort the cosmic microwave background. The data require that such domains of matter and antimatter are separated by a length scale l_B with, conservatively,

$$l_B \gtrsim 3 \text{ kpc} \tag{4.2}$$

the radius of our galaxy, and probably [2, 3]

$$l_B \gtrsim 10 \text{ kpc} \tag{4.3}$$

the scale of the Virgo cluster.

The asymmetry between baryons (b) and antibaryons ($\bar{b}$) may be quantified by the difference in their number densities $n_B \equiv n_b - n_{\bar{b}}$. However, the expansion of the universe dilutes both n_b and $n_{\bar{b}}$ and, hence, their difference, since, as explained in section 1.4, each scales as $R(t)^{-3}$, where $R(t)$ is the cosmological scale factor. It is, therefore, customary to use the ratio

$$\eta \equiv \frac{n_B}{n_\gamma} \tag{4.4}$$

to measure the asymmetry. n_γ is the photon number density given by the Boltzmann distribution (see section 5.1)

$$n_\gamma = 2\frac{\zeta(3)}{\pi^2}T^3 \tag{4.5}$$

when the temperature is T. From the measured microwave background [4], $T = T_0 = 2.725 \pm 0.002$ K at present and this gives

$$n_\gamma \approx 411 \text{ cm}^{-3}. \tag{4.6}$$

The present net baryon number density may be written in terms of the current critical density, defined in (1.37),

$$\rho_c = \frac{3}{8\pi^2}m_P^2 H_0^2 = 1.88h^2 \times 10^{-29} \text{ g cm}^{-3} \tag{4.7}$$

where $h \equiv H_0/100$ km s^{-1} Mpc^{-1} measures the present Hubble constant, $m_P \equiv G_N^{-1/2} = 1.22 \times 10^{19}$ GeV is the Planck mass and [5]

$$h = 0.0.71^{+0.04}_{-0.03}. \tag{4.8}$$

Then

$$n_B = \frac{\Omega_B}{m_B}\rho_c = 1.1 \times 10^{-5}\Omega_B h^2 \text{ cm}^{-3} \tag{4.9}$$

and

$$\eta = 2.65 \times 10^{-8}\Omega_B h^2 \tag{4.10}$$

where $\Omega_B \equiv \rho_B/\rho_c$ measures the baryon energy density as a fraction of the critical density. The measured primordial deuterium and hydrogen abundances require [5]

$$\Omega_B h^2 = 0.024 \pm 0.001 \tag{4.11}$$

which gives

$$\eta = (6.36 \pm 0.26) \times 10^{-10}. \tag{4.12}$$

The conservation of entropy in a comoving volume, when the universe is in local thermal equilibrium, means that the entropy density s also scales as $R(t)^{-3}$. Thus, the baryon asymmetry may alternatively be measured by

$$\eta_B \equiv \frac{n_B}{s} \tag{4.13}$$

where

$$s = \frac{2\pi^2}{45} g_{*S,T} T^3 \tag{4.14}$$

with

$$g_{*S,T} = \sum_{\text{bosons}} g_i \left(\frac{T_i}{T}\right)^3 + \frac{7}{8} \sum_{\text{fermions}} g_i \left(\frac{T_i}{T}\right)^3 \tag{4.15}$$

counting the total effective number of massless degrees of freedom at the temperature T, $g_i = 1$ for a real scalar, $g_i = 2$ for a real (massless) gauge field, $g_i = 4$ for a spin-$\frac{1}{2}$ Dirac field and $g_i = 2$ for a Weyl (chiral) field. We are allowing the possibility that different species are at different temperatures. When all $T_i = T$, $g_{*S,T} = N_*$ given in equation (1.104), and (4.14) reduces to (2.21). This is an excellent approximation until $t \sim 1$ s (or $T \sim 1$ MeV). However, as noted in section 1.8, it is *not* true today. The advantage of using η_B as a measure of the baryon asymmetry is that it is conserved, as long as baryon-number-violating interactions occur very slowly. The relationship between s and n_γ is

$$s = \frac{\pi^4}{45\zeta(3)} g_{*S,T} n_\gamma \simeq 1.8 g_{*S,T} n_\gamma \tag{4.16}$$

so

$$\eta = 1.8 g_{*S,T} \eta_B. \tag{4.17}$$

Thus, η is *not* constant in time, since $g_{*S,T}$ changes as the temperature drops and the number of effective massless modes decreases. The present entropy $s_0 = 7.0394 n_{\gamma,0}$ and the same data (4.12) give

$$\eta_B = (9.03 \pm 0.37) \times 10^{-11}. \tag{4.18}$$

So the challenge confronting theorists is to explain this small, non-zero number. The natural assumption is that 'originally' there was zero asymmetry. In equilibrium at a temperature $T \lesssim 1$ GeV, the nucleon and antinucleon densities are

$$n_N = n_{\bar{N}} = 2\left(\frac{m_N T}{2\pi}\right)^{3/2} e^{-m_N/T}. \tag{4.19}$$

As the universe cools, the nucleons and antinucleons annihilate with a rate

$$\Gamma_{\text{ann}} = n_N \langle \sigma_{\text{ann}} v \rangle \tag{4.20}$$

where $\langle \ldots \rangle$ denotes thermal averaging, σ_{ann} is the annihilation cross section and v is the relative velocity. The annihilation continues so long as the rate is larger than the expansion rate H of the universe:

$$H = \left(\frac{8\pi\rho}{3m_P^2}\right)^{1/2} = \frac{2\pi}{3} \left(\frac{\pi g_{*,T}}{5}\right)^{1/2} \frac{T^2}{m_P} \tag{4.21}$$

assuming that the energy density $\rho = (\pi^2/30)g_{*,T}T^4$ is dominated by relativistic particles. Here

$$g_{*,T} = \underbrace{\sum_{\text{bosons}} g_i \left(\frac{T_i}{T}\right)^4}_{} + \frac{7}{8} \underbrace{\sum_{\text{fermions}} g_i \left(\frac{T_i}{T}\right)^4}_{} \tag{4.22}$$

satisfies $g_{*,T} = g_{*S,T} = N_*$ (with N_* defined in (1.104)) when all particle species i are at the same temperature $T_i = T$. (See section 5.1.) The thermal average

$$\langle \sigma_{\text{ann}} v \rangle \simeq m_\pi^{-2} \tag{4.23}$$

and, at $T = T_f \simeq 20$ MeV, the annihilation rate falls below the expansion rate, nucleons and antinucleons are so dilute that they cannot annihilate further and their number densities become frozen at

$$\frac{n_N}{n_\gamma} = \frac{n_{\overline{N}}}{n_\gamma} = \frac{2\pi^2}{\zeta(3)}\left(\frac{m_N}{2\pi T_f}\right)^{3/2} e^{-m_N/T_f} \tag{4.24}$$

$$\simeq 10^{-18} \tag{4.25}$$

using (4.6). This is far smaller than the value (4.12) which derives from the measured primordial abundances of the light nuclei. Thus, the assumed zero initial asymmetry is inconsistent with the nucleosynthesis data.

Of course, statistical fluctuations can generate a non-zero initial asymmetry. At present our galaxy contains about 10^{79} photons and 10^{69} nucleons. When $T \gtrsim 1$ GeV, however, the comoving volume containing our galaxy contained about 10^{79} baryons and antibaryons. Thus, statistical fluctuations might generate an asymmetry

$$N_B - N_{\overline{B}} \sim \sqrt{N} \tag{4.26}$$

so that, instead of (4.19), we have

$$n_N - n_{\overline{N}} \sim \frac{1}{\sqrt{N}}n_N \sim 10^{-39.5}n_N \tag{4.27}$$

which again is far too small to explain the nucleosynthesis data.

The conclusion is that the initial baryon asymmetry must be non-zero to explain the size of the asymmetry we observe today. Of course, the required value may be input by hand as an initial condition but aesthetically this is unattractive. The consensus is that the asymmetry derives from new physics in the early universe. We turn next to the three necessary conditions for baryogenesis, first derived by Sakharov [6].

4.2 Conditions for baryogenesis

If we start from a universe with a net baryon number B of zero and evolve to one with a non-zero value, it is clear that baryon number is not conserved. Thus the first condition is that *there are baryon-number non-conserving interactions in nature.* Aside from the baryon-number asymmetry itself, there is no direct experimental evidence of such interactions and any theory which contains them is constrained by the current lower bound on the lifetime (τ_p) of the proton [7]

$$\tau_p \gtrsim 10^{31}\text{--}10^{33} \text{ yr.} \tag{4.28}$$

The generation of a non-zero baryon number for the universe, or 'baryogenesis', also requires that *there are C- and CP-violating interactions in nature.* To see this, suppose that a process $i \to f$, with initial state i and final state f, violates baryon-number conservation, so $B_i - B_f \neq 0$. If charge conjugation C were an exact symmetry, then the process $\bar{i} \to \bar{f}$, where $\bar{i}$ is obtained from i by replacing all particles by their antiparticles and similarly for $\bar{f}$, would occur at the same rate as the former. Since $B_{\bar{i}} = -B_i$ and $B_{\bar{f}} = -B_f$, the net baryon number produced by the two processes $B_i - B_f + B_{\bar{i}} + B_{\bar{f}}$ is zero. A similar argument applies if CP-invariance were exact; parity reversal P reverses the momenta of all participating particles but when these are integrated over the (identical) allowed phase space the net baryon number produced by the two processes is again zero. The TCP-invariance of any particle physics model (T is time reversal) ensures that if there is CP-violation, then there is also T-violation and it is easy to see that if T-invariance were an exact symmetry, then baryon number would be conserved. Of course, we have long known that C-invariance is violated by weak interactions and that CP-violation occurs at the milliweak level in kaon decays [8]. There is also limited evidence for T-violation in kaon decays [9]. Thus there is no *a priori* need for new physics from this condition.

The final condition is that *the baryon-number non-conserving processes occur when the universe is not in thermal equilibrium.* To see this, consider a particle X with non-zero baryon number in thermal equilibrium at a temperature $T \ll m_X$. The number density n_X of X particles is given by

$$n_X \simeq g_X (m_X T)^{3/2} e^{(-m_X + \mu_X)T} \tag{4.29}$$

where μ_X is the chemical potential. Likewise, in thermal equilibrium the number density $n_{\bar{X}}$ of the antiparticles $\bar{X}$ is

$$n_{\bar{X}} \simeq g_X (m_X T)^{3/2} e^{(-m_X + \mu_{\bar{X}})T}. \tag{4.30}$$

If $X, \bar{X}$ participate in baryon-number non-conserving processes, as required by the first condition, then the process

$$XX \to \bar{X}\bar{X} \tag{4.31}$$

is allowed and, in equilibrium, this requires

$$2\mu_X = 2\mu_{\bar{X}}. \tag{4.32}$$

Then the asymmetry vanishes, since

$$\eta \propto n_X - n_{\bar{X}} = 0. \tag{4.33}$$

Only a departure from thermal equilibrium will permit an asymmetry. Such a departure can arise, for example, during a phase transition in which gauge symmetry breaking occurs. It can also arise due to the expansion of the universe during the decay of a heavy particle.

4.3   Out-of-equilibrium decay of heavy particles

Given enough time any particle, no matter how weakly it interacts, will reach thermal equilibrium. However, in an expanding universe, it becomes increasingly difficult for any given species X of particle to remain in thermal equilibrium. This is because the expansion dilutes the densities of all particles with which X interacts and thereby inhibits the rate of the interactions needed to maintain equilibrium; and also because the rate of decay Γ_X of the X particles eventually falls below the expansion rate H of the universe and the decays are unable to reduce the numbers of X particles to the levels required to stay in equilibrium.

Suppose X is a superheavy boson, having non-zero baryon number, which decays to lighter fermions f in a baryon-number non-conserving process. Gauge vector bosons and Higgs scalar particles with these properties arise naturally in grand unified theories (GUTs) which unify the strong and electroweak interactions [10, 11]. At high temperatures $T \gg m_X$, we assume that all particles are in equilibrium, and that the net baryon number B is zero. The number densities of the X particles and their antiparticles are

$$n_X = n_{\bar{X}} = \frac{\zeta(3)}{\pi^2} g_X T^3 \tag{4.34}$$

with $g_X = 2, 1$ corresponding respectively to X being a vector, scalar particle. Thus, using (4.5),

$$\frac{n_X}{n_\gamma} = \frac{n_{\bar{X}}}{n_\gamma} = \frac{1}{2} g_X. \tag{4.35}$$

To maintain equilibrium densities as the universe expands, the X and $\bar{X}$ particles must reduce their numbers sufficiently and, so long as they are able to do so, the net baryon number remains zero. For $T \lesssim m_X$, the equilibrium densities will then reduce relative to that of the photons. From the analogue for X particles of (4.19) and (4.5), the relative densities are given by

$$\frac{n_X^{\text{eq}}}{n_\gamma} = \frac{n_{\bar{X}}^{\text{eq}}}{n_\gamma} = \frac{g_X \pi^2}{2\zeta(3)} \left(\frac{m_X}{2\pi T}\right)^{3/2} e^{-m_X/T}. \tag{4.36}$$

For baryogenesis, the most important quantity in determining whether thermal equilibrium can be maintained when $T \sim m_X$ is the decay rate Γ_X of the

X particle, which controls the numbers of X and $\bar{X}$ particles. Unless this is sufficiently rapid, thermal equilibrium densities cannot be maintained as T falls. If at some temperature $T \gtrsim m_X$ the X bosons cannot decay in the time scale H^{-1} associated with the expansion of the universe, then they decouple from the thermal bath while they are still relativistic and their densities satisfy (4.35). Thus at a lower temperature $T \lesssim m_X$, their abundance is much larger than the equilibrium densities satisfying (4.36). The condition for this to happen is that

$$\Gamma_X \lesssim H|_{T=m_X} . \tag{4.37}$$

Using (4.21), this requires

$$m_X^2 \gtrsim \frac{3}{2\pi} \left(\frac{5}{\pi} \right)^{1/2} N_*^{-1/2} \Gamma_X m_P \tag{4.38}$$

assuming that all (relativistic) particle species are at the same temperature, so $g_{*,T} = N_*$. The overabundance which occurs when this condition is satisfied allows the possibility of baryogenesis. Whether or not it *is* satisfied depends upon the particular GUT model in which the X particles arise.

If X is a superheavy gauge boson, for example,

$$\Gamma_X \sim \alpha_G m_X \tag{4.39}$$

where $\alpha_G = g_G^2/4\pi$ is the GUT 'fine structure constant'. Then (4.38) gives

$$m_X \gtrsim N_*^{-1/2} \alpha_G m_P. \tag{4.40}$$

For the non-supersymmetric $SU(5)$ GUT (which, incidentally, does not satisfy the constraint (4.28) on the proton lifetime),

$$\alpha_G \sim \tfrac{1}{42} \qquad m_X \sim 10^{15} \text{ GeV} \qquad N_* = \tfrac{427}{4} \tag{4.41}$$

provided that the colour triplet, electroweak Higgs particles are superheavy, the constraint (4.40) is not obviously satisfied. However, for the supersymmetric $SU(5)$ GUT,

$$\alpha_G \sim \tfrac{1}{25} \qquad m_X \sim 2 \times 10^{16} \text{ GeV} \qquad N_* = \tfrac{915}{4} \tag{4.42}$$

the larger values of N_* and m_X outweigh the larger value of α_G and the non-equilibrium condition (4.38) is marginally satisfied.

However, if X is a superheavy Higgs particle, its decay width

$$\Gamma_X \sim \left(\frac{m_f}{m_W} \right)^2 \alpha_G m_X \tag{4.43}$$

can be much smaller than that of the superheavy gauge boson, because the Yukawa coupling is suppressed (unless f is a top quark) by a factor m_f/m_W relative to the gauge coupling [10, 11]. Then the condition (4.38) can be easily satisfied, since, in this case, we require only that

$$m_X \gtrsim N_*^{-1/2} \left(\frac{m_f}{m_W} \right)^2 \alpha_G m_P. \tag{4.44}$$

In supergravity GUTs, baryon number non-conserving interactions can also arise via hidden-sector effects [12]. Then X can be an observable-sector gauge-singlet scalar which is coupled only gravitationally to observable-sector fermions. In this case,

$$\Gamma_X \sim \frac{m_X^3}{m_P^2} \tag{4.45}$$

and (4.38) gives

$$m_X \lesssim N_*^{1/2} m_P \tag{4.46}$$

which is always satisfied. Thus, the general conclusion is that the decay of superheavy *scalar* particles in a supersymmetric or supergravity GUT affords the best opportunity for the out-of-equilibrium decays necessary for baryogenesis.

As soon as the age H^{-1} of the universe becomes equal to the lifetime Γ_X^{-1} of the X, $\bar{X}$ particles they begin to decay and generate a non-zero net baryon number. Using (4.37), this occurs at a temperature T_{dec} satisfying

$$H|_{T=T_{dec}} = \Gamma_X \lesssim H|_{T=m_X}. \tag{4.47}$$

Thus, from (4.21),

$$T_{dec} < m_X. \tag{4.48}$$

Suppose that the X particle has decay channels $X \to f_n$ to a final state f_n producing baryon number B_n. Then the $\bar{X}$ has decay channels $\bar{X} \to \bar{f}_n$ producing baryon number $-B_n$, and the net baryon number produced by all of these decays is

$$\Delta B = \Gamma_X^{-1} \sum_n B_n [\Gamma(X \to f_n) - \Gamma(\bar{X} \to \bar{f}_n)]. \tag{4.49}$$

This gives a net baryon number density arising from the decays of

$$n_B = n_X \Delta B \simeq n_\gamma \Delta B \tag{4.50}$$

using (4.35). Thus, the baryon asymmetry (4.4) is

$$\eta \equiv \frac{n_B}{n_\gamma} \simeq \frac{1}{2} g_X \Delta B \tag{4.51}$$

or

$$\eta_B \equiv \frac{n_B}{s} \simeq \frac{45\zeta(3)}{2\pi^4} \frac{g_X}{N_*} \Delta B. \tag{4.52}$$

As anticipated, ΔB and, hence, the baryon asymmetry, vanishes if none of the X decays produces a baryon number and if there is no C- or CP-violation. Further, if thermal equilibrium were maintained any net baryon number produced by the decays is cancelled by inverse decay.

The foregoing analysis presumed that the X and $\bar{X}$ decays release no entropy, which is a poor approximation if $T_{\text{dec}} \ll m_X$. In this case, the energy density ρ of the universe is dominated by X particles. If this is converted entirely into radiation at a reheating temperature T_R given by

$$\rho \simeq \rho_X \simeq n_X m_X = \frac{\pi^2}{30} N_* T_R^4 = \frac{3}{8\pi} m_P^2 \Gamma_X^2 \qquad (4.53)$$

using (4.47), then

$$\frac{n_X}{s} = \frac{3}{4} \frac{T_R}{m_X} = \frac{3}{4} \left(\frac{45 m_P^2 \Gamma_X^2}{4\pi^3 N_* m_X^4} \right)^{1/4} \qquad (4.54)$$

and the baryon asymmetry becomes

$$\eta_B = \frac{3}{4} \frac{T_R}{m_X} \Delta B \qquad (4.55)$$

instead of (4.52). Either way, it seems that an encouragingly small amount of CP-violation is entailed to generate an asymmetry on the scale (4.18) observed. To see whether it is, we need to calculate ΔB in the various models containing baryon number and CP non-conservation.

4.4 Baryogenesis in GUTs

Grand unified theories (GUTs) seek to unify the three separate gauge groups $SU(3)$, $SU(2)$ and $U(1)$ of the standard model in a simple group G:

$$G \supset SU(3) \times SU(2) \times U(1). \qquad (4.56)$$

(See [13] for a review.) The GUT hypothesis is that above some high energy (GUT) scale M_G,

$$M_G \gtrsim 10^{15} \text{ GeV} \qquad (4.57)$$

G is an exact symmetry, which is spontaneously broken at the GUT scale to the standard model, which is itself spontaneously broken at the electroweak scale. In this way the (low-energy) gauge coupling strengths $(\alpha_1, \alpha_2, \alpha_3)$ of the standard model are all determined from the unknown (high-energy) coupling strength α_G of G, by using the renormalization group equations to 'run' between the GUT and the electroweak energy scales. We have discussed this in some detail in [10] but the essential point is that the evolution of the coupling strengths depends upon the matter content of the low-energy theory. Since neither α_G nor m_G is known *a priori*, the GUT hypothesis can only be tested by starting from the values of the coupling strengths measured at the electroweak scale and running to high energies to see whether they converge to a single value (α_G).

The known matter content of the standard model consists of three generations of

$$Q_L = (3, 2, \tfrac{1}{6}) \qquad B = \tfrac{1}{3} \qquad L = 0$$
$$u_L^c = (\bar{3}, 1, -\tfrac{2}{3}) \qquad B = -\tfrac{1}{3} \qquad L = 0$$
$$d_L^c = (\bar{3}, 1, \tfrac{1}{3}) \qquad B = -\tfrac{1}{3} \qquad L = 0 \qquad (4.58)$$
$$L_L = (1, 2, -\tfrac{1}{2}) \qquad B = 0 \qquad L = 1$$
$$e_L^c = (1, 1, 1) \qquad B = 0 \qquad L = -1$$

using the notation (n_3, n_2, Y), where n_3 specifies the colour $SU(3)$ representation, n_2 the weak $SU(2)$ representation and Y is the weak hypercharge. B and L are the baryon and lepton numbers (the superfix ^{c} indicates the charge conjugate particle). In addition, the electroweak Higgs

$$h_1 = (1, 2, \tfrac{1}{2}) \qquad (4.59)$$

is an essential ingredient of the standard model, whose discovery is currently awaited, hopefully at the LHC. If we assume just this matter content, besides the 12 gauge vector bosons of the three gauge groups, it is found that the coupling strengths converge and reach a point of closest approach but not coincidence, at the energy scale and coupling strength given in (4.41).

It is remarkable that the couplings come as close as they do and this in itself lends general support to the GUT hypothesis. However, the failure to converge precisely to a common value shows that if the GUT hypothesis is correct, then there must be matter additional to that of the standard model. Remarkably, the supersymmetric standard model, in which all of the matter particles (4.58) have supersymmetric (bosonic) partners (sparticles), all of the gauge bosons have (fermionic) supersymmetric partners (gauginos) and the Higgs doublet h_1 has a (fermionic) higgsino partner, *does* produce the convergence sought [14]. The calculated unification scale and coupling constant are given in (4.42). (Supersymmetry requires an additional Higgs doublet

$$h_2 = (1, 2, -\tfrac{1}{2}) \qquad (4.60)$$

plus its superpartner.)

This convergence represents the best evidence we have both for the GUT hypothesis and for low-energy supersymmetry and it is, therefore, natural to wonder whether a GUT with this matter content produces baryogenesis at the level needed to produce the observed asymmetry (4.12) or (4.18). In a general GUT, the matter content (4.58) (and Higgs fields) of the standard model, or its supersymmetric extension, constitute partial or complete representations R of the

GUT gauge group G. The coupling to the gauge bosons A_μ^A of the fermionic matter has the standard form

$$\mathcal{L} = \sum_R \bar{R}\gamma^\mu (i\partial_\mu - g_G A_\mu^A t^A)R \tag{4.61}$$

where t^A are the matrix representations of G corresponding to the representation R to which the fermions belong. For the $SU(5)$ GUT, each generation of (4.58) belongs to two irreducible representations $\bar{5}$ and 10 of the group, which decompose into representations of $SU(3)_c \times SU(2)_L \times U(1)_Y$ as follows:

$$\bar{5} = (\bar{3}, 1, \tfrac{1}{3}) + (1, 2, -\tfrac{1}{2}) = [d_L^c, L_L] \tag{4.62}$$

$$10 = (3, 2, \tfrac{1}{6}) + (\bar{3}, 1, -\tfrac{2}{3}) + (1, 1, 1) = [Q_L, u_L^c, e_L^c]. \tag{4.63}$$

Evidently the matrix t^A couples the gauge boson A_μ^A to fermions in the representations R and $\bar{R}$, where $\bar{R}$ contains the complex conjugate representations, to those given in (4.58), i.e.

$$
\begin{array}{lll}
\bar{Q}_L = (\bar{3}, 2, -\tfrac{1}{6}) & B = -\tfrac{1}{3} & L = 0 \\[4pt]
\bar{u}_L^c = (3, 1, \tfrac{2}{3}) & B = \tfrac{1}{3} & L = 0 \\[4pt]
\bar{d}_L^c = (3, 1, -\tfrac{1}{3}) & B = \tfrac{1}{3} & L = 0 \\[4pt]
\bar{L}_L = (1, 2, \tfrac{1}{2}) & B = 0 & L = -1 \\[4pt]
\bar{e}_L^c = (1, 1, -1) & B = 0 & L = 1.
\end{array} \tag{4.64}
$$

For baryogenesis, we are concerned with those gauge bosons which are coupled to fermions with a net non-zero baryon number. In the case of the $SU(5)$ GUT with the gauge bosons in the adjoint 24 representation, all of the 12 gauge bosons additional to the 12 of the standard model have this property. They transform as

$$(3, 2, -\tfrac{5}{6}) + (\bar{3}, 2, \tfrac{5}{6}) \tag{4.65}$$

representations of $SU(3) \times SU(2) \times U(1)$. We denote the (colour triplet) $SU(2)$ doublet by (X, Y), and the $SU(3) \times SU(2)$ symmetry requires that

$$m_X = m_Y \simeq M_G. \tag{4.66}$$

The allowed decay modes

$$
\begin{array}{llll}
X \rightarrow dv, & ul, & d^c u^c & \tag{4.67} \\
Y \rightarrow dl, & u^c u^c &
\end{array}
$$

are shown in figure 4.1.

All violate baryon number conservation. In all cases the difference in the baryon number B and the lepton number L of the final state is

$$B - L = -\tfrac{2}{3} \tag{4.68}$$

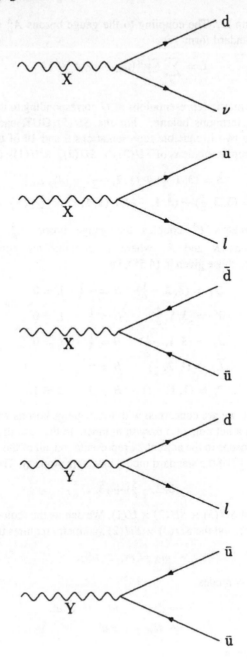

Figure 4.1. Baryon-number non-conserving decays of SU(5) GUT gauge bosons.

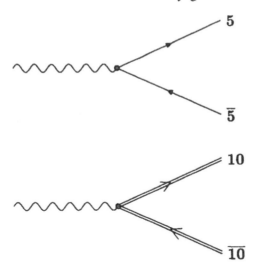

Figure 4.2. Gauge vector boson vertices in the minimal SU(5) GUT.

so we may consistently assign (X, Y) this value of $B - L$, which shows that $B - L$ is conserved in all gauge-boson-mediated processes. B, however, is not separately conserved. Indeed couplings (4.67) and their complex conjugates induce the proton decay modes

$$p \rightarrow \pi^0 e^+, \qquad \pi^+ \nu \qquad (4.69)$$

at a rate which, in the non-supersymmetric model, is calculated [15] to be about one hundred times the measured upper bound implied by (4.28). All of these couplings arise from the Feynman diagram vertices shown in figure 4.2.

Higgs bosons are needed in a GUT to generate both the superheavy masses needed for the non-standard model gauge bosons and Higgs particles, as well as electroweak scale masses for the $W^\pm$ and Z gauge bosons. In the minimal non-supersymmetric $SU(5)$ GUT, the electroweak Higgs doublet is accommodated in a **5** representation H, which in addition includes colour triplet scalars H_3 transforming as $(\mathbf{3}, \mathbf{1}, -\frac{1}{3})$ of $SU(3) \times SU(2) \times U(1)$. The Yukawa couplings have the form

$$\mathcal{L}_Y = \chi_{[IJ]} h_D \psi^I \overline{H}^J + \varepsilon^{IJKLM} \chi_{[IJ]} h_U \chi_{[KL]} H_M + \text{h.c.} \qquad (4.70)$$

where ψ^I, $\chi_{[IJ]}$ are the $\overline{\mathbf{5}}$, **10** representations of $SU(5)$ which include the matter content (4.58) of the standard model. h_U, h_D are complex matrices $(h_{U,D})_{fg}$ acting on the (undisplayed) generation-space labels of ψ^I and $\chi_{[IJ]}$. Then the colour triplet Higgs particles have decay modes similar to those of the X-boson

$$H_3 \rightarrow u^c d^c, \qquad ul, \qquad d\nu \qquad (4.71)$$

all of which violate baryon-number conservation. These are shown in figure 4.3.

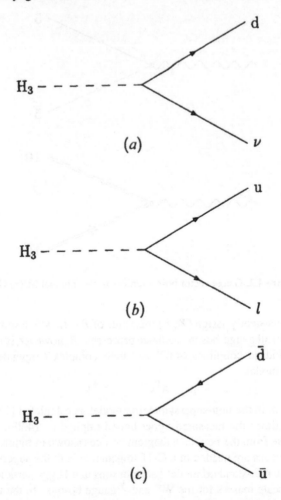

Figure 4.3. Baryon-number non-conserving decays of SU(5) GUT colour triplet Higgs bosons.

As before, in (4.68), the difference in the baryon number B and the lepton number L is the same for all decays:

$$B - L = -\tfrac{2}{3} \tag{4.72}$$

so $B - L$ is conserved in all Higgs mediated processes. CP-violation arises from complex phases which cannot be absorbed by field redefinitions but, as we shall see, there is no contribution to ΔB at tree level. These couplings arise from the Feynman vertices shown in figure 4.4.

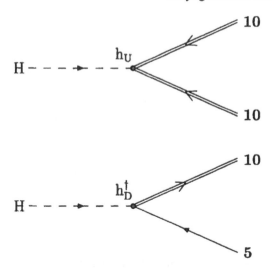

Figure 4.4. Higgs (5) boson vertices in the minimal SU(5) GUT. Single lines represent 5 representations, double lines 10 representations.

We have already noted that, in the minimal $SU(5)$ GUT, the requirements (4.40), (4.44) necessary for departure from thermal equilibrium are more likely to be satisfied by the massive, colour-triplet Higgs scalar H_3 than by the massive gauge bosons, so X, Y decays will not contribute significantly to the baryon asymmetry of the universe. Nevertheless, for completeness, we consider the contributions to ΔB defined in (4.49) from both sources. First we note that the tree-level contribution shown in figures 4.1 and 4.2 is zero. This is clear because in the Born approximation the process $X \rightarrow d\nu$, for example, and $\bar{X} \rightarrow \bar{d}\bar{\nu}$ derive from terms in the Lagrangian which are Hermitian conjugate to each other, so their amplitudes are complex conjugates. Since the kinematics of the two processes is identical,

$$\Gamma(X \rightarrow d\nu)|_{\text{Born}} = \Gamma(\bar{X} \rightarrow \bar{d}\bar{\nu})|_{\text{Born}} \tag{4.73}$$

and the contribution to ΔB given in (4.49) is zero. The same argument applies to the decays of Y and H_3.

At the next order, we need to include radiative corrections and look for (CP-violating) contributions to ΔB arising from the interference between the Born terms and these single-loop radiative corrections. For example, consider the radiative correction to the decay

$$H_3 \rightarrow u_f \ell_g \tag{4.74}$$

shown in figure 4.5 (f, g are generation labels).

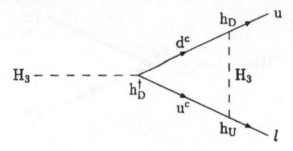

Figure 4.5. Radiative correction to $H_3 \to ul$.

The matrix element has the form

$$\Delta \mathcal{M} \sim (h_D h_D^\dagger h_U)_{fg} I_H \tag{4.75}$$

where I_H is the Feynman loop integral involved. Since the mass of the colour-triplet Higgs satisfies

$$m_{H_3} >> m_u + m_\ell \tag{4.76}$$

I_H is complex. To one-loop order, the square of the total matrix element $\mathcal{M}$ satisfies

$$|\mathcal{M}|^2 - |\mathcal{M}_0|^2 \simeq 2\,\mathrm{Re}[\Delta \mathcal{M} \mathcal{M}_0^\dagger]$$
$$\propto 2\,\mathrm{Re}[(h_D h_D^\dagger h_U)_{fg}(h_U^\dagger)_{gf} I_H] \tag{4.77}$$

(no summation). $\mathcal{M}_0$ is the amplitude for the Born approximation, shown in figure 4.3(b). For the corresponding antiparticle decay, we just replace all coupling constants by their complex conjugates and the difference between the rates is given by

$$\Gamma(H_3 \to u_f \ell_g) - \Gamma(\bar{H}_3 \to \bar{u}_f \bar{\ell}_g) \propto \mathrm{Im}[(h_D h_D^\dagger h_U)_{fg}(h_U^\dagger)_{gf}]\,\mathrm{Im}(I_H). \tag{4.78}$$

Thus, when we sum over the generation labels the contributions cancel, since

$$\mathrm{tr}(h_D h_D^\dagger h_U h_U^\dagger) = \mathrm{real}. \tag{4.79}$$

In fact, all such one-loop interference terms are the absorptive parts of one or other of the two-loop diagrams shown in figure 4.6; the absorptive part is obtained by cutting the internal (fermion) loop and placing the cut fermions on the mass shell. The contribution just discussed arises from the absorptive part of the diagram in figure 4.6(g). It is clear that all of the other diagrams also make zero contribution to the baryon asymmetry [16]: the gauge couplings are real and the scalar couplings enter only in the combinations

$$\mathrm{tr}(h_U h_U^\dagger) = \mathrm{real}$$
$$\mathrm{tr}(h_D h_D^\dagger) = \mathrm{real}. \tag{4.80}$$

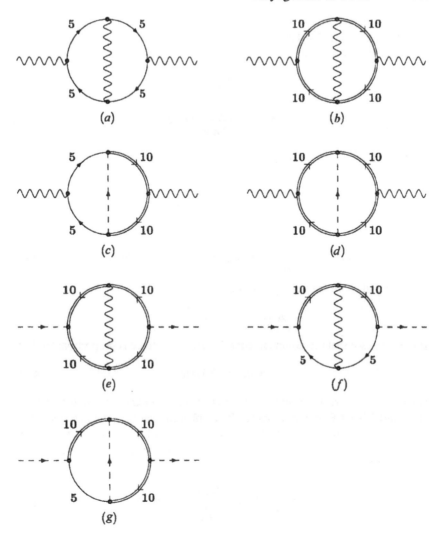

Figure 4.6. Two-loop diagrams giving one-loop radiative corrections to gauge boson and Higgs boson decays.

The first non-zero contributions arise only from the three-loop radiative corrections to H_3 decays, and the four-loop corrections to X,Y decays. For example, there are three-loop radiative corrections to H_3 decay whose contributions to the difference between the decay widths of H_3 and $\bar{H}_3$ is proportional to $\mathrm{Im}T$, where

$$T \equiv \mathrm{tr}(h_D^\dagger h_U h_U^\dagger h_U h_D^\dagger h_D h_U^\dagger h_D) \tag{4.81}$$

which, in general, is non-zero [17–19]. The Yukawa couplings $h_{U,D}$ determine the fermion masses. The fermions in the three families of 10 and $\bar{5}$ representations are unitarily related to the mass eigenstates, the connection being given by

$$h_U = \frac{8G}{\sqrt{2}m_W} P m_U Q \tag{4.82}$$

$$h_D = \frac{8G}{\sqrt{2}m_W} R m_D S \tag{4.83}$$

where P, Q, R, S are unitary 3×3 matrices and

$$m_U = \text{diag}(m_t, m_c, m_u) \tag{4.84}$$

$$m_D = \text{diag}(m_b, m_s, m_d). \tag{4.85}$$

Then

$$2\,\text{Im}\,T = \frac{g^8}{16m_W^8}\,\text{tr}(m_U^2[m_U A m_D^2 A^\dagger, m_U B m_D^2 B^\dagger]) \tag{4.86}$$

where A and B are the unitary matrices

$$A = P^\dagger R \qquad B = Q S^\dagger. \tag{4.87}$$

It is easy to see that the dominant contribution to the trace is proportional to [20]

$$m_b^4 m_t^3 m_c f(\theta) \sin\delta \tag{4.88}$$

where $f(\theta)$ is a real function of the mixing angles characterizing the matrices A, B and δ is a CP-violating phase. Remembering that the total decay width of the (colour-triplet) Higgs scalar is given by (4.43) with $m_f = m_t$ the heaviest fermion, we conclude that the baryon asymmetry deriving from the minimal $SU(5)$ GUT satisfies

$$\Delta B \lesssim \left(\frac{\alpha_G}{2\pi}\right)^3 \frac{m_b^4 m_t m_c}{m_W^6} \sim 10^{-15}. \tag{4.89}$$

However, using (4.52) and the observational data (4.18), we require that

$$\Delta B \simeq 6N_* \times 10^{-10} \gtrsim 10^{-7} \tag{4.90}$$

so there is no doubt that this mechanism cannot explain the measured asymmetry. In any case, we have already noted that this minimal theory gives an unacceptably high proton decay rate.

The foregoing discussion suggests that to increase the predicted value of ΔB, we need to arrange that the asymmetry can arise via one-loop corrections to the Higgs decays. This entails enlarging the Higgs content. The simplest method is to include a second 5 of Higgs scalars H', with a different mass or lifetime, whose couplings are of the same form (4.70) but with the coupling matrices $h_{U,D}$

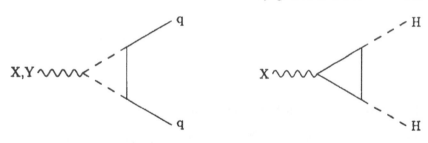

Figure 4.7. CP-violating, baryon-number non-conserving decays of X,Y bosons.

replaced by $h'_{U,D}$. Then, besides the radiative correction to H_3 decay shown in figure 4.5, there will be a similar diagram involving H'_3 exchange. The difference between the decay rates now satisfies

$$\Gamma(H_3 \to u_f \ell_g) - \Gamma(\bar{H}_3 \to \bar{u}_f \bar{\ell}_g) \propto \text{Im}[h'_D h^\dagger_D h'_U h^\dagger_U] \, \text{Im}(I_{H'}) \qquad (4.91)$$

instead of (4.78). In general, this is non-zero [18].

Unfortunately, such a model is unsatisfactory in other respects. It will have flavour-changing, Higgs-mediated neutral current decays [21] and, in addition, like the model with a single **5** of Higgs, it continues to possess the strong CP problem. (See section 5.3.1.) We shall see later that the most attactive solution to the latter problem utilizes a Peccei–Quinn $U(1)$ symmetry [22]. This *requires* additional Higgs doublets, so that one doublet H is coupled only to d_R quarks and the other doublet H' only to u_R quarks. This is realized in an $SU(5)$ GUT by coupling one Higgs **5** to the matter $(\bar{\mathbf{5}})(\mathbf{10})$ and the other to the $(\mathbf{10})(\mathbf{10})$ fields [23]. Thus, instead of (4.70), we have

$$\mathcal{L}_Y = \chi_{[IJ]} h_D \bar{\psi}^I \bar{H}^J + \chi_{[IJ]} h'_U \chi_{[KL]} H'_M \epsilon^{IJKLM} + \text{h.c.} \qquad (4.92)$$

This looks as though the one-loop radiative correction shown in figure 4.5 is now forbidden, since the exchanged Higgs has to have both D-type and U-type couplings. However, it *is* allowed, since the Higgs mass eigenstates mix H and H'. Nevertheless, there is no CP-violation and, hence, no contibution to the baryon asymmetry, for the reason (4.79) given earlier. Instead, in this model, the (CP-violating) baryon asymmetry arises from the decays of X, Y gauge bosons with a pair of Higgs bosons in the intermediate or final state [24,25] (see figure 4.7), and a sufficient asymmetry arises for a wide range of parameters [17,26].

Another variant of the $SU(5)$ minimal model is to introduce a Higgs multiplet belonging to a different representation, i.e. not a **5**. Since, in $SU(5)$,

$$\bar{\mathbf{5}} \times \bar{\mathbf{5}} = \overline{\mathbf{10}} + \overline{\mathbf{15}}$$
$$\bar{\mathbf{5}} \times \mathbf{10} = \mathbf{5} + \mathbf{45}$$
$$\mathbf{10} \times \mathbf{10} = \bar{\mathbf{5}} + \overline{\mathbf{45}} + \overline{\mathbf{50}} \qquad (4.93)$$

Higgs belonging to **10**, **15**, **45** or **50** representations can be coupled to the matter fermions. In particular, the introduction of a **45** Higgs representation also allows the generation of sufficient baryon asymmetry for a wide range of parameters [17].

4.5   Baryogenesis in SO(10) GUTs

We have discussed the $SU(5)$ GUT at some length, despite its inadequacies. There are, however, other GUTs, based on larger (higher-rank) groups, which some consider to be more attractive, and which, in any case, possess features which are not present in the $SU(5)$ GUTs. In particular, as a vector-like theory, $SO(10)$ has the desirable feature of being automatically anomaly-free. Another attraction is the fact that it includes all of the fermions of one generation within a single irreducible **16**-dimensional (spinor) representation of the group. In terms of the representations of $SU(5) \times U(1) \subset SO(10)$, it decomposes as

$$16 = (1, -5) + (\bar{5}, 3) + (10, -1) \tag{4.94}$$

with the second number specifying the $U(1)$ charge of the $SU(5)$ irreducible representation. The $\bar{5}$ and **10** irreducible representations are precisely those to which the matter fermions are assigned in $SU(5)$, shown in (4.62),(4.63), and the extra $SU(5)$ singlet state, which *must* exist if $SO(10)$ is correct, could be a right-chiral neutrino $\nu_R \sim \nu_L^c$: this latter possibility looks feasible as well as attractive in the light of evidence [27] for neutrino oscillations which can, of course, only arise if neutrinos have (different) masses and, therefore, possibly right-chiral components. (The existence of such a state also allows for the possibility of a Majorana mass term.) The inclusion of both of the chiral components of all fermions within the same GUT representation means that charge conjugation (C) and parity reversal (P) are naturally symmetries of the GUT and are spontaneously broken when the GUT breaks to a gauge group which does not have this property. This occurs at a high scale if, as it might, $SO(10)$ breaks to $SU(5)$ or to $SU(5) \times U(1)$. This happens if, for instance, the GUT symmetry-breaking Higgs are also in a **16** representation.

However, there are other possibilities. Clearly

$$SO(10) \supset SO(6) \times SO(4) \tag{4.95}$$

so, since

$$SO(6) \cong SU(4)$$
$$SO(4) \cong SU(2) \times SU(2) \tag{4.96}$$

one possibility is that the $SO(10)$ breaks at the GUT scale as

$$SO(10) \xrightarrow{M_G} SU(4) \times SU(2)_L \times SU(2)_R \equiv G_{422}. \tag{4.97}$$

This happens if, for instance, the GUT symmetry-breaking Higgs transforms as a **54**-dimensional representation. Then the **16**-dimensional fermion representation is given by

$$16 = (4, 2, 1) + (\bar{4}, 1, 2)$$

$$= \left[\begin{pmatrix} u_i \\ d_i \end{pmatrix}_L, \begin{pmatrix} v_e \\ e \end{pmatrix}_L \right] + \left[\begin{pmatrix} d_i^c \\ -u_i^c \end{pmatrix}_L, \begin{pmatrix} e^c \\ -v_e^c \end{pmatrix}_L \right] \qquad (4.98)$$

for the first generation. At later stages, the $SU(4)$ breaks at a scale M_c and the $SU(2)_R$ breaks at a scale M_R as

$$SU(4) \xrightarrow{M_c} SU(3)_c \times U(1)' \qquad (4.99)$$

and

$$SU(2)_R \xrightarrow{M_R} U(1)_{T_R^3}. \qquad (4.100)$$

So either

$$G_{422} \xrightarrow{M_c} G_{3122} \xrightarrow{M_R} G_{3121} \xrightarrow{M_1} G_{sm} \qquad (4.101)$$

or

$$G_{422} \xrightarrow{M_R} G_{421} \xrightarrow{M_c} G_{3121} \xrightarrow{M_1} G_{sm} \qquad (4.102)$$

where

$$G_{3122} \equiv SU(3)_c \times U(1)' \times SU(2)_L \times SU(2)_R \qquad (4.103)$$
$$G_{3121} \equiv SU(3)_c \times U(1)' \times SU(2)_L \times U(1)_{T_R^3} \qquad (4.104)$$
$$G_{421} \equiv SU(4) \times SU(2)_L \times U(1)_{T_R^3} \qquad (4.105)$$

and $G_{sm} = SU(3)_c \times SU(2)_L \times U(1)_Y$ is the standard model gauge group.

The content of the **4** representation of $SU(4)$ under the decomposition (4.99) is

$$4 = (3, \tfrac{1}{3}) + (1, -1) \qquad (4.106)$$

since $U(1)'$ is a (traceless) generator of $SU(4)$. We have used a normalization of the hypercharge Y' which shows that, for fermions *only*,

$$Y' = B - L. \qquad (4.107)$$

This was first noted by Pati and Salam [28] and, for this reason $U(1)'$ is sometimes denoted $U(1)_{B-L}$. At any rate, unlike the $SU(5)$ GUTs, it *is* possible to break $B - L$ conservation in $SO(10)$ models. As previously noted, the scale M_1 at which $U(1)'$ and, therefore, $B - L$ conservation is broken is not necessarily the scale at which $SO(10)$ is broken.

In fact, the $SO(10)$ group has an element D which interchanges the charge conjugate doublets within the **16** representation (4.98). To see this, we choose the decomposition (4.95) so that the Cartan subalgebra of $SO(6)$ is generated by the $SO(10)$ generators M^{12}, M^{34}, M^{56} and that of $SO(4)$ by $M^{78}, M^{9,10}$. The Cartan subalgebras of $SU(2)_{L,R}$ are then generated by $T_{L,R}^3 = \frac{1}{2}(M^{78} \pm M^{9,10})$. We first note that

$$D \equiv M^{23} M^{67} \tag{4.108}$$

where

$$(M^{ab})_{ij} \equiv \delta_i^a \delta_j^b - \delta_j^a \delta_i^b \tag{4.109}$$

is an element of $SO(10)$. The action of D on the underlying 10-dimensional space is to reflect the coordinates $x^a \leftrightarrow -x^a$ ($a = 2, 3, 6, 7$) leaving the remainder invariant. Thus, the effect of D on the Cartan subalgebra of $SU(4) \cong SO(6)$ is to reverse the signs of the generators

$$D : T^{3,8,15} \rightarrow -T^{3,8,15} \tag{4.110}$$

while on that of $SU(2)_L \times SU(2)_R$

$$D : T_L^3 \leftrightarrow T_R^3. \tag{4.111}$$

Thus,

$$D : (\mathbf{4}, \mathbf{2}, \mathbf{1}) \leftrightarrow (\bar{\mathbf{4}}, \mathbf{1}, \mathbf{2}) \tag{4.112}$$

as asserted.

Because of this, unlike the $SU(5)$ GUTs, a baryon asymmetry is often not generated at the GUT-breaking scale but rather at the lower scale M_1 at which $U(1)'$ is broken. Whether or not this happens, of course, depends upon the Higgs fields responsible for this symmetry breaking and their coupling to the matter fields. The Higgs fields that couple to matter must be in one or more of the $SO(10)$ representations occurring in the product

$$\mathbf{16} \times \mathbf{16} = \mathbf{10}_s + \mathbf{120}_a + \mathbf{126}_s. \tag{4.113}$$

For example, with the colour-triplet Higgs particles belonging to the **10**-dimensional representation, the symmetry (4.112) ensures that

$$\Gamma(H_3 \rightarrow \ell Q) = \Gamma(\bar{H}_3 \rightarrow \bar{\ell}\bar{Q}) \tag{4.114}$$

at one-loop level, so no baryon asymmetry results, similarly in the QQ channels [29]. However, if the Higgs content is enlarged to include a **45**-dimensional, a **126**-dimensional and an additional **10**-dimensional representations, then with a suitable choice of $\langle 45 \rangle$ we can break

$$SO(10) \rightarrow G_{3122} \tag{4.115}$$

in a way which breaks the D-symmetry (G_{3122} is defined in (4.103)). This is done by ensuring that the non-zero VEVs are odd under D. In addition, the extra Higgs content splits g_L and g_R, the coupling constants of the $SU(2)_{L,R}$ groups, *and* ensures that the colour-triplet Higgs mass eigenstates are *complex* superpositions of those coming from the two **10**s. This is sufficient to generate a baryon asymmetry at the GUT symmetry-breaking scale [29].

The lesson to be learnt from these considerations is that a baryon asymmetry *can* arise in such theories but that the energy scale at which it arises may be much lower than the GUT scale. The magnitude of any such asymmetry depends sensitively on the details of the particular model but, in many models, there is ample room in parameter space to accommodate the observed asymmetry.

4.6   Status of GUT baryogenesis

The discussion in the two previous sections of baryogenesis using the baryon-number non-conserving interactions of a GUT was based upon the *assumption* that the superheavy GUT gauge bosons or Higgs particles whose decays produce the desired baryon asymmetry are, in the first place, in thermal equilibrium and then, as the universe expanded and the temperature dropped, came out of equilibrium when the condition (4.37) was satisfied. It is at least debatable whether this assumption is well founded.

We shall see in chapter 7 that there are strong theoretical reasons, and some support from observational data, for believing that the universe went through a period of 'inflation', during which the scale factor grew by a factor of order 10^{27}, so that the observable universe evolved from a single Hubble volume. It is, therefore, essential that the baryon asymmetry we observe was generated *after* inflation: any asymmetry generated earlier would be so diluted by the inflation as to render it utterly unobservable at the present time. This is the source of the difficulty with the scenario envisaged hitherto. To generate the required amount of inflation requires the inflaton potential to be rather flat: (the 'inflaton' (ϕ) is the presumed field whose evolution determines how much inflation actually occurs). This means that the mass of the inflaton is relatively low [30], in the range

$$m_\phi \lesssim 10^{13}\text{--}10^{15} \text{ GeV} \tag{4.116}$$

to account for the observed flatness and homogeneity of the universe and to solve the horizon problem[1]. When it reaches the minimum of its potential, the inflaton oscillates about its value at the minimum. As it does so, somehow, this low-entropy, cold universe evolves into a hot universe dominated by radiation. The key question is: What is the temperature of this 'reheated' universe? This is a vital question because the bound (4.116) means that, in some cases,

$$m_\phi < 2m_X \tag{4.117}$$

where m_X is the mass of the particle whose decays generate the baryon asymmetry. This is the case in the minimal $SU(5)$ GUT, for example, where

---

[1] This is because $m_\phi^2 \sim V''(\phi)$, and the double derivative is constrained by the condition (7.45) with $V(\phi)$ satisfying (7.117).

$X = H_3$ is the colour-triplet Higgs particle with

$$m_{H_3} \gtrsim 10^{14} \text{ GeV}. \tag{4.118}$$

In these circumstances, the decay

$$\phi \to X\bar{X} \tag{4.119}$$

is kinematically forbidden, so if X particles are created, they must be created by thermal production in the reheated universe. (This is why our comments in section 4.5 about the possibility in other GUTs of baryon asymmetry arising at a scale well below M_G are pertinent.) So the next question is: What is the abundance of the out-of-equilibrium X particles thus created? It is beyond our scope to discuss here the calculation of the reheating temperature, the abundance, and other questions which arise, in any detail. The interested reader is referred to [31] and references therein. We shall content ourselves with noting the conclusions which have been reached from these studies. A mechanism called 'parametric resonance', deriving from nonlinear quantum effects, leads to a phenomenon called 'preheating' during which copious production of X particles occurs even though these are heavier than the inflaton. It appears that the out-of-equilibrium scenario, upon which our analysis was predicated, arises naturally and that the right amount of baryon asymmetry is produced for a very wide range of decay widths of the X particles.

Notwithstanding this highly welcome outcome, one is naturally led to wonder whether the observed baryon asymmetry might not have a different origin. Although, as we have noted, there is evidence of (supersymmetric) unification of coupling strengths, this does not necessarily entail the existence of a GUT. It is quite conceivable, in the context of string theory for example, that there is no GUT. If so, the observed baryon asymmetry must have a different origin. This is the topic to which we now turn.

4.7 Baryon-number non-conservation in the Standard Model

It is easy to see that the standard model Lagrangian, having the local gauge symmetry group $SU(3)_c \times SU(2)_L \times U(1)_Y$, is also invariant under the (classical) global $U(1)$ transformations associated with the baryon number (B) and lepton numbers (N_ℓ, $\ell = e, \mu, \tau$) in which fermion fields $\psi(x)$ transform as

$$\psi(x) \to e^{iB\theta}\psi(x) \tag{4.120}$$

$$\psi(x) \to e^{iN_\ell\theta}\psi(x). \tag{4.121}$$

When θ is local, the first of these (4.120), for example, applied to the kinetic term produces a change δS in the action

$$\delta S = -\int d^4x \, (\bar{\psi}\gamma^\mu B_\psi \psi)\partial_\mu\theta$$

$$= \int d^4x \, \theta \partial_\mu (\bar\psi \gamma^\mu B_\psi \psi) \tag{4.122}$$

where B_ψ is the baryon number of ψ. Since the remainder of the Lagrangian density is invariant under the (global and therefore the local) transformation, the (Noether) current

$$j_\mu^{(B)} \equiv \sum_\psi \bar\psi \gamma_\mu B_\psi \psi \tag{4.123}$$

associated with baryon number, is divergenceless

$$\partial^\mu j_\mu^{(B)} = 0 \tag{4.124}$$

and B is conserved at every order of perturbation theory. Similarly, for the lepton currents,

$$j_\mu^{(L_\ell)} \equiv \sum_\psi \bar\psi \gamma_\mu L_{\ell\psi} \psi \tag{4.125}$$

However, classical symmetries such as these are generally not preserved at the quantum level [32–34]. In particular, in a chiral theory, in which the left- and right-chiral fermion states are coupled differently, as occurs in electroweak theory, there are chiral anomalies resulting from the non-invariance of the field measure $\mathcal{D}\psi \mathcal{D}\bar\psi$ in the functional integral determining the generating function [34]. This non-invariance is equivalent to a further change $\delta S'$ in the action

$$\delta S' = \frac{1}{32\pi^2} \int d^4x \, \theta(x) [\text{tr}[\{F_{\mu\nu}^{(L)}, \tilde{F}^{\mu\nu(L)}\}B] - \text{tr}[\{F_{\mu\nu}^{(R)}, \tilde{F}^{\mu\nu(R)}\}B]] \tag{4.126}$$

where $F_{\mu\nu}^{(L)} \equiv t^a F_{\mu\nu}^{a(L)}$ is the gauge field strength coupled to the left-chiral component $\psi_L \equiv \frac{1}{2}(1 - \gamma_5)\psi$ of ψ; t^a is the generally matrix-valued coupling of the representation of the gauge group to which the left-chiral component ψ_L of the fermion belongs, $\tilde{F}_{\mu\nu}^{(L)} \equiv \frac{1}{2}\epsilon_{\mu\nu\rho\sigma} F^{\rho\sigma(L)}$ is the dual field strength and $\{ , \}$ denotes anticommutator. The trace is over all fermions ψ. Similarly for $F_{\mu\nu}^{(R)}$. Thus, combining this with (4.122), we see that quantum effects require that

$$\partial^\mu j_\mu^{(B)} = -\frac{1}{32\pi^2} \text{tr}[\{F_{\mu\nu}^{(L)}, \tilde{F}^{\mu\nu(L)}\}B - \{F_{\mu\nu}^{(R)}, \tilde{F}^{\mu\nu(R)}\}B]. \tag{4.127}$$

For the standard model, we note that there is no contribution to the divergence from the $SU(3)_c$ group, because, for each quark flavour,

$$F_{\mu\nu}^{(L)} = F_{\mu\nu}^{(R)}. \tag{4.128}$$

The left-chiral quark states Q_L of the first generation are coupled both to the $SU(2)_L$ field strength $W_{\mu\nu}^a$ ($a = 1, 2, 3$) with $t^a = g_2 \frac{1}{2}\tau^a$ (τ^a are the 2 × 2 Pauli matrices) and to the $U(1)_Y$ field strength $B_{\mu\nu}$ with strength $g_1 Y_L = g_1 \frac{1}{6}$. The right-chiral quark fields u_R, d_R are coupled only to $B_{\mu\nu}$ with strength $g_1 Y_R = g_1 \frac{2}{3}, g_1 \frac{-1}{3}$ respectively. Thus the baryon number current

$$j_\mu^{(B)} \equiv \tfrac{1}{3} \sum_q \bar{q} \gamma_\mu q \qquad (4.129)$$

satisfies

$$\partial^\mu j_\mu^{(B)} = -\frac{N_G}{32\pi^2}[g_2^2 W_{\mu\nu}^a \tilde{W}^{a\mu\nu} - g_1^2 B_{\mu\nu} \tilde{B}^{\mu\nu}] \qquad (4.130)$$

where $N_G = 3$ is the number of fermion generations. (It is important to remember that each quark flavour q occurs in three colours.) The meaning of this important equation is that the quantum average, the expectation value, of the divergence is equal to the expression on the right-hand side in a fixed background field given by the gauge fields $W_\mu^a(x)$ and $B_\mu(x)$ [35].

In the same way, it is easily seen that the lepton number currents

$$j_\mu^{(L_\ell)} \equiv \bar{\nu}_\ell \gamma_\mu \nu_\ell + \bar{\ell} \gamma_\mu \ell \qquad (\ell = e, \mu, \tau) \qquad (4.131)$$

satisfy

$$\partial^\mu j_\mu^{(L_\ell)} = \frac{1}{N_G} \partial^\mu j_\mu^{(B)} \qquad (4.132)$$

and it follows that $(1/N_G)B - L_\ell$ is conserved for each ℓ. Also, defining the total lepton number

$$L \equiv \sum_\ell L_\ell \qquad (4.133)$$

we see that $B - L$ is conserved, even though neither B nor L is.

Both of the field-strength factors on the right-hand side of (4.130) are expressible as total divergences:

$$B_{\mu\nu}\tilde{B}^{\mu\nu} = \partial_\mu k^\mu \qquad (4.134)$$

where

$$k^\mu = \epsilon^{\mu\nu\rho\sigma} B_{\nu\rho} B_\sigma \qquad (4.135)$$

and

$$W_{\mu\nu}^a \tilde{W}^{a\mu\nu} = \partial_\mu K^\mu \qquad (4.136)$$

where

$$\begin{aligned} K^\mu &= \epsilon^{\mu\nu\rho\sigma}[W_{\nu\rho}^a W_\sigma^a + \tfrac{1}{3} g_2 \epsilon^{abc} W_\nu^a W_\rho^b W_\sigma^c] \\ &= 2\epsilon^{\mu\nu\rho\sigma} \text{tr}[W_{\nu\rho} W_\sigma - \mathrm{i}\tfrac{2}{3} g_2 W_\nu W_\rho W_\sigma]. \end{aligned} \qquad (4.137)$$

Note that, although the divergences $\partial_\mu k^\mu$ and $\partial_\mu K^\mu$ are gauge invariant, the individual currents are *not*. In the first instance, let us consider the (classical) background vacuum fields in the $SU(2)_L \times U(1)_Y$ theory. The standard solution is to take all fields to be zero except for the Higgs doublet for which

$$\phi^{\text{vac}} = \frac{1}{\sqrt{2}} \begin{pmatrix} 0 \\ v \end{pmatrix} \equiv \phi_0 \tag{4.138}$$

where v is constant. Both K^μ and k^μ are then zero, of course. Because of the $SU(2)_L \times U(1)_Y$ gauge invariance of the Lagrangian, we could as well choose a gauge transformation of this solution. Then,

$$\phi^{\text{vac}}(x) = U(x)U_1(x)\phi_0 \tag{4.139}$$

where $U_1(x)$ is a general element of $U(1)_Y$ and $U(x)$ is likewise a general element of $SU(2)_L$, so we may write

$$U(x) = a(x)I_2 + i\tau \cdot b(x) \tag{4.140}$$

with

$$\det U(x) = a(x)^2 + b(x) \cdot b(x) = 1. \tag{4.141}$$

Thus, $U(x)$ can be regarded as a mapping from spacetime into the three-sphere (S^3) which is the group space of $SU(2)$. In this gauge, the $U(1)$ vector potential is non-zero:

$$B_\mu^{\text{vac}} = \frac{i}{g_1}(\partial_\mu U_1)U_1^{-1} \tag{4.142}$$

but, of course, the (gauge-invariant) field strength $B_{\mu\nu}^{\text{vac}}$ is still zero and k^μ remains zero:

$$k_\mu^{\text{vac}} = 0. \tag{4.143}$$

Similarly, in this gauge, the $SU(2)$ gauge field becomes

$$W_\mu^{\text{vac}} = \frac{i}{g_2}(\partial_\mu U)U^{-1} \equiv \frac{1}{2}\tau \cdot W_\mu \tag{4.144}$$

and, with the parametrization (4.140), this gives

$$W_\mu^{\text{vac}} = -\frac{2}{g_2}(a\partial_\mu b - b\partial_\mu a + \partial_\mu b \times b) \tag{4.145}$$

For future reference, we note that, using (4.141),

$$W_\mu^{\text{vac}} \cdot W_\nu^{\text{vac}} = \frac{4}{g_2^2}[(\partial_\mu a)(\partial_\nu a) + (\partial_\mu b) \cdot (\partial_\nu b)]$$

$$\equiv \frac{4}{g_2^2}\gamma_{\mu\nu} \tag{4.146}$$

where $\gamma_{\mu\nu}$ is the metric on S^3 for the spacetime coordinates. The vacuum state described by (4.144) may be taken to be time-independent and such that $U \to I_2$

as $|x| \to \infty$. Then U effectively maps the S^3 of real space, obtained by identifying points at infinity, into the S^3 which is the $SU(2)$ group space [36]. Such mappings are characterized by their homotopy class. Since

$$\pi_3(S^3) = Z, \tag{4.147}$$

where Z is the group of integers, any such mapping is associated with an integer which counts the number of times the spacetime S^3 is wrapped around the (unit) S^3 of the internal $SU(2)$ space.

Now, the field strength $W_{\mu\nu}^{vac}$ constructed from W_{μ}^{vac} is zero but the current K^{μ} (4.137) is no longer zero:

$$K^{\mu vac} = -\frac{4ig_2}{3}\epsilon^{\mu\nu\rho\sigma}\, \text{tr}[W_{\nu}^{vac}W_{\rho}^{vac}W_{\sigma}^{vac}]. \tag{4.148}$$

So

$$\begin{aligned} K^{0vac} &= \frac{g_2}{3}\epsilon^{0ijk}\epsilon^{abc}(W_i^{avac}W_j^{bvac}W_k^{cvac}) \\ &= -2g_2 \det W^{vac}. \end{aligned} \tag{4.149}$$

Also

$$\det W^{vac} = \sqrt{\det G} \tag{4.150}$$

where

$$G_{ij} = W_i^{vac} \cdot W_j^{vac} = \frac{4}{g_2^2}\gamma_{ij} \tag{4.151}$$

and γ_{ij} is the metric (4.146) on S^3 for the spatial coordinates. Thus, if we define

$$K \equiv \int d^3x \, K^{0vac} \tag{4.152}$$

we have that

$$\begin{aligned} K &= -\frac{16}{g_2^2}\int d^3x \sqrt{\det \gamma} \\ &= -\frac{16}{g_2^2}\int_{S^3} dV \\ &= -\frac{32\pi^2}{g_2^2}n \end{aligned} \tag{4.153}$$

where n is an integer which counts the number of wrappings of the internal S^3 provided by the mapping U; the volume of the unit S^3 is $2\pi^2$. For a general (non-vacuum) field configuration,

$$K = \frac{32\pi^2}{g_2^2}N_{CS} \tag{4.154}$$

where N_{CS} is the Chern–Simons number. It derives from the existence of a 3-form $K_{(3)}$ which arises because the 4-form $\mathrm{tr}(W_{(2)} \wedge W_{(2)})$ is exact; $W_{(2)} \equiv W_{\mu\nu} \, \mathrm{d}x^\mu \wedge \mathrm{d}x^\nu$ is the field strength 2-form. Thus, in our (vacuum) case, N_{CS} is just (minus) the winding number n.

Evidently electroweak theory, unlike QED, has an infinity of topologically distinct vacua, which may be labelled with their Chern–Simons number. These vacua are not *physically* distinct [36]: they are gauge transformations of each other, as we have emphasized. Now consider the total baryon number

$$B \equiv \int \mathrm{d}^3x \, j^{0(B)} \tag{4.155}$$

and let the gauge and Higgs fields be general and have time dependence. Then the baryon number B will also be time-dependent. In a time interval (t_i, t_f), the change is

$$\Delta B = \int_{t_i}^{t_f} \mathrm{d}t \, \partial_0 B = \int \mathrm{d}^4x \, \partial_\mu j^{\mu(B)} \tag{4.156}$$

assuming that $j^{(B)}$ vanishes at spatial infinity. Note that using (4.130) this *is* gauge invariant. Using (4.134),(4.136) and (4.143), we see that

$$\begin{aligned} \Delta B &= -\frac{N_G g_2^2}{32\pi^2} \int \mathrm{d}^4x \, \partial_\mu K^\mu \\ &= -\frac{N_G g_2^2}{32\pi^2} \Delta K. \end{aligned} \tag{4.157}$$

Now suppose that in this time interval the gauge and Higgs fields traverse a non-contractible loop in the field configuration space, starting and finishing in a *vacuum* configuration. Then the change ΔK is given just by the vacuum formula (4.153), which gives

$$\Delta B = -N_G \Delta n = N_G \Delta N_{CS}. \tag{4.158}$$

In other words, in the standard model, baryon-number non-conservation arises in integer multiples of N_G whenever the initial and final vacuum states are topologically distinct.

It is instructive to consider what is happening using Dirac's picture of the vacuum as the state with all negative-energy fermion levels filled [37]. We start (at t_i) and finish (at t_f) in this state. In the presence of non-zero field strengths and Higgs fields, we expect the energy levels to be displaced at intermediate times. In the (trivial) case of QED, for example, although the levels are perturbed from their values at t_i, they each return to their original level at t_f, because the electromagnetic field interacts with the left- and right-chiral fermion components with equal strength. However, the baryogenesis with which we are concerned derives from the chiral nature of electroweak theory: only left-chiral fermions interact with the $SU(2)$ gauge bosons. Depending on the field configurations, this

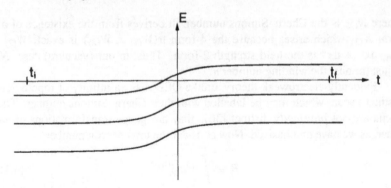

Figure 4.8. Fermion energy-level crossing in electroweak theory.

chiral property allows the energy levels of *all* of the (infinite number of) negative energy states to be raised by one (or more) level such that at t_f all of the negative energy levels remain occupied. In the process, one (or more) of the *positive* energy levels is occupied and we see one (or more) fermions produced. (See figure 4.8.)

Of course, if just one fermion is produced, angular momentum is not conserved, and, in fact, electroweak theory with just one doublet *is* inconsistent: it has a chiral anomaly. In the realistic case, each generation has four doublets: three (colours of) quark doublet and one lepton doublet. Thus, with $n_G = 3$, there are 12 doublets in all and with the minimum of just one level crossing, 12 fermions are created, nine quarks and three leptons. An allowed process, which has total charge zero, so charge is conserved but has baryon number 3 and each lepton number 1, so that $N_\ell - \frac{1}{3}B$ is conserved might create from the vacuum

$$uudeuddv_\mu uddv_\tau \tag{4.159}$$

or, equivalently,

$$pn \to \bar{n}e^+\bar{v}_\mu\bar{v}_\tau. \tag{4.160}$$

4.8   Sphaleron-induced baryogenesis

At intermediate times between t_i and t_f, there are non-vacuum field configurations, which necessarily have higher energy associated with them. The situation is, therefore, analogous to a particle moving in a one-dimensional periodic potential $V(x)$, in which a potential barrier separates adjacent minima. If the particle has energy E less than the barrier height, classically it remains trapped in one of the valleys of the potential, oscillating between the turning points where the total energy $E = V(x)$. However, quantum mechanically there is a non-zero probability of penetrating the barrier. In the semi-classical approximation, the amplitude for transmission is given by the WKB formula in which the amplitude is suppressed by an exponential factor

$$T = \exp\left(-\int_{x_0}^{x_1} p \, dx\right) \qquad (4.161)$$

where $p(x) = \sqrt{2m[V(x) - E]}$ and the limits of integration x_0 and x_1 are the two points at which the kinetic energy of the classical particle is zero, so that $V(x_0) = E = V(x_1)$. It is convenient to choose the zero of energy such that $E = 0$. Suppose that x_0 is a minimum of $V(x)$, so that we are considering the probability of a particle in classical equilibrium tunnelling through the potential barrier. In several dimensions, (minus) the exponent in (4.161) generalizes [38] to

$$b \equiv \int_{x_0}^{x_1} ds \, \sqrt{2mV(x)} \qquad (4.162)$$

(where $ds^2 = dx \cdot dx$) and the integral is to be evaluated along the path for which b is a minimum. The required path $x(\tau)$, therefore, satisfies

$$m\frac{d^2x}{d\tau^2} = \nabla V \qquad (4.163)$$

with

$$\frac{1}{2}m\frac{dx}{d\tau} \cdot \frac{dx}{d\tau} - V(x) = 0. \qquad (4.164)$$

Equation (4.163) is just the Euler–Lagrange equation for the imaginary-(or Euclidean-)time version of Hamilton's principle, in which the formal substitution $\tau = it$ is made. In other words, it minimizes the Euclidean action

$$S_E = \int d\tau \, L_E \qquad (4.165)$$

where

$$L_E \equiv \frac{1}{2}m\frac{dx}{d\tau} \cdot \frac{dx}{d\tau} + V(x). \qquad (4.166)$$

Equivalently, it describes the motion of a particle in time τ moving in the inverted potential $-V(x)$. It is clear then that the classical equilibrium point x_0 can only be reached asymptotically, as $\tau \to -\infty$

$$\lim_{\tau \to -\infty} x = x_0. \qquad (4.167)$$

We can choose the time at which the particle reaches x_1, where $dx/d\tau = 0$ next, to be $\tau = 0$. Then the exponent b may be written as

$$b = \int_{x_0}^{x_1} ds \, \sqrt{2mV(x)} = \int_{-\infty}^{0} d\tau \, L_E. \qquad (4.168)$$

For positive τ, the motion is just the reverse of the motion for negative τ, so that

$$b = \tfrac{1}{2} \int_{-\infty}^{\infty} \mathrm{d}\tau \, L_E = \tfrac{1}{2} S_E. \tag{4.169}$$

Thus, the quantum-mechanical transition *rate* Γ at which the particle tunnels through the potential barrier is exponentially suppressed as [39, 40]

$$\Gamma \propto e^{-S_E}. \tag{4.170}$$

In a quantum field theory, to find the transition rate between adjacent vacua we need a field configuration which interpolates between them in Euclidean spacetime, as in section 2.9. We can get a feel for what is involved using an observation made by Belavin *et al* [41] and 't Hooft [42]. For a Euclidean spacetime,

$$\int \mathrm{d}^4 x \, (W_{\mu\nu}^a - \tilde{W}_{\mu\nu}^a)^2 \geq 0. \tag{4.171}$$

So, ignoring the Higgs contribution, the action is

$$\begin{aligned}
S_E &= \tfrac{1}{4} \int \mathrm{d}^4 x \, W_{\mu\nu}^a W_{\mu\nu}^a = \tfrac{1}{4} \int \mathrm{d}^4 x \, \tilde{W}_{\mu\nu}^a \tilde{W}_{\mu\nu}^a \\
&\geq \tfrac{1}{4} \int \mathrm{d}^4 x \, W_{\mu\nu}^a \tilde{W}_{\mu\nu}^a \\
&= \tfrac{1}{4} \int \mathrm{d}^4 x \, \partial_\mu K^\mu \\
&= \frac{8\pi^2}{g_2^2} \Delta N_{CS}
\end{aligned} \tag{4.172}$$

where we have used (4.136) and (4.154). So a gauge field configuration which interpolates between vacua with $\Delta N_{CS} = 1$ will have Euclidean action

$$S_E(1) \geq \frac{2\pi}{\alpha_2}. \tag{4.173}$$

Already we can see that the tunnelling probability is likely to be incredibly small, since

$$\exp[-S_E(1)] \sim 10^{-80} \tag{4.174}$$

using

$$\alpha_2^{-1} \simeq \alpha_{\text{em}}^{-1} \sin^2 \theta_W \simeq 30 \tag{4.175}$$

as suggested by current data. Of course, (at zero temperature) in a pure Yang–Mills theory, such as this, nothing sets the overall scale, so we may not yet write down the tunnelling rate per unit volume. To do this, we need to consider the spontaneously broken theory.

A related problem is to determine the energy scale of the potential barrier separating adjacent minima. The schematic diagram in figure 4.9 suggests that

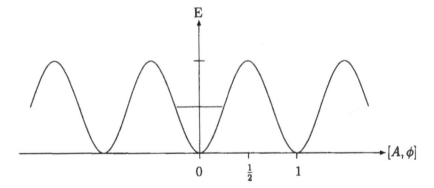

Figure 4.9. Schematic picture of the dependence of the static energy of the gauge-Higgs system upon the field configuration $[A, \phi]$. The minima are topologically distinct vacua, labelled by the integer-valued Chern–Simons numbers. The maxima, on this slice, are unstable sphaleron configurations.

the barrier height will be the energy associated with a field configuration having Chern–Simons number $N_{CS} = \frac{1}{2}$. (Remember that N_{CS} is only integral in a vacuum field configuration.) Manton [36] and Klinkhamer and Manton [43] constructed just such an object. They assumed a static, spherically symmetric ansatz in which the Higgs field

$$\phi(r) = h(m_W r)U(r)\phi_0 \tag{4.176}$$

with

$$\phi^0 = \frac{1}{\sqrt{2}} \begin{pmatrix} 0 \\ v \end{pmatrix} \tag{4.177}$$

as before, $m_W = \frac{1}{2}g_2 v$ and

$$W_i = \frac{i}{g_2} f(m_W r)(\partial_i U)U^{-1} \tag{4.178}$$

where U has the form (4.140)

$$U = \frac{1}{r}[z I_2 + i y \tau_1 + i x \tau_2] \tag{4.179}$$

and

$$r = |r| = (x^2 + y^2 + z^2)^{1/2}. \tag{4.180}$$

h and f are functions to be determined. The energy of the field configuration is given by

$$E = \int d^3x \left[\frac{1}{2} \text{tr}(W_{ij} W_{ij}) + \frac{1}{4} B_{ij} B_{ij} + (D_i \phi)^\dagger (D_i \phi) + V(\phi)\right] \tag{4.181}$$

where $W_{\mu\nu}$, $B_{\mu\nu}$ are the usual field strengths, $D_i\phi$ is the gauge covariant derivative

$$D_i\phi = \partial_i\phi + ig_2 W_i\phi + i\tfrac{1}{2}g_1 B_i\phi \qquad (4.182)$$

and

$$V(\phi) = \lambda(\phi^\dagger\phi - \tfrac{1}{2}v^2)^2. \qquad (4.183)$$

In the first instance, the $U(1)$ fields are ignored, i.e. $\theta_W = 0$. Substituting the ansatz gives the energy in the form

$$E = \frac{2m_W}{\alpha_2} \int_0^\infty d\xi\, F\left(\xi, f, f', h, h'; \frac{m_H}{m_W}\right) \qquad (4.184)$$

where $\xi = m_W r$, so that f and h are functions of ξ alone, and $m_H^2 = 2\lambda v^2$ is the Higgs mass. The Euler–Lagrange equations which follow from requiring E to be minimized may be solved approximately or numerically, with f and h required to approach zero as $\xi \to \infty$. Substituting the solutions back into E gives the minimum energy in the form

$$E_{\min} = \frac{2m_W}{\alpha_2}\mathcal{E}\left(\frac{m_H}{m_W}\right). \qquad (4.185)$$

The scale is set by the prefactor

$$\frac{2m_W}{\alpha_2} \simeq 5\,\text{TeV} \qquad (4.186)$$

and the function $\mathcal{E}$ is rather slowly varying: it increases only from 1.5 to 2.7 as m_H increases from zero to infinity, taking the value 2.1 when $m_H = 2.8m_W$. In the range 25 GeV $< m_H <$ 250 GeV, $\mathcal{E}$ is well approximated by

$$\mathcal{E}(x) = 1.58 + 0.32x - 0.05x^2. \qquad (4.187)$$

Allowing $\theta_W \neq 0$ changes $E_{\min}$ by about 100 GeV.

Although the configuration has the minimum energy among those satisfying the ansatz, it is nevertheless unstable against perturbations which do not satisfy the ansatz. This was to be expected too from the schematic diagram in figure 4.9, in which the top of the potential barrier is evidently a maximum. Because of this instability, the configuration was named a 'sphaleron', from the Greek meaning 'ready to fall'. The energy of the sphaleron satisfies

$$8\,\text{TeV} < E_{\text{sph}} < 14\,\text{TeV} \qquad (4.188)$$

and measures the height of the saddle point in configuration space over which the vacuum must be 'pushed' to reach a topologically distinct vacuum. This suggests that the exponentially small tunnelling rate (4.174) might be evaded by supplying O(10 TeV) of energy, for example in a pn collision. Then some baryon number non-conserving transition, such as (4.160), might be observed. Such energies will soon be available at the LHC. However, it is not just the potential barrier which has to be surmounted. The energy scale (4.188) suggests that about 30 weak gauge bosons must be created and assembled into the highly coherent sphaleron configuration, so that the process is highly suppressed by phase-space considerations unless the energy is considerably in excess of the sphaleron mass.

The other possibility, and the one which is of primary concern to us, is that, at temperatures $k_B T \sim E_{\text{sph}}$, thermal effects might give baryon number non-conservation at an appreciable rate. To calculate the thermal transition rate we must remember that the effective potential becomes temperature dependent [44] and, in consequence, so does the the Higgs field expectation value $v(T)$ and, hence, the W-boson mass $m_W(T)$. In fact, using (2.55), since λ in (2.93) is $\frac{1}{4}$ that in (2.42),

$$m_W(T) = \frac{1}{2}gv(T) = m_W\left[\frac{CT}{2\sqrt{\lambda}|m|} + \left(1 - \frac{T^2}{T_1^2}\right)^{1/2}\right] \qquad (4.189)$$

where

$$|m| = \sqrt{\lambda}v = \frac{m_H}{\sqrt{2}} \qquad (4.190)$$

at tree level,

$$C = \frac{3e^3(1 + 2\cos^3\theta_W)}{4\pi\sin^3 2\theta_W} \qquad (4.191)$$

in electroweak theory and, from (2.54),

$$T_1 = T_0\left(1 + \frac{C^2 T_0^2}{4\lambda m^2}\right)^{-1/2} \qquad (4.192)$$

with T_0 given by (2.95)

$$T_0 = |m|\left[\frac{\lambda}{2} + \frac{e^2(1 + 2\cos^2\theta_W)}{4\sin^2 2\theta_W} + \frac{h_t^2}{4}\right]^{-1/2}. \qquad (4.193)$$

The measured values of the parameters give

$$133 \text{ GeV} \lesssim T_0 \lesssim 360 \text{ GeV} \qquad (4.194)$$

with the lower bound deriving from the current lower bound [7]

$$m_H \gtrsim 114.3 \text{ GeV} \qquad \text{at 95\% CL} \qquad (4.195)$$

on the Higgs mass. Then

$$\frac{T_1 - T_0}{T_0} \lesssim 2.3 \times 10^{-3} \qquad (4.196)$$

with the upper bound saturated when m_H achieves its lower bound. The temperature-dependent sphaleron energy satisfies

$$E_{\text{sph}}(T) = \frac{2m_W(T)}{\alpha_2}\mathcal{E} \tag{4.197}$$

with

$$1.9 < \mathcal{E} < 2.7 \tag{4.198}$$

and the associated Boltzmann factor $\exp[-E_{\text{sph}}(T)/T]$ means that the baryon-number non-conservation is unsuppressed when

$$T > E_{\text{sph}}(T). \tag{4.199}$$

However, this is satisfied only when T is practically at the critical temperature at which the energy barrier disappears in any case. Careful analytical estimates of the baryon-number changing rates, valid in the range

$$2m_W(T) \ll T \ll \frac{2m_W(T)}{\alpha_2} \tag{4.200}$$

have been made by Arnold and McLerran [45]. They find a transition rate per unit volume

$$\frac{\Gamma_{\text{sph}}}{V} = \kappa(2m_W(T))^4\left(\frac{2m_W(T)}{\alpha_2 T}\right)^3 e^{-E_{\text{sph}}(T)/T} \tag{4.201}$$

where the numerical factor $\kappa \sim 11$, although a (non-perturbative) numerical 'measurement' of the diffusion rate of the Chern–Simons number over the barrier [46] suggests that this may be too large by a factor of order 10.

At high temperatures where $T > T_1$, the Higgs field VEV $v(T)$ is zero and there is no sphaleron. The gauge fields, however, can still generate baryon-number non-conservation. The energy E_b of such a configuration is controlled by the *a priori* length scale ℓ of the configuration which changes the Chern–Simons number N_{CS}. Presumably the scale of the action is set by (4.173), so

$$E_b\ell \sim \frac{2\pi}{\alpha_2}. \tag{4.202}$$

To avoid Boltzmann suppression, we require

$$E_b \sim \frac{2\pi}{\alpha_2\ell} < T \tag{4.203}$$

so

$$\ell \gtrsim \frac{2\pi}{\alpha_2 T}. \tag{4.204}$$

Thus, the transition rate per unit volume is

$$\frac{\Gamma_b}{V} \sim \frac{1}{\ell^3 t} \sim \left(\frac{\alpha_2 T}{2\pi}\right)^4 \tag{4.205}$$

if we assume $\ell \sim t$. However, the scale (4.204) is also the scale at which perturbation theory breaks down in a hot plasma and it has been argued [47] that plasma damping effects increase the time scale, so that

$$t \sim \frac{\ell}{\alpha_2} \tag{4.206}$$

basically because there are fewer ways large field configurations can cross the barrier than smaller ones. This means that

$$\frac{\Gamma_b}{V} \sim \alpha_2 \left(\frac{\alpha_2 T}{2\pi} \right)^4 \tag{4.207}$$

and indeed a lattice simulation [48] gives

$$\frac{\Gamma_b}{V} = (29 \pm 6)\alpha_2^5 T^4. \tag{4.208}$$

Numerically there is no difference between the two expressions.

At any rate, it is clear that there is no Boltzmann suppression at high temperatures, so we *do* have a source of baryon-number non-conservation. Whether or not it can explain the observed baryon asymmetry of the universe, of course, depends upon the other two Sakharov criteria: the amount of CP-violation and whether the system is out of thermal equilibrium. The picture we have in mind is that, in some region of space, there is a non-trivial gauge and Higgs field configuration, of the type we have discussed, which leads to the non-conservation of baryon number. To generate a baryon asymmetry we require CP-violating interactions involving the quark fields. Then the transition rate for the sphaleron-induced process with $\Delta N_{CS} = +1$ will differ (slightly) from that with $\Delta N_{CS} = -1$. Provided that the system is not in thermal equilibrium, there is then the possibility of a net non-zero baryon-number asymmetry.

4.9 CP-violation in electroweak theory

In electroweak theory the sole source of CP-violation is via the Cabibbo–Kobayashi–Maskawa (CKM) matrix, which derives a CP-violating phase from the unremovable phases in the Yukawa interactions of the quarks when we transform to the mass eigenstates. The form of these interactions is

$$\mathcal{L}_Y = \phi^\dagger \bar{d}_R h_D Q_L + \psi^\dagger \bar{u}_R h_U Q_L + \text{h.c.} \tag{4.209}$$

where Q_L is the quark doublet and u_R, d_R the singlets, ϕ is the Higgs scalar doublet and $\psi = i\tau_2 \phi^*$. As in equation (4.70), h_U and h_D are complex matrices acting on the undisplayed generation indices of Q_L, u_R, d_R. Also, as in equation (4.78), we need to construct a diagram with non-vanishing imaginary parts in both the loop integral and the trace over generation indices. It is easy to see that (4.81)

does not derive from a diagram allowed by the vertices in (4.209). In fact, the first
contribution arises at the *12th* order in perturbation theory, from the trace

$$T_{12} \equiv \text{tr}(h_U^\dagger h_U h_U^\dagger h_U h_D^\dagger h_D h_D^\dagger h_D h_U^\dagger h_U h_D^\dagger h_D) + \text{h.c.} \tag{4.210}$$

Again we may diagonalize h_U and h_D as in (4.82), (4.83), (4.84) and (4.85).
Then [49, 50],

$$T_{12} = \frac{g_2^{12}}{64m_W^{12}} \text{tr}(|m_U|^4 V |m_D|^4 V^\dagger |m_U|^2 V |m_D|^2 V^\dagger) \tag{4.211}$$

where

$$V = QS^\dagger \tag{4.212}$$

is the CKM matrix and

$$|m_U|^4 = \text{diag}(m_t^4, m_c^4, m_u^4) \tag{4.213}$$

and similarly for $|m_D|^4$, $|m_U|^2$ and $|m_D|^2$. Then

$$\text{Im}\, T_{12} = \frac{g_2^{12}}{64m_W^{12}} \sum_{i,j,k,l} (m_U^4)_i (m_D^4)_j (m_U^2)_k (m_D^2)_l \,\text{Im}[V_{ij} V_{kl} V_{il}^* V_{kj}^*]. \tag{4.214}$$

Now

$$\text{Im}[V_{ij} V_{kl} V_{il}^* V_{kj}^*] = J \sum_{p,q} \epsilon_{ikp} \epsilon_{jlq} \tag{4.215}$$

where J is the Jarlskog invariant [49]. In the standard CKM and Particle Data
Group parametrizations respectively, it is given by

$$\begin{aligned} J &= \sin^2\theta_1 \sin\theta_2 \sin\theta_3 \cos\theta_1 \cos\theta_2 \cos\theta_3 \sin\delta \\ &= \sin\theta_{12} \sin\theta_{13} \sin\theta_{23} \cos\theta_{12} \cos\theta_{13}^2 \cos\theta_{23} \sin\delta_{13}. \end{aligned} \tag{4.216}$$

Then

$$\begin{aligned} \text{Im}\, T_{12} &= \frac{g_2^{12}}{64m_W^{12}} J (m_t^2 - m_c^2)(m_c^2 - m_u^2)(m_u^2 - m_t^2)(m_b^2 - m_s^2) \\ &\quad \times (m_s^2 - m_d^2)(m_d^2 - m_b^2) \\ &\simeq \frac{g_2^{12}}{64m_W^{12}} J m_t^4 m_b^4 m_c^2 m_s^2. \end{aligned} \tag{4.217}$$

So we expect the scale of CP-violation in the standard model to be controlled by
the parameter

$$\delta_{sm} \simeq A \left(\frac{\alpha_2}{2\pi}\right)^6 J \left(\frac{m_t^2 m_b^2 m_c m_s}{m_W^6}\right)^2. \tag{4.218}$$

With $A \sim 10^3$ perhaps to allow for the large number of Feynman diagrams, we still have

$$\delta_{sm} \lesssim 10^{-25} \qquad (4.219)$$

using the current data [7]. This tiny number scales the difference in free energies and, hence, the difference between the rates of the $\Delta N_{CS} = \pm 1$ processes. By itself, it is sufficient to exclude the possibility of explaining the observed baryon number asymmetry of the universe in the context of the standard model electroweak theory, so extra sources of CP-violation are clearly needed. Such sources arise naturally in the supersymmetric version of the theory which we shall discuss shortly. Before doing so, however, we shall see that there are further reasons why the standard model electroweak theory cannot yield the measured asymmetry.

4.10 Phase transitions and electroweak baryogenesis

We have seen that electroweak theory possesses (sphaleron-induced) baryon-number non-conserving processes, as well as CP-violation via the CKM matrix. However, there is a rather general argument that these cannot generate the baryon asymmetry of the universe in the absence of a phase transition or if the phase transition is second order [51].

Suppose the latter and consider the universe at temperature T satisfying

$$m_W < T < T_c \qquad (4.220)$$

with T_c the critical temperature of the phase transition. For a second-order phase transition

$$m_W(T) = m_W \left(1 - \frac{T^2}{T_c^2} \right)^{1/2}. \qquad (4.221)$$

We can see this from (4.189) by setting C to zero. Then $v(T)$ and, hence, $m_W(T)$, approach zero *continuously* as T approaches T_c from below; also, when $C = 0$, $T_0 = T_1 = T_c$. As in the case of GUT baryogenesis (4.37), we now require that the rate Γ_{sph} of baryon-number non-conserving processes is smaller than the (Hubble) rate $H(T)$ associated with the expansion of the universe, so that the baryons are decoupled from the thermal bath. The total sphaleron rate Γ_{sph} is obtained by scaling the rate per unit volume Γ_{sph}/V, given in (4.201), with $R^3(t)$, where $R(t)$ is the scale factor, proportional to T^{-1} in the radiation-dominated era. Thus, roughly,

$$\Gamma_{sph} \sim kT \exp\left[-\frac{E_{sph}(T)}{T} \right] \qquad (4.222)$$

where $E_{sph}(T)$ is given in (4.197). With $H(T)$ given by (4.21), decoupling only happens when the temperature T drops below T^* with

$$\frac{T^*}{T_c} \simeq \left[1 + \left(\frac{T_c}{m_W} \frac{\alpha_2}{2\mathcal{E}} \ln \frac{m_P}{T^*} \right)^2 \right]^{-1/2}. \qquad (4.223)$$

This gives

$$0.71 < \frac{T^*}{T_c} < 0.88 \qquad (4.224)$$

corresponding to the bounds (4.194). Thus the sphaleron-induced baryon-number non-conserving processes with which we are concerned are not decoupled until the temperature is well below T_c but still well above m_W [51] and, at these temperatures, the exponential suppression essentially turns off the baryon-number production. At the higher temperatures where the Boltzmann suppression is evaded, the rate exceeds the universal expansion rate and the resulting thermal equilibrium washes out any baryon asymmetry. So the picture we have is as follows: As the temperature drops below T_c, the symmetry breaks and the field ϕ develops a non-zero VEV at the minimum of the potential, as in figure 2.1. At each point in space thermal fluctuations perturb the field ϕ which then 'rolls' classically to the new value at the global minimum. If this phase transition is second order or a continuous crossover, the slow rolling to the new minimum means that the departure from thermal equilibrium is too small until T drops below T^* but by then the sphaleron-induced baryon production has been turned off by the Boltzmann exponential suppression.

Thus, the only possibility is that the phase transition is *first* order. In this case, as discussed in section 2.9, for temperatures $T > T_1$, the only minimum of V_{eff} is with the system in the symmetric, unbroken phase characterized by zero VEV. As the temperature falls below T_1, a local minimum of V_{eff} develops, separated by a potential barrier from that at zero VEV, and below the critical temperature $(T < T_c)$ this broken phase becomes the global minimum, see figure 2.2. Nucleation of the broken phase proceeds by the formation of bubbles of this true vacuum in the sea of false (symmetric phase) vacuum. At some supercooled temperature below T_c, the size of the bubbles becomes large enough for them to overcome the surface tension effects and they expand to fill the whole of space and complete the phase transition. Finally, for temperatures below T_0, the local minimum at zero VEV disappears and only the broken phase is stable.

A baryon asymmetry may be generated as the wall of an expanding bubble passes through a region containing particles in the unbroken phase. The Higgs field changes rapidly because of the wall motion, as do other fields, and these interact with the particles giving concentrations quite far from equilibrium. If the baryon-number non-conserving processes *and* the CP-violating processes both occur in or near the wall, a net non-zero baryon asymmetry can result: this scenario is called *local* baryogenesis [52, 53]. After the wall has passed the region we are discussing is in the true (broken phase) vacuum with $v(T) \neq 0$, so it is important that, in this phase, the sphaleron-induced, baryon-number non-conserving processes are turned off by the Boltzmann suppression, so that any baryon asymmetry produced during the non-equilibrium era is frozen in. The condition for this to happen is [50]

$$\frac{v(T_c)}{T} \gtrsim 1. \qquad (4.225)$$

This is called the 'sphaleron washout condition'. In *non-local* baryogenesis, CP-violating interactions of the particles with the bubble wall produce an asymmetry in a quantum number *other* than the baryon number and the resulting particles carry this asymmetry into the unbroken phase, away from the wall. Then baryon-number non-conserving effects convert *this* asymmetry into a baryon asymmetry, (some of) which is frozen in the broken phase after the bubble wall has passed. If the speed of the wall is greater than the sound speed in the plasma, the former process dominates [54]; otherwise the latter does but, in general, both may occur and the total baryon asymmetry is the sum of that generated by the two processes.

We have already noted that the phase transition *is* (weakly) first order in electroweak theory and that at the phase transition (2.61) gives

$$\phi_c = v(T_c) = \frac{2CT_c}{3\lambda} \qquad (4.226)$$

so

$$\frac{v(T_c)}{T_c} = \frac{2C}{3\lambda} \lesssim 0.17 \qquad (4.227)$$

which does *not* satisfy (4.225). In fact, *requiring* that (4.225) is satisfied would require

$$m_H \lesssim 47 \text{ GeV} \qquad (4.228)$$

in clear contradiction to the current lower bound (4.195).

The perturbative calculations of the corrections to the effective potential are not *a priori* reliable, because of the so called 'infrared problem' that afflicts finite-temperature field theory [55, 56]. It derives from the existence of an expansion parameter of the form

$$\epsilon = \frac{g_2^2}{e^{-m/T} - 1} \simeq \frac{g_2^2 T}{m} \qquad \text{when } m \ll T \qquad (4.229)$$

where m is some bosonic mass appearing in the propagators. Then light modes, those with $m \ll g_2^2 T$, interacting with the Higgs are a problem that should be treated *non*-perturbatively. The direct method of carrying out a four-dimensional finite temperature lattice simulation is difficult because the weak coupling entails the existence of multiple length scales which are difficult to fit simultaneously on a finite lattice. Also, in practice, chiral fermions cannot be handled efficiently. However, all of these problems can be overcome by using a finite-temperature effective field theory [57–59], obtained by integrating out (perturbatively) all non-zero Matsubara modes, which includes, in particular, all fermions. The resulting effective theory is then three-dimensional and involves only the surviving infrared modes, the Higgs and the spatial components of the $SU(2)$ and $U(1)$ gauge fields. This theory is ideally suited for lattice simulations. It is found that, in the m_H–T_c plane, there is a line of first-order phase transitions that end at a critical point after which there is only a crossover transition. The endpoint is known to high precision and is at

$$m_{H,c} = 72.3 \text{ GeV} \qquad T_c = 109.2 \text{ GeV}. \qquad (4.230)$$

Further, the condition (4.225) is satisfied only for

$$m_H \lesssim 10 \text{ GeV}. \qquad (4.231)$$

Thus, the more accurate calculation of the phase transition shows that the current bound (4.195) is far from allowing a first-order phase transition, let alone baryogenesis.

This analysis also indicates how the situation might be improved. The strength of the phase transition, as measured by $v(T_c)/T_c$, can be increased by substantially decreasing the effective (three-dimensional) scalar self-coupling λ_3, as is suggested by (4.227). This requires a new, non-perturbative degree of freedom and this happens when there are extra *scalar* degrees of freedom, as occurs naturally in the supersymmetric version of electroweak theory to which we now turn.

In summary, the standard model electroweak theory does not explain the observed baryon asymmetry for two reasons: (i) because there is insufficient CP-violation; and (ii) because the phase transition is too weakly first order (in the sense described earlier) to suppress the erasure of any baryon number asymmetry produced in the symmetric phase.

4.11   Supersymmetric electroweak baryogenesis

We have already noted that the extra matter contained in the minimal supersymmetric standard model (MSSM) might allow the alleviation of the problematic features of the non-supersymmetric theory that preclude baryogenesis at the level observed in nature. Supersymmetry entails the existence of new fermions. In particular, there are charginos and neutralinos, mass eigenstates that are generically superpositions of the charged or neutral weak gauginos and Higgsinos. There are also new bosons and we shall be specifically concerned with top squarks. Diagonalization of the mass matrices of all of these states generally leads to new sources of CP-violation. As we shall see, it is the existence of new particles (and thereby of new sources of CP-violation), rather than the supersymmetry itself, that might allow the MSSM to explain the observed baryon asymmetry.

Now consider an expanding bubble of the broken phase, with the bubble wall propagating through the hot plasma into the symmetric phase, perturbing the (quasi-)particle distributions from equilibrium. Inside the bubble, baryon-number non-conservation is small, because of exponential suppression by the sphaleron's Boltzmann factor, provided that the sphaleron washout condition (4.225) is satisfied. Effectively, baryon number is conserved inside the bubble. However, outside the bubble, anomalous baryon-number non-conservation is rapid. One way to see how baryogenesis occurs [60] is to think of the wall of the expanding bubble feeling a 'wind' of particles in the symmetric phase. These particles may pass through the wall into the broken phase or be reflected back into the symmetric phase region. The latter will interact and be slowed by other particles approaching the wall, before eventually passing through the wall. Thus, an accumulation of particles develops in front of the bubble wall. As a result of the CP violation, there is a non-zero difference between the transmission coefficients of these particles and their antiparticles across the wall of the bubble and a similar difference between the reflection coefficients. These differences, in turn, generate local source terms for the net number densities associated with the particles and, in particular, the Higgs number and axial top number, which appear in the coupled Boltzmann equations, tending to pull the system away from equilibrium. These particles are chosen because they participate in particle-number-changing transitions in the wall that are fast compared with relevant time scales but they carry charges that are approximately conserved in the symmetric phase. Transport effects then generate a local excess (or deficit) of left-chiral charginos (say) over their antiparticles ahead of the advancing bubble wall. Unsuppressed baryon-number non-conservation in the symmetric phase then converts these densities into a net baryon asymmetry, which is frozen as the bubble wall sweeps through, provided that the sphaleron washout condition is satisfied.

All calculations [60–64], of n_B in the symmetric phase are done using coupled diffusion equations for the relevant number densities, which include contributions arising from source currents generated at the bubble wall, scattering processes involving the top quark Yukawa coupling, as well as Higgs-number and axial-top-number-violating processes in the bubble wall and broken phase. Higgs-number and quark-number diffusion terms are also included. There is general agreement on the equations to be used (see (4.232)) but little on how to determine the source currents.

We have also noted that the MSSM affords new mechanisms for satisfying the washout condition (4.225) that are unavailable to the standard model. Specifically, the possible existence of a *light $SU(2)$ scalar top squark $\tilde{t}_R$* (a 'stop') that interacts strongly with the Higgs field might drive the necessary reduction in the effective three-dimensional scalar self-coupling λ_3 needed to generate a sufficiently strong first-order phase transition with a Higgs mass satisfying the current bounds [65]. Any such scalar gives a negative contribution to λ_3 at one-loop level but a light *left* stop $\tilde{t}_L$ is inconsistent with electroweak precision measurements. Numerical calculations [65] have confirmed two-loop estimates and shown that these are even somewhat conservative. The conclusion is that there *are* parts of the MSSM parameter space not excluded by experiment where the electroweak phase transition is strong enough to allow baryogenesis. However, besides needing a light stop $m_{\tilde{t}_R} \lesssim m_t$, there must either be a much heavier stop $m_{\tilde{t}_L} \gtrsim 10 m_t$ or else two independent light Higgs particles $m_{h,A} \lesssim 120\,\text{GeV}$. ($h$ is the scalar Higgs and A the pseudoscalar.)

The next question then is whether in this region of parameter space the additional sources of CP-violation in the MSSM can generate source terms for the various particle densities that are strong enough to induce sufficient baryon asymmetry in the symmetric phase, which is then frozen in as the bubble wall passes through. The most direct method of baryogenesis would, of course, be to use the left-chiral quarks themselves, since any CP-violating contributions to their distributions would directly bias the sphaleron interactions and thereby generate a baryon asymmetry. However, in the MSSM (but not in *all* two-Higgs doublet models) the Higgs potential is real at tree level and CP-violating contributions to quark masses arise only at one-loop order. Moreover, such contributions are potentially suppressed by the GIM mechanism, as in the standard model. We are, therefore, forced to consider CP-violating perturbations in other particle densities. Specifically, we consider squarks, which couple to quarks strongly via the strong supergauge interactions, and charginos, which couple strongly to third-generation quarks via Yukawa interactions. Neutralinos also contribute but their coupling to fermions is weaker than that of the charginos, so the transport of any asymmetry to the quark sector is much less efficient and we neglect them. The coupled diffusion equations that control the densities with CP-violating sources have the general form

$$D_i \xi_i' + v_w \xi_i' + \Gamma_i(\xi_i + \xi_j + \cdots) = S_i \qquad (4.232)$$

where $\xi_i \equiv \mu_i/T$ with μ_i the chemical potential of the ith species, D_i is a diffusion constant, v_w is the velocity of the bubble wall, Γ_i is the inelastic rate converting species i into other species $j, \ldots$, and S_i is the CP-violating source term created at the bubble wall; the prime denotes differentiation with respect to the z-direction in which the bubble wall propagates. Different approaches have been proposed for calculating the source terms [31, 63, 66], and it is unclear whether they agree. We shall use the classical force method [63, 64, 67, 68], in which the particles move in the plasma under the influence of a classical force exerted on them by the spatially varying Higgs field. Because of CP-violation, particles and antiparticles experience (slightly) different forces. The source terms in the diffusion equations are proportional to the thermal average of this CP-violating component of the force.

For illustrative purposes, we shall follow the treatment of Cline *et al* [69]. The set of coupled diffusion equations can be simplified considerably by taking account of the hierarchy of inelastic reaction rates Γ_i that change the particle species i into other species $j, \ldots$. The electroweak sphaleron rate Γ_b, of order $\alpha_2^4 T$ (see (4.205)), is the slowest and can be ignored until we are ready to compute the actual baryon asymmetry. In contrast, the various gauge interaction rates, of order $\alpha_a T$, are fast and can be taken to be in equilibrium on the time scale for particles to diffuse in front of the bubble wall:

$$\alpha_a T \gg \frac{D_i}{v_w^2}. \qquad (4.233)$$

Then, in particular, the chemical potentials of the weak bosons are zero, so the chemical potentials of quarks in the same doublet are equal:

$$\xi_{t_L} = \xi_{b_L} \equiv \xi_{q_3}. \qquad (4.234)$$

Similarly, the assumption that supergauge interactions are in equilibrium implies that

$$\xi_p = \xi_{\tilde{p}} \tag{4.235}$$

for any particle p and associated sparticle $\tilde{p}$.   Intermediate between these interaction rates are those of various inelastic processes, including those associated with the interaction Lagrangian and 'strong sphaleron' processes. We shall see in the following chapter that, although baryon number is conserved by QCD, the axial baryon-number current

$$J_\mu^{(5)} \equiv \sum_q \bar{q}\gamma_\mu\gamma_5 q \tag{4.236}$$

is anomalous and there are strong sphaleron solutions, analogous to the electroweak sphalerons, that connect different topological sectors.   Their contribution to the coupled diffusion equations conveys the CP-violation to all quark flavours.   The rate Γ_{ss} of such processes is of order $\alpha_3^4 T$ and comparable with those arising from the top Yukawa coupling, so they must be included. Although neither of the foregoing assumptions is a particularly good approximation, it is believed, or at least hoped, that they lead to multiplicative errors of at most of order unity in the predicted baryon asymmetry.

In the analysis of [69], the sources S_i of the CP-violating asymmetry arise solely from the chargino sector. The squark sector is ineffective because, for bosons, the CP-violating source arises only at second order in the gradient expansion of the CP-violating mass. Charginos are mass eigenstates arising from Higgsino–Wino mixing. However, any asymmetry from the Wino component can only be transported to a chiral asymmetry in quarks and squarks via mixing effects, whereas a Higgsino asymmetry is transported directly via strong Yukawa interactions.   Thus, the set of diffusion equations to be considered can be simplified by neglecting chargino mixing and dropping any Wino contributions. With the foregoing approximations, the number of coupled diffusion equations is reduced to those for the two Higgsino densities $\xi_{\tilde{h}_1}$ and $\xi_{\tilde{h}_2}$, which have CP-violating sources $S_{\tilde{h}_1}$ and $S_{\tilde{h}_2}$, plus those for the third quark generation doublet ξ_{q_3} and the right-chiral top ξ_{t_R}, and those for first- and second-generation doublets $\xi_{q_{1,2}}$ and right-chiral singlets ξ_{q_R}, which are coupled to the first four equations only by strong-sphaleron interactions. The equations may be solved numerically and the last step is to calculate the baryon asymmetry induced by weak sphaleron effects on the CP-violating asymmetries ξ_i.

Since the weak sphaleron derives entirely from the $SU(2)_L$ component of the electroweak gauge group, the baryon asymmetry results from left-chiral quark and lepton asymmetries in front of the bubble wall. The latter are essentially zero in this approach and the former enter the baryon-number-violating rate equation

$$\frac{dn_B}{dt} = \frac{3}{2}\Gamma_b\left(\xi_{q_L} - A\frac{n_B}{T^2}\right) \tag{4.237}$$

via

$$\xi_{qL} \equiv 3(\xi_{q_1} + \xi_{q_2} + \xi_{q_3}) \tag{4.238}$$

because each doublet occurs in three colours. Here Γ_b is the weak-sphaleron baryon-number non-conserving rate (4.205) or (4.208). The second term on the right-hand side is the Boltzmann term by which the baryon number would relax to zero if the sphaleron processes had time to equilibrate in front of the bubble wall. The n_B here is, therefore, related to the quark and lepton asymmetries, μ_q and μ_ℓ, that result from equilibriating all flavour-changing interactions that are faster than the sphaleron rate Γ_b in the symmetric phase. Thus, A is given by

$$A\frac{n_B}{T^2} = 9\mu_q + \sum_\ell \mu_\ell \tag{4.239}$$

since each sphaleron creates nine quarks and three leptons. All quarks have the same chemical potential, because of efficient mixing, but lepton mixing might be weak. The calculation of these chemical potentials depends on which interactions equilibriate on the relevant timescale. The asymmetry n_A for any particle species a is given by

$$n_A \equiv n_a - n_{\bar{a}} = \kappa_a \frac{\mu_a T^2}{6} \tag{4.240}$$

where $\kappa_a = 1, 2$, respectively, when a is a fermion or boson. In electroweak baryogenesis, the relevant time scale is Γ_b^{-1} and, on this scale, both chiralities of all six quark flavours in three colours do equilibrate and we include a number N_{sq} of light squarks. Thus, from the quarks and squarks, we have

$$n_B = \frac{1}{3}(n_Q + n_{sQ}) = \frac{\mu_q}{18}[6 \times 3 \times 2 + 2 \times 3N_{sq}]T^2 \tag{4.241}$$

so that

$$\mu_q = \frac{n_B}{2T^2}\left(1 + \frac{N_{sq}}{6}\right)^{-1}. \tag{4.242}$$

Similarly, since only the left-chiral leptons equilibrate, but not the right,

$$n_\ell = \tfrac{1}{3}\mu_\ell T^2 \tag{4.243}$$

and

$$\sum_\ell \mu_\ell = 3\sum_\ell \frac{n_\ell}{T^2} = 3\frac{n_B}{T^2}. \tag{4.244}$$

Thus,

$$A = \frac{9}{2}\left(1 + \frac{N_{sq}}{6}\right)^{-1} + 3. \tag{4.245}$$

The solution of (4.237) is found by transforming to the wall frame in which $\partial_t \to -v_w \partial_z$. Then

$$n_B = \frac{3\Gamma_{sph}}{2v_w} \int_0^\infty dz\, \xi_{qL} e^{-bz} \tag{4.246}$$

where

$$b \equiv \frac{3A\Gamma_{\mathrm{sph}}}{2v_w T^3}.$$ (4.247)

Consequently, the quark asymmetry ξ_{q_L}, found from solving the coupled diffusion equations in terms of the Higgsino sources $S_{\tilde{h}_1}(z)$ and $S_{\tilde{h}_2}(z)$, determines the baryon asymmetry n_B and, hence, the baryon-to-photon and baryon-to-entropy ratios $\eta_B \equiv n_B/n_\gamma \simeq 7n_B/s$.

The conclusion is that the MSSM *can* explain the observed value (4.18) but several independent parameters must be optimally tuned to do so. First, the CP-violating phase in the chargino mass matrix must be close to maximal. In order to accommodate experimental data on electric dipole moments, especially that of mercury, this requires that the lower generation squarks must have masses of order 10 TeV. Actually this is necessary to maximize the chiral quark asymmetry. Also $m_{\tilde{t}_L} \sim 10$ TeV is required to give sufficiently large radiative corrections to m_h, given the already noted need for a light $\tilde{t}_R$ ($m_{\tilde{t}_R} \lesssim m_t$) to satisfy the washout condition. In addition, $\tan\beta \equiv v_u/v_d \lesssim 3$ is required, the wall velocity v_w must be close to its optimal value of 0.02 and the walls should be as thin as they can be for the validity of the classical force method, about $6/T$. ($v_{u,d}$ are, respectively, the VEVs of the (two) Higgs doublets h_1 in (4.59), used to give masses to the uplike quarks, and h_2 in (4.60), used to give masses to the downlike quarks.) Similar conclusions have been reached by Huber and Schmidt [70].

Although the MSSM can explain the observed baryon asymmetry, it is evidently not generic. The region of parameter space in which it does so is very constrained, and might well be excluded by future experiments that set new bounds on sparticle masses, for example. We therefore comment briefly on alternatives that have been proposed but which, however, have not been as fully studied as the other methods we have described.

4.12 Affleck–Dine baryogenesis

The discussion of baryogenesis in the previous section made hardly any use of the *supersymmetry* of the MSSM. Rather, the MSSM supplied new fields that strengthened the first-order phase transition and which also developed the chiral asymmetries that were subsequently converted to a baryon asymmetry. In contrast, the Affleck–Dine mechanism [71] uses a generic feature of any supersymmetric theory, namely the existence of 'flat' directions, to generate a large VEV for a field carrying non-zero $B - L$ in the early universe; at temperatures higher than the electroweak phase transition, we have already noted that weak sphaleron processes are in equilibrium, so that any $B + L$ asymmetry is erased. In fact, perturbative baryon-number conservation in the MSSM is achieved by imposing a discrete R-symmetry that has the effect of excluding certain dimension-four operators from the superpotential that would otherwise explicitly generate baryon-number non-conservation, see [11] for example. Even so, there remains the possibility of dimension-five, and higher, non-renormalizable operators, whose effects are suppressed by at least one power of a (hopefully) superheavy scale M, e.g. the Planck mass m_P. The alternative approach to baryogenesis, proposed by Affleck–Dine utilizes both of these features.

First, recall that in a supersymmetric theory there are scalar fields with non-zero baryon or lepton number. Before supersymmetry breaking, there are generally many (D- and F-) flat directions. These are directions (in field configuration space) along which the potential is a constant (i.e. 'flat'). These include directions that allow gauge invariant combinations of squark and/or slepton fields to develop non-zero VEVs. For example [72], the MSSM superpotential (see [11])

$$W = \lambda_u Q H_u u^c + \lambda_d Q H_d d^c + \lambda_e L H_d e^c \qquad (4.248)$$

has an F-flat direction parametrized by the complex field ϕ as follows:

$$Q_1^\alpha = \begin{pmatrix} \phi \\ 0 \end{pmatrix} \qquad L_1 = \begin{pmatrix} 0 \\ \phi \end{pmatrix} \qquad d_2^\alpha = \phi \qquad (4.249)$$

with all other fields zero; subscripts label the generations, and superscripts are colour labels. The auxiliary F-terms ($F_\phi \equiv \partial W/\partial \Phi$, as in section 2.7) of all chiral superfields Φ vanish in this direction, and the fact that ϕ is complex shows that there is global $U(1)$ symmetry associated with it. This direction is also D-flat. In other words, the D-terms ($D^a \equiv \sum_\Phi \Phi^\dagger t^a \Phi$, as in section 2.7) for all three gauge groups also vanish. It follows that the scalar potential

$$V = \frac{1}{2} \sum_i g_i^2 D_i^a D_i^a + \sum_\Phi F_\Phi^* F_\Phi \qquad (4.250)$$

is zero and, therefore, flat in this direction (the summed index a runs over the adjoint representation of the corresponding group.) Thus, the scalar particle associated with the field ϕ is massless. Fields such as this, associated with flat directions, are called 'moduli' fields and the massless particles associated with them raise cosmological questions that we shall discuss later. In our example, the combination of non-zero fields associated with the flat direction

$$X = Q_1^\alpha L_1 d_2^{c,\alpha} \qquad (4.251)$$

is gauge invariant and has $B - L = -1$. In general, the gauge-invariant combination X is proportional to a power of the field parametrizing the flat direction:

$$X \propto \phi^m \qquad (4.252)$$

In our example, $m = 3$. As detailed later, various effects lift the flatness and may allow ϕ to develop a VEV. If these VEVs are large, the subsequent evolution of the universe can develop a substantial baryon asymmetry.

To see how, recall that the superpotential W is expected to have higher-order, non-renormalizable terms. These must be gauge and R-parity invariant. Thus, focusing on a single flat direction, in general

$$W_{nr} = \frac{\lambda}{nM^{n-3}} X^k = \frac{\lambda}{nM^{n-3}} \phi^n \qquad (4.253)$$

where $n = mk$; k is even if X has odd R-parity, as in the previous example, and M is some large mass. This gives F-terms that are *non*-zero in the specifed direction. Note, however, that although the superpotential W_{nr} does not conserve $U(1)_{B-L}$, the scalar potential V derived from it *does*:

$$V_{nr} = \frac{|\lambda|^2}{M^{2n-6}} |\phi|^{2n-2} \qquad (4.254)$$

since it is proportional to $(\phi^*\phi)^{kn-1}$.

We shall see later, in chapter 7, that there is good reason for believing that in the early universe there was a period of 'inflation', during which the Hubble constant H was approximately constant. The second, and most important, point is that global supersymmetry is necessarily broken during inflation. Since the vacuum energy

$$\langle V \rangle = \frac{3}{8\pi} H^2 m_P^2 \qquad (4.255)$$

is non-zero, there must be non-zero F and/or D components for some matter fields, and a soft potential develops. This includes a mass term for the (erstwhile massless) moduli field ϕ, of order H, the (instantaneous) Hubble constant [72]. In fact, $H \sim 10^{13-15}$ GeV during inflation, as can be seen from (7.41) with $V(\phi)$ given by (7.115) in order to explain the observed cosmic microwave background. A mass term too conserves $B - L$. The supersymmetry breaking also induces (soft) A-terms in the scalar potential. These have the same form as the superpotential W, so, in our example, they have the form

$$V_A = \frac{A}{M^{n-3}} \phi^n + \text{h.c.} \qquad (4.256)$$

Like the mass term, the scale of A is of order the current Hubble constant H. Note that such terms do *not* conserve $B - L$ and this is the source of the $(B - L)$-violation necessary to generate a net $B - L$ in the evolution of the flat direction. The Sakharov criteria, necessary for baryogenesis, also require CP-violation and we shall see later that a CP-violating phase difference between this A-term and that arising from the hidden-sector supersymmetry breaking is essential to getting a non-zero baryon asymmetry at the end.

The equation of motion for the field ϕ is

$$\ddot{\phi} + 3H\dot{\phi} + \frac{\partial V}{\partial \phi^*} = 0 \qquad (4.257)$$

where V is the potential obtained by combining the contributions from V_{nr} in (4.254), V_A in (4.256) and the mass term for ϕ. It has the form

$$V = -cH_I^2|\phi|^2 + \left(\frac{a\lambda H_I\phi^n}{nM^{n-3}} + \text{h.c.}\right) + |\lambda|^2\frac{|\phi|^{2n-2}}{M^{2n-6}} \tag{4.258}$$

where H_I is the (approximately constant) value of the Hubble parameter during inflation and a, c are constants of O(1). This, of course, is just the equation of a damped oscillator and the important point is that, during inflation, it is close to being critically damped. If $c < 0$, which corresponds to positive mass-squared for ϕ, V has a minimum at $\phi = 0$. In this case, the average value of the field evolves exponentially to $\phi = 0$ and the large value at the end of inflation needed to get a baryon asymmetry is not achieved [72]. However, if $c > 0$, which, in general, requires non-minimal Kähler terms, V has a single minimum at ϕ_0, where

$$|\phi_0| = \left(\frac{\beta H_I M^{n-3}}{\lambda}\right)^{1/(n-2)} \tag{4.259}$$

with β a numerical constant which depends on a, c, n. Thus, $|\phi_0|$ is parametrically between H_I and M, which can easily be large. In the angular direction, the potential varies as $\cos(\arg a + \arg\lambda + n\arg\phi)$ and has n degenerate minima. Further, again because of the near critical damping, the field evolves rapidly to one of these minima, provided c is not too small [72]. Thus, at the end of inflation, the average value of the field has a large value with a well-defined phase, which is constant over scales large compared with the horizon.

This sets the boundary condition for the next era. After inflation the universe enters a matter-dominated era in which the Hubble constant is explicitly time-dependent:

$$H = \frac{2}{3t} \tag{4.260}$$

(see section 1.4). The equation of motion is still given by (4.257), with V of the form (4.258) but with H now given by (4.260). As t increases, H decreases, so that the instantaneous minimum of V also decreases. Solving the equation of motion reveals that ϕ tracks just behind this decreasing minimum.

This evolution continues until $H \sim m_{3/2} \sim 1$ TeV, where $m_{3/2}$ is the gravitino mass. At that point, the soft supersymmetry-breaking terms from the hidden sector become comparable with those arising from inflation and, at later times, dominate the evolution. The hidden-sector terms contribute a *positive* mass-squared term for ϕ as well as an A-term, both having scales determined by $m_{3/2}$. Thus, the additional contribution to the potential V has the form

$$V_{\text{hs}} = m_\phi^2|\phi|^2 + \left(\frac{Am_{3/2}\lambda\phi^n}{nM^{n-3}} + \text{h.c.}\right) \tag{4.261}$$

where $m_\phi \sim m_{3/2}$ and $A = \text{O}(1)$. Consequently the equation of motion (4.257) becomes *underdamped* as H decreases below $m_{3/2}$. Two important effects now come in to play. First, the positive mass-squared term dominates the inflationary contribution, so that ϕ begins to oscillate undamped about $\phi = 0$, with initial condition $\phi = \phi_0(t)$ at $t \sim m_{3/2}^{-1}$. The oscillation of ϕ is the coherent condensate with number density

$$n_\phi = \frac{\rho_\phi}{m_\phi} = m_\phi |\phi|^2 \tag{4.262}$$

where ρ_ϕ is the energy density in the condensate. Second, when the (CP-violating and $(B - L)$-violating) hidden-sector A-term dominates the inflationary term, the potential in the angular direction varies as $\cos(\arg A + \arg \lambda + n \arg \phi)$. Thus, if $\arg A \neq \arg a$, a non-zero 'torque' is created and a non-zero $\dot\theta$ develops. This is precisely what is needed to create a baryon/lepton asymmetry. The number density for the $U(1)_{B-L}$ charge of the condensate is

$$n_{B-L} = i(\phi^* \partial_0 \phi - \phi \partial_0 \phi^*) = 2|\phi|^2 \dot\theta. \tag{4.263}$$

The equation of motion has been integrated numerically by Dine *et al* [72], who find that, at late times ($t \gg m_{3/2}$), the ratio n_{B-L}/n_ϕ generically evolves to a constant of order unity. At late times, the potential is dominated by the (positive) mass term which, of course, conserves $B - L$. Thus, the $B - L$ created during the time when $H \sim m_{3/2}$ is conserved.

It remains only to convert this O(1) ratio to the physically relevant baryon to entropy ratio n_{B-L}/s. When $H \sim m_{3/2}$, the energy density of the condensate $\rho_\phi \sim m_{3/2}^2 |\phi|^2$ is much smaller than the energy density ρ_I associated with the coherent oscillations of the inflaton, $\rho_I \sim \frac{3}{8\pi} H^2 m_P^2$. Using (4.259), we see that

$$\frac{\rho_\phi}{\rho_I} \simeq \left(\frac{m_{3/2} M^{n-3}}{\lambda m_P^{n-2}} \right)^{2/(n-2)}. \tag{4.264}$$

This ratio remains approximately constant until the inflaton decays at some time when $H < m_{3/2}$. Provided that the inflaton decays dominate, the entropy density is given by

$$s \simeq \frac{\rho_I}{T_R} \tag{4.265}$$

where T_R is the reheat temperature after inflaton decay. Thus

$$\frac{n_{B-L}}{s} = \frac{n_{B-L}}{n_\phi} \frac{T_R}{m_\phi} \frac{\rho_\phi}{\rho_I}. \tag{4.266}$$

The ratio (4.264) is very sensitive to n. For $n > 4$, $M \sim m_P$ and a reasonable T_R, the ratio n_{B-L}/s is generally too large. For example, to get the observed value of $n_{B-L}/s \sim 10^{-10}$ with $n = 6$ requires T_R to be of order the electroweak scale. In contrast, $n = 4$ gives

$$\frac{n_{B-L}}{s} \sim 10^{-10} \left(\frac{T_R}{10^6 \text{ GeV}} \right) \left(\frac{10^{-3} M}{\lambda m_P} \right) \tag{4.267}$$

which gives an acceptable ratio for a reasonable range of the parameters.

As long as the ϕ condensate decays via $(B - L)$-conserving decay processes *after* the inflaton, the previous estimate of the ratio is insensitive to the details of the decay. However, even though most of the inflaton energy is not converted to radiation until $H = H_R < m_{3/2}$, the condensate might also decay via thermal scattering as soon as ϕ is small enough, which is when $g\phi < T$, where g is the gauge or Yukawa coupling constant. This must be unimportant when $H \sim m_{3/2}$. However [72], the value of ϕ/T is rather sensitive to the integer n. For $n = 6$, the required condition is comfortably satisfied but the $n = 4$ case is borderline.

The only flat direction having $B - L$ non-zero and having $n = 4$ corresponds to

$$W = \frac{\lambda}{M}(LH_u)^2 \qquad (4.268)$$

which might arise directly at the string scale or be generated from a GUT by integrating out a heavy standard-model singlet field N with coupling gLH_uN. At low energies, this operator generates neutrino masses

$$m_\nu \sim \frac{\lambda}{2M}v^2 \qquad (4.269)$$

where v is the Higgs VEV. Then

$$\frac{n_{B-L}}{s} \sim 10^{-10} \left(\frac{T_R}{10^6 \text{ GeV}}\right)\left(\frac{10^{-8}\text{eV}}{m_\nu}\right) \qquad (4.270)$$

where m_ν is the *lightest* neutrino mass. Taking $T_R < 10^9$ GeV then requires at least one neutrino to be lighter than about 10^{-5} eV, with an even stronger bound of 10^{-8} eV if the constraint $T_R < 10^6$ GeV is enforced to ensure that thermal scattering is unimportant when $H \sim m_{3/2}$. Interestingly, in this case, non-minimal Kähler terms are not needed to ensure a negative mass-squared for ϕ. This is because such terms can arise for the Higgs field via radiative corrections.

4.13 Exercises

1. Show that the nucleon–antinucleon annihilation rate falls below the expansion rate when the temperature $T \simeq 20$ MeV.
2. Show that in a time-reversal invariant theory baryon number is conserved.
3. Verify that for the $SU(5)$ GUT N_* has the value given in equation (4.41), and that for the supersymmetric theory, it has the value given in (4.42).
4. Show that the superheavy gauge bosons in the $SU(5)$ GUT have the decay modes given in equation (4.67) and that the colour-triplet Higgs has the decay modes given in (4.71).
5. Verify that in the $SU(5)$ GUT there are no two-loop contributions to $\Gamma(H_3) - \Gamma(\bar{H}_3)$ and that the three-loop contribution given in equation (4.81) does arise in the theory.

6. Verify equation (4.86).
7. If the $SO(10)$ GUT symmetry-breaking Higgs transforms as the **16**-dimensional representation, show that the gauge group breaks to $SU(5) \times U(1)$, and if the Higgs transforms as the **54**-dimensioinal representation, then the gauge group breaks to G_{422} given in equation (4.97).
8. Verify equations (4.130) and (4.131).
9. Verify equation (4.136).
10. Show that in the standard model the communication of CP-violation to the baryon-number non-conserving processes, via the phase in the CKM matrix, arises first in 12th order perturbation theory, as given in equation (4.210). Verify equation (4.214).
11. Verify that decoupling of the baryon-number non-conserving interactions occurs at a temperature T^* given by equation (4.223).
12. Verify equation (4.240) relating the asymmetry n_A to the chemical potential μ_a.
13. Show that in the direction (4.249) all of the F-terms derived from the superpotential (4.248) are zero.
14. Show that in the direction (4.249) all of the D-terms are also zero.
15. Verify that the potential (4.258) has a minimum of the form given in (4.259).
16. Verify (4.264).
17. Show that in the MSSM the subspace of field directions in which all D-terms vanish is 37-dimensional [72].

4.14 General references

- Riotto A 1998 *Trieste 1998, High Energy Physics and Cosmology* pp 326–436, arXiv:hep-ph/9807454
- Trodden M 1999 *Rev. Mod. Phys.* **71** 1463, arXiv:hep-ph/9803479
- Cline J M, Joyce M and Kainulainen K 2000 *JHEP* **0007** 018, arXiv:hep-ph/0006119

Bibliography

[1] Saeki T *et al* 1998 *Phys. Lett.* B **422** 319
[2] Stecker F W, Morgan D L and Bredekamp J 1971 *Phys. Rev. Lett.* **27** 1469
Stecker F W 1985 *Nucl. Phys.* B **252** 25
Mohanty A K and Stecker F W 1984 *Phys. Lett.* B **143** 351
Gao Y T, Stecker F W, Gleiser M and Klein D B 1990 *Astrophys. J. Lett.* **37** 361
[3] Steigman G 1976 *Ann. Rev. Astron. Astrophys.* **14** 334
[4] Mather J C 1999 *et al Astrophys. J* **512** 511
[5] Spergel D N *et al* 2003 *Astrophys. J. Suppl.* **148** 175, arXiv:astro-ph/0302209

[6] Sakharov A D 1967 *Pis'ma Z. Eksp. Teor. Fiz* **5** 32
   English translation 1967 *JETP Lett.* **5** 24
[7] Particle Data Group 2002 *Phys. Rev.* D **66** 010001
[8] Christenson J H, Cronin J W, Fitch V L and Turlay R 1964 *Phys. Rev. Lett.* **13** 138
[9] Schubert K R *et al* 1970 *Phys. Lett.* B **31** 662
   Angelopoulos A *et al* 1998 *Phys. Lett.* B **444** 43
[10] See, for example, Bailin D and Love A 1994 *Introduction to Gauge Field Theory* (revised edition) (Bristol: IOP)
[11] See, for example, Bailin D and Love A 1994 *Supersymmetric Gauge Field Theory and String Theory* (Bristol: IOP)
[12] Nilles H P 1984 *Phys. Rep.* **110** 1
[13] Langacker P 1981 *Phys. Rep.* C **72** 185
[14] Ellis J, Kelley S and Nanopoulos D V 1991 *Phys. Lett.* B **260** 131
   Amaldi V, de Boer W and Furstenau H 1991 *Phys. Lett.* B **260** 447
[15] Tomozawa Y 1981 *Phys. Rev. Lett.* **46** 463
   Berezenskii V, Joffe B and Kogan Ya 1981 *Phys. Lett.* B **105** 33
   Donoghue J and Golowich E 1982 *Phys. Rev.* D **26** 2888
   Lucha W 1983 *Phys. Lett.* B **122** 381
   Isgur N and Wise M 1982 *Phys. Lett.* B **118** 179
[16] Nanopoulos D V and Weinberg S 1979 *Phys. Rev.* D **20** 2484
[17] Harvey J A, Kolb E W, Reiss D B and Wolfram S 1982 *Nucl. Phys.* B **201** 16
[18] Ellis J, Nanopoulos D V and Gaillard M K 1979 *Phys. Lett.* B **80** 350
   Ellis J, Nanopoulos D V and Gaillard M K 1979 *Phys. Lett.* B **82** 464 (erratum)
[19] Barr S, Segrè G and Weldon H A 1979 *Phys. Rev.* D **20** 2494
[20] Kolb E W and Turner M S 1983 *Ann. Rev. Nucl. Part. Sci.* **33** 645
[21] Masiero A and Segrè G 1982 *Phys. Lett.* B **109** 349
[22] Peccei R and Quinn H R 1977 *Phys. Rev. Lett.* **38** 1440
   1977 *Phys. Rev.* D **16** 1791
[23] Wise M, Georgi H and Glashow S L 1981 *Phys. Rev. Lett.* **47** 402
[24] Yanagida T and Yoshimura M 1980 *Nucl. Phys.* B **168** 534
[25] Masiero A and Yanagida T 1982 *Phys. Lett.* B **109** 353
[26] Harvey J A, Kolb E W, Reiss D B and Wolfram S 1973 *Phys. Rev. Lett.* **31** 661
[27] Super-Kamiokande collaboration 1998 *Phys. Rev. Lett.* **81** 1562
[28] Pati J C and Salam A 1973 *Phys. Rev.* D **8** 1240
   Pati J C and Salam A 1973 *Phys. Rev. Lett.* **31** 661
   Pati J C and Salam A 1974 *Phys. Rev.* D **10** 275
[29] Chang D, Mohapatra R N and Parida M K 1984 *Phys. Lett.* B **142** 55
[30] Lyth D H and Riotto A 1999 *Phys. Rep.* **314** 1
[31] Riotto A 1998 *Theories of Baryogenesis* Lectures delivered at the Summer School on High Energy Physics and Cosmology, Miaramare, Trieste, arXiv: hep-ph/9807054
[32] Adler S L 1969 *Phys. Rev.* **177** 2426
[33] Bell J S and Jackiw R 1969 *Nuovo Cimento* A **60** 47
[34] Fujikawa K 1979 *Phys. Rev. Lett.* **42** 1195
[35] Weinberg S 1996 *The Quantum Theory of Fields* vol II (Cambridge: Cambridge University Press) p 366
[36] Manton N S 1983 *Phys. Rev.* D **28** 2019
[37] Shaposhnikov M E 1998 *Contemp. Phys.* **39** 177
[38] Banks T, Bender C and Wu T T 1973 *Phys. Rev.* D **8** 3346

Banks T, Bender C and Wu T T 1973 **8** 3366
[39] Coleman S 1977 *Phys. Rev.* D **15** 683
Coleman S 1985 *Aspects of Symmetry* (Cambridge: Cambridge University Press) ch 7
[40] Coleman S 1977 *Phys. Rev.* D **15** 2929
Coleman S 1977 *Phys. Rev.* D **16** 1248 (erratum)
Callan C G and Coleman S 1977 *Phys. Rev.* D **16** 1762
[41] Belavin A A, Polyakov A M, Schwartz A S and Tyupkin Yu S 1975 *Phys. Lett.* B **59** 85
[42] 't Hooft G 1976 *Phys. Rev. Lett.* **37** 8
't Hooft G 1976 *Phys. Rev.* D **14** 3432
[43] Klinkhamer F R and Manton N S 1984 *Phys. Rev.* D **30** 2212
[44] Weinberg S 1974 *Phys. Rev.* D **9** 3357
[45] Arnold P and McLerran L 1987 *Phys. Rev.* D **36** 581
[46] Moore G D 1998 *Phys. Lett.* B **439** 357
[47] Arnold P, Son D and Yaffe L G 1997 *Phys. Rev.* D **55** 6264
Arnold P 1997 *Phys. Rev.* D **55** 7781
[48] Moore G D, Hu C and Muller B arXiv:hep-ph/9710436
[49] Jarlskog C 1985 *Phys. Rev. Lett.* **55** 1039
[50] Shaposhnikov M E 1986 *JETP Lett.* **44** 465
[51] Kuzmin V A, Rubakov V A and Shaposhnikov M E 1985 *Phys. Lett.* B **155** 36
[52] Turok N and Zarozny T 1990 *Phys. Rev. Lett.* **65** 2331
Turok N and Zarozny T 1991 *Nucl. Phys.* B **349** 727
McLerran L, Shaposhnikov M E, Turok N and Voloshin M 1991 *Phys. Lett.* B **258** 451
[53] Dine M, Huet P, Singleton R and Susskind L 1991 *Phys. Lett.* B **257** 351
Dine M, Huet P and Singleton R 1992 *Nucl. Phys.* B **375** 625
[54] Lue A, Rajagopal K and Trodden M 1997 *Phys. Rev.* D **55** 1250, arXiv:hep-ph/9612282
[55] Linde A D 1980 *Phys. Lett.* B **96** 289
[56] Gross D J, Pisarski R D and Yaffe L G 1981 *Rev. Mod. Phys.* **53** 43
[57] Shaposhnikov M E 1996 *Erice 1996, Effective Theories and Fundamental Interactions* pp 360–83, arXiv:hep-ph/9610247
[58] Nieto A 1997 *Int. J. Mod. Phys.* A **12** 1431, arXiv:hep-ph/9612291
[59] Laine M 1997 *Eger 1997, Strong and Electroweak Matter '97* pp 160–77, arXiv:hep-ph/9707415
[60] Huet P and Nelson A E 1996 *Phys. Rev.* D **53** 4578
[61] Cohen A G and Nelson A E 1992 *Phys. Lett.* B **297** 111
[62] Cohen A G, Kaplan D B and Nelson A E 1994 *Phys. Lett.* B **336** 41
[63] Joyce M, Prokopec T and Turok N 1994 *Phys. Lett.* B **338** 269
Joyce M, Prokopec T and Turok N 1995 *Phys. Rev. Lett.* **75** 1695 erratum 3375
Joyce M, Prokopec T and Turok N 1996 *Phys. Rev.* D **53** 2598
[64] Cline J M, Joyce M and Kainulainen K 1998 *Phys. Lett.* B **417** 79
Cline J M, Joyce M and Kainulainen K 1999 *Phys. Lett.* B **448** 321 (erratum)
[65] See Laine M arXiv:hep-ph/0010275, and references therein
[66] Huet P and Sather E 1995 *Phys. Rev.* D **51** 379
[67] Cline J M and Kainulainen K 2000 *Phys. Rev. Lett.* **85** 5519, arXiv:hep-ph/0002272
[68] Kainulainen K, arXiv:hep-ph 0002273
[69] Cline J M, Joyce M and Kainulainen K 2000 *JHEP* 07 018

Cline J M, Joyce M and Kainulainen K 2001 arXiv:hep-ph/0110031 (erratum)
[70] Huber S J and Schmidt M G 2001 *Nucl. Phys.* B 606 183, arXiv:hep-ph 0003122
[71] Affleck I and Dine M 1985 *Nucl. Phys.* B 249 361
[72] Dine M, Randall L and Thomas S 1996 *Nucl. Phys.* B 458 291, arXiv:hep-ph/9507453

Chapter 5

Relic neutrinos and axions

5.1 Introduction

We saw in chapter 1 that, for much of the time, the constituents of the early universe were in approximate thermal equilibrium. This is because the rates for the interactions of these constituents were large compared with the expansion rate H. However, this thermal equilibrium was not maintained *all* of the time. If it were, the current state of the universe would be entirely determined by its temperature and we noted in the previous chapter the huge disparity between, for example, the equilibrium baryon abundance and that actually observed. Departures from equilibrium are, therefore, extremely important in determining the abundance of the relics that can be observed today.

The equilibrium number density $n_{X,\text{eq}}$ of a species X is given by

$$n_{X,\text{eq}} = \frac{g}{(2\pi)^3} \int d^3 p \, \frac{1}{e^{E(p)/T} \pm 1} \tag{5.1}$$

where g is the number of internal degrees of freedom, $E(p) = \sqrt{(|p|^2 + m_X^2)}$ and $+1$ relates to fermions and -1 to bosons. In the relativistic limit $T \gg m_X$, this gives for *bosons*

$$n_{X,\text{eq}} = \frac{\zeta(3)}{\pi^2} g T^3 \tag{5.2}$$

and for *fermions*

$$n_{X,\text{eq}} = \frac{3\zeta(3)}{4\pi^2} g T^3 \tag{5.3}$$

where $\zeta(3) = 1.202\,06$. A similar calculation shows that both the energy density

$$\rho_{X,\text{eq}} = \frac{g}{(2\pi)^3} \int d^3 p \, E(p) \frac{1}{e^{E(p)/T} \pm 1} \tag{5.4}$$

and the pressure

$$p_{X,\text{eq}} = \frac{g}{(2\pi)^3} \int d^3 p \, \frac{|p|^2}{3E(p)} \frac{1}{e^{E(p)/T} \pm 1} \tag{5.5}$$

scale as T^4 in the relativistic limit and the total (relativistic) energy density is

$$\rho = 3p = \frac{\pi^2}{30} g_{*,T} T^4 \tag{5.6}$$

where

$$g_{*,T} = \sum_{\text{bosons}} g_i \left(\frac{T_i}{T}\right)^4 + \frac{7}{8} \sum_{\text{fermions}} g_i \left(\frac{T_i}{T}\right)^4 \tag{5.7}$$

and we are allowing for the possibility that different particle species i may be at different temperatures T_i.

In thermal equilibrium, the entropy per comoving volume

$$s \equiv \frac{S}{V} \approx \frac{\rho + p}{T} \tag{5.8}$$

is dominated by the contribution of relativistic particles and, to a good approximation, it is given by

$$s = \frac{2\pi^2}{45} g_{*S,T} T^3 \tag{5.9}$$

where

$$g_{*S,T} = \sum_{\text{bosons}} g_i \left(\frac{T_i}{T}\right)^3 + \frac{7}{8} \sum_{\text{fermions}} g_i \left(\frac{T_i}{T}\right)^3. \tag{5.10}$$

Comparing (5.7) and (5.10), we see that when $T_i = T$, so that all particle species are at the same temperature, $g_{*,T} = g_{*S,T} = N_*$, with N_* defined in (1.104). However, in general, they differ.

Since the entropy per comoving volume is conserved, it is useful to measure the abundance of a species X by scaling its number density with the entropy density. We therefore define

$$Y_X \equiv \frac{n_X}{s}. \tag{5.11}$$

Then, using (5.2),(5.3) and (5.9), the equilibrium abundance in the relativistic limit is

$$Y_{X,\text{eq},T} = \frac{45\zeta(3)}{2\pi^4} \frac{g_{\text{eff}}}{g_{*S,T}} = 0.278 \frac{g_{\text{eff}}}{g_{*S,T}} \tag{5.12}$$

where for *bosons* $g_{\text{eff}} \equiv g$, and for *fermions* $g_{\text{eff}} \equiv 3g/4$.

All cosmological relics contribute to the current total energy density ρ_0 of the universe, and it is customary to scale these densities with the critical density

$$\rho_c \equiv \frac{3H_0^2}{8\pi G_N} = 10.54 h^2 \text{ keV cm}^{-3} \tag{5.13}$$

where $H_0 = 100h \text{ km s}^{-1} \text{ Mpc}^{-1}$ is the present Hubble constant and

$$h = 0.71^{+0.04}_{-0.03}. \tag{5.14}$$

The dimensionless measure of the total energy density is then defined by

$$\Omega_0 \equiv \frac{\rho_0}{\rho_c} \tag{5.15}$$

and, similarly, for a relic species X whose current energy density is $\rho_{X,0}$, we define

$$\Omega_X \equiv \frac{\rho_{X,0}}{\rho_c}. \tag{5.16}$$

Thus,

$$\sum_X \Omega_{X,0} = \Omega_0 \tag{5.17}$$

and the energy density for any one relic species must be less than the total energy density, so

$$\Omega_{X,0} < \Omega_0 \tag{5.18}$$

where data from the latest microwave anisotropy probe (WMAP) [1] give

$$\Omega_0 = 1.02 \pm 0.02. \tag{5.19}$$

The current energy density of a species X is given in terms of the current abundance by

$$\rho_{X,0} = n_{X,0} m_X = Y_{X,0} s_0 m_X \tag{5.20}$$

so provided we can calculate the current abundance $Y_{X,0}$, and the current entropy density s_0, a bound on the mass m_X of any relic species may be obtained:

$$m_X < \frac{\rho_c \Omega_0}{s_0 Y_{X,0}}. \tag{5.21}$$

A stronger bound may be obtained by replacing Ω_0 by Ω_m where the latter derives from the total *matter* content in the universe. The WMAP analysis gives

$$\Omega_m h^2 = 0.135^{+0.008}_{-0.009} \tag{5.22}$$

with h given by (5.14).

In the next section, we shall apply the foregoing considerations to neutrinos, in order to see what can be inferred about their masses. In section 5.3 we shall attempt a similar analysis for 'axions', hypothetical particles that are required to exist if the 'strong CP problem' of the standard $SU(3) \times SU(2) \times U(1)$ model of particle physics is solved by the 'Peccei–Quinn' mechanism, currently the only known solution of this problem. Axions must be very light, like the neutrinos. If they have survived until the present, their mass too is strongly constrained by various astrophysical and cosmological data.

5.2 Relic neutrinos

Uniquely among elementary particles, neutrinos participate only in weak (and gravitational) interactions. In the early universe, scattering processes, such as $\nu e \leftrightarrow \nu e$, and annihilation processes, such as $\nu\bar{\nu} \leftrightarrow e\bar{e}$, kept the neutrinos in thermal equilibrium. The total cross section for such processes is $\sigma \sim G_F^2 T^2$, just on dimensional grounds, since the weak (Fermi) coupling constant $G_F \sim 10^{-5} m_N^{-2}$ has dimensions $[M^{-2}]$. Since from (5.3) the relativistic number density $n_{\nu,\text{eq}}$ is proportional to T^3, the total interaction rate $\Gamma_{\text{int}} \sim \sigma \nu n_{\nu,\text{eq}} \sim G_F^2 T^5$. When this is large compared with the Hubble expansion rate

$$H = \sqrt{\frac{8\pi G_N \rho}{3}} = \sqrt{\frac{4\pi^3 G_N g_{*,T}}{45}} T^2 \tag{5.23}$$

there is thermal equilibrium. However, when $T \sim 1$ MeV, the two rates are comparable: $\Gamma_{\text{int}} \sim H$. Below this temperature, the Hubble expansion dominates and thermal equilibrium is not maintained. The neutrinos are, therefore, decoupled or 'frozen out'. Their abundance is frozen at the value obtained at the decoupling temperature $T_{\text{dec}} \sim 1$ MeV. Thus, the present abundance

$$Y_{\nu,0} = Y_{\nu,\text{eq},T_{\text{dec}}} \tag{5.24}$$

where, using (5.12),

$$Y_{\nu,\text{eq},T_{\text{dec}}} = \frac{n_\nu}{s} = 0.278 \frac{g_{\text{eff}}}{g_{*S,T_{\text{dec}}}}. \tag{5.25}$$

For a single (left-)chiral neutrino species $g_{\text{eff}} = 3/2$ (including the antineutrino) and, since $T_{\text{dec}} \sim 1$ MeV,

$$g_{*S,T_{\text{dec}}} = 2 + \tfrac{7}{8}(4 + 3 \times 2) = \tfrac{43}{4} \tag{5.26}$$

keeping only the electron and three families of chiral neutrinos as relativistic at this temperature.

In order to determine the bound (5.21), we first need to calculate the present entropy density

$$s_0 = \frac{2\pi^2}{45} g_{*S,T_0} T_0^3 \tag{5.27}$$

where g_{*S,T_0} is given by (5.10). At $T = T_0$, the (relativistic) species contributing to s_0 are the photons, having $g = 2$, and the three families of neutrinos, also with $g = 2$ (including the antineutrinos). Thus,

$$g_{*S,T_0} = 2 + \frac{7}{8} \times 3 \times 2 \left(\frac{T_\nu}{T_0}\right)^3. \tag{5.28}$$

The temperature of the neutrinos T_ν differs from T_0 because after neutrino decoupling, when the temperature falls below $T = m_e \sim 0.5$ MeV, electrons and positrons annihilate via $e^+e^- \rightarrow \gamma\gamma$ and the entropy in the $e^\pm$ pairs is transferred to the photons but *not* to the neutrinos which are already decoupled. For $T_{dec} > T > m_e$, the species in thermal equilibrium with the photons are the photons ($g = 2$) and the electrons and positrons ($g = 4$), so that $g_* = 2 + \frac{7}{8}4 = \frac{11}{2}$. When $T \ll m_e$, only the photons are in equilibrium, so that $g_* = 2$. Conservation of entropy, which is proportional to g_*sT^3, therefore requires that the photon temperature increases by a factor $(\frac{11}{4})^{1/3}$ following the pairs' annihilation, while the temperature of the neutrinos is unaffected. Thus,

$$\frac{T_\nu}{T_0} = \left(\frac{4}{11}\right)^{1/3} \tag{5.29}$$

$$g_{*S,T_0} = \frac{43}{11} \tag{5.30}$$

and, with $T_0 = 2.725$ K, equation (5.27) gives

$$s_0 = 2889 \text{ cm}^{-3}. \tag{5.31}$$

Finally, we note that from (5.19) and (5.14), the WMAP data give

$$\Omega_0 h^2 = 0.51 \pm 0.04. \tag{5.32}$$

Putting all of this together, we find that

$$\Omega_{\nu\bar{\nu}} H_0^2 = \frac{8\zeta(3)}{3\pi} \frac{g_{eff} g_{*S,T_0}}{g_{*S,T_{dec}}} G_N T_0^3 m_\nu \tag{5.33}$$

where

$$\Omega_{\nu\bar{\nu}} \equiv \frac{\rho_{\nu,0}}{\rho_c}. \tag{5.34}$$

Thus,

$$m_\nu = \Omega_{\nu\bar{\nu}} h^2 (94.1 \text{ eV}) \tag{5.35}$$

and (5.21) gives the Cowsik–McClelland bound [2, 3]

$$m_\nu < 48 \text{ eV} \tag{5.36}$$

or, if we impose the stronger constraint deriving from (5.22),

$$m_\nu < 12.7 \text{ eV}. \tag{5.37}$$

5.3   Axions

5.3.1   Introduction: the strong CP problem and the axion solution

We have already alluded in section 4.7 to the infinity of topologically distinct vacua in electroweak theory that derive from the non-trivial (third) homotopy class of the electroweak $SU(2)$ gauge group, as noted in (4.147). Since $SU(2)$ is a subgroup of $SU(3)$, similar conclusions apply to QCD and

$$\pi_3(SU(3)) = Z. \tag{5.38}$$

Indeed the pure $SU(3)$ gauge theory has the well-known 'instanton' solutions [4], which approach these vacua as $|x| \to \infty$. These have (Euclidean) action S_E satisfying

$$S_E = \frac{8\pi^2 |q|}{g_3^2}. \tag{5.39}$$

Here q is the Pontryagin index and is given by

$$\frac{g_3^2}{16\pi^2} \int d^4x \ \mathrm{tr}(G^{\mu\nu}\tilde{G}^{\mu\nu}) \tag{5.40}$$

and it counts the number of wrappings of the S^3, that is the $SU(2)$ group manifold, by the unitary matrix $U_3(x)$ specifying the (pure gauge transformation) vacuum at infinity: $G_a^{\mu\nu}$ is the gluon field strength. See, for example, [5]. The consequence of this is that the true QCD vacuum, the so-called 'θ-vacuum', is a superposition of these states

$$|\theta\rangle = \sum_q e^{-iq\theta} |q\rangle \tag{5.41}$$

where $|q\rangle$ is the 'vacuum' with Pontryagin index q. Then, if we define V_1 as the operator that changes the winding number by one unit, so that

$$V_1|q\rangle \equiv |q+1\rangle \tag{5.42}$$

we see that the θ-vacuum is an eigenstate of V_1 with eigenvalue $e^{i\theta}$. This means that the effective Lagrangian has an additional piece (a so-called 'θ-term')

$$\mathcal{L}_{\mathrm{eff}} = \mathcal{L} + \frac{\theta g_3^2}{32\pi^2} G_{\mu\nu}^a \tilde{G}^{a\mu\nu} \tag{5.43}$$

which is parity (P), time-reversal (T) and CP non-invariant.

A similar additional term also arises when an axial $U(1)$ transformation is performed on all of the quark fields:

$$U(1)_A : q \to e^{i\alpha\gamma_5} q. \tag{5.44}$$

The axial current $j_\mu^{(5)}$, defined by

$$\begin{aligned}
j_\mu^{(5)} &= \sum_q \bar{q}\gamma_\mu\gamma_5 q \\
&= \sum_q [\bar{q}_R\gamma_\mu q_R - \bar{q}_L\gamma_\mu q_L]
\end{aligned} \tag{5.45}$$

where q_R and q_L are the chiral components of q, is anomalous [6] (see also chapter 4 [33, 34]) (see [5] for an account of this). In fact, for massless quarks,

$$\partial^\mu j_\mu^{(5)} = \frac{N_f g_3^2}{16\pi^2} G_{\mu\nu}^a \tilde{G}^{a\mu\nu} \tag{5.46}$$

where N_f is the number of flavours ($= 2N_G$). Thus, as in (4.126), such a θ-term can be *removed* by performing the $U(1)_A$ transformation (5.44) with

$$\alpha = -\frac{\theta}{2N_f}. \tag{5.47}$$

With all quarks massless, the θ-term is not physical, since it can be removed by an (unobservable) $U(1)_A$ transformation. However, this is not the end of the problem, since quarks are *not* massless. The contribution of the mass terms to the QCD Lagrangian is

$$\mathcal{L}_m = -\bar{q}_{Li} M_{ij} q_{Rj} - \bar{q}_{Ri} M_{ij}^\dagger q_{Lj} \tag{5.48}$$

where $i, j = 1, \ldots, N_f$ label the quark flavours, and M_{ij} is the mass matrix. The effect of the $U(1)_A$ transformation (5.44) on M is

$$U(1)_A \; : \; M \to e^{2i\alpha} M$$
$$M^\dagger \to e^{-2i\alpha} M^\dagger. \tag{5.49}$$

Thus, if M was initially Hermitian, so that there are no γ_5s in the mass terms, it is no longer so after the transformation and the transformed Lagrangian has reacquired the P and T non-conserving interactions which the transformation (5.44) with α satisfying (5.47) sought to remove. The quantity $\bar{\theta}$ defined by

$$\bar{\theta} = \theta + 2N_f \arg(\det M) \tag{5.50}$$

is invariant under $U(1)_A$ transformations and parametrizes the T-violation in the (strong) QCD Lagrangian: $\bar{\theta}$ is the effective QCD vacuum angle in the basis where all quark masses are real, positive and γ_5 free. If non-zero, it contributes to the neutron electric dipole moment d_n and the measured upper bound on this requires [7]

$$\bar{\theta} \lesssim 10^{-10}. \tag{5.51}$$

The outstanding question, then, is why $\bar{\theta}$ is so small. This is the 'strong CP problem'.

For each value of the parameter $\bar{\theta}$, we have a different QCD theory and there is no *a priori* reason why one (very small or zero) value is preferred over another. A possible escape from this is that $\bar{\theta}$ is the expectation value of a field $\bar{\theta}(x)$, whose VEV is determined dynamically by an effective potential, as happens when the electroweak symmetry is spontaneously broken by the Higgs-doublet field. Then it is conceivable that, at the minimum of the effective potential, $\bar{\theta} = 0$.

Before addressing the solution of the strong CP problem, we should note that CP-violation might also arise in electroweak theory from terms analogous to the QCD θ-term (5.43), namely

$$\mathcal{L}_\theta = \frac{\theta_2 g_2^2}{32\pi^2} W_{\mu\nu}^a \tilde{W}^{a\mu\nu} + \frac{\theta_1 g_1^2}{32\pi^2} B_{\mu\nu} \tilde{B}^{\mu\nu}. \tag{5.52}$$

Each of the θ-terms is a total divergence (see (4.134) and (4.136)) and so can only contribute surface terms to the action. However, if we consider Euclidean path integral configurations with finite action, the field strengths $F_{\mu\nu}$ must fall off faster than $1/r^2$ as $r \to \infty$, where r is the Euclidean distance. For the $U(1)$ case, this requires that the vector potential B_μ falls off faster than $1/r$ and then the surface integral is negligible as $r \to \infty$. The basic reason is that the θ_1-term has a trivial topological structure. We therefore drop it henceforth. In contrast, the θ_2-term is necessitated by the non-trivial topological structure (4.147) of SU(2). It contributes non-zero surface terms to the action, even though it is a total divergence (4.136). In this case, there are Euclidean path integral configurations with finite action in which the field strengths $W_{\mu\nu} \sim 1/r^4$ as $r \to \infty$ but $W_\mu \sim 1/r$. For these configurations, the surface terms are *not* negligible. However, we have already noted that of the four global $U(1)$ symmetries of the standard model, associated with baryon number (B) and the lepton numbers ($L_\ell, \quad \ell = e, \mu, \tau$), three, namely $\frac{1}{3}B - L_\ell$, are exactly conserved, and the remaining one, say $U(1)_B$, is anomalous, as noted in (4.130). Thus, in a manner directly analogous to that previously described for $U(1)_A$ (when the quarks were massless), we may perform a $U(1)_B$ transformation of the quark fields which removes the θ_2 term. So θ_2 is evidently not an observable parameter and the only observable θ parameter is that associated with QCD.

This observation indicates that a possible solution of the θ problem is to introduce a *further* $U(1)$ symmetry, designed to allow the removal of the θ-term by an appropriate transformation which changes $\arg \det M$. This is the solution proposed by Peccei and Quinn [8] in which the symmetry group of the standard model is augmented by an additional global, chiral $U(1)$ symmetry, known universally now as the $U(1)_{PQ}$ symmetry. However, we cannot do this with just the minimal Higgs content of the standard model. This is because if we use the $U(1)$ to rephase the Higgs doublet by a phase factor $e^{i\delta}$, say, the down-type masses are rephased by $e^{i\delta}$, but the up-type by $e^{-i\delta}$, so that $\arg(\det M)$ is unaltered. Thus we *must* introduce additional scalars. Further, the scalars must not have equal and opposite $U(1)$ charges, since otherwise the previous objection still applies. It follows that the $U(1)$ cannot be the existing $U(1)_Y$ of the standard model, so a new $U(1)_{PQ}$ is required. Additional global $U(1)$ symmetries arise quite commonly in semi-realistic compactifications of heterotic and type I/II string theories. Thus, their introduction to solve the θ problem seems less unattractive now than when it was first proposed. Generically, at the minimum of the effective potential, *both* the local gauge symmetries and the global symmetry are spontaneously broken. Then, by Goldstone's theorem (see [5] for a discussion), there is a scalar (Goldstone) boson having zero mass at the Lagrangian level. This is the axion $a(x)$. It is associated with the phase of the $U(1)_{PQ}$ transformation and, under a $U(1)_{PQ}$ transformation parametrized by α, it transforms according to

$$\frac{a(x)}{v_{PQ}} \rightarrow \frac{a(x)}{v_{PQ}} + \alpha \tag{5.53}$$

where v_{PQ} is the VEV associated with the spontaneous breaking of the $U(1)_{PQ}$ global symmetry. Under the same transformation, a chiral fermion (f) transforms as

$$f(x) \rightarrow e^{-ix_f\alpha} f(x) \tag{5.54}$$

where x_f is the Peccei–Quinn (PQ) charge of f. Then the (PQ) current associated with the symmetry is

$$j^{(PQ)\mu} \equiv \frac{\partial \mathcal{L}}{\partial(\partial_\mu \alpha)} = v_{PQ}\partial^\mu a + \sum_f x_f \bar{f}\gamma^\mu f \tag{5.55}$$

and this is conserved at the classical level, because of the $U(1)_{PQ}$ symmetry. However, because it is a *chiral* symmetry, the symmetry is anomalous, just as $U(1)_A$ is. The anomaly has a form similar to that in (5.46):

$$\partial^\mu j_\mu^{(PQ)} = \xi_3 \frac{g_3^2}{32\pi^2} G_{\mu\nu}^a \tilde{G}^{a\mu\nu} + \xi_2 \frac{g_2^2}{32\pi^2} W_{\mu\nu}^a \tilde{W}^{a\mu\nu} + \xi_1 \frac{g_1^2}{32\pi^2} B_{\mu\nu}\tilde{B}^{\mu\nu} \tag{5.56}$$

where the parameters ξ_i ($i = 1, 2, 3$) are model-dependent constants determined by the $U(1)_{PQ}$ charges of the (chiral) fermion states.

The axion field $a(x)$ appears explicitly in the Yukawa couplings of the fermions to the scalar fields: it is these couplings which generate fermion mass terms when the gauge symmetry is spontaneously broken. We now make a *local* transformation of the fermion fields:

$$f(x) \rightarrow \exp\left[\frac{-ia(x)x_f}{v_{PQ}}\right] f(x) \tag{5.57}$$

chosen so that the axion field is removed from the Yukawa terms. Because it is a *local* transformation, the fermion kinetic terms generate (derivative) interactions with the axion field

$$\bar{f}\gamma^\mu i\partial_\mu f \rightarrow \bar{f}\gamma^\mu i\partial_\mu f + \frac{x_f}{v_{PQ}}(\partial_\mu a)\bar{f}\gamma^\mu f \tag{5.58}$$

and because $U(1)_{PQ}$ is anomalous (see equation (5.56)), extra non-derivative axion interactions are generated:

$$\mathcal{L}_{anom} = \frac{a(x)}{v_{PQ}}\left[\xi_3 \frac{g_3^2}{32\pi^2} G_{\mu\nu}^a \tilde{G}^{a\mu\nu} + \xi_2 \frac{g_2^2}{32\pi^2} W_{\mu\nu}^a \tilde{W}^{a\mu\nu} + \xi_1 \frac{g_1^2}{32\pi^2} B_{\mu\nu}\tilde{B}^{\mu\nu}\right]. \tag{5.59}$$

Thus, the effective Lagrangian is

$$\mathcal{L}_{\text{eff}} = \mathcal{L}_{sm} + \left[\bar{\theta} + \frac{\xi_3}{v_{PQ}}a(x)\right]\frac{g_3^2}{32\pi^2}G_{\mu\nu}^a\tilde{G}^{a\mu\nu}$$

$$+ \left[\theta_2 + \frac{\xi_2}{v_{PQ}}a(x)\right]\frac{g_2^2}{32\pi^2}W_{\mu\nu}^a\tilde{W}^{a\mu\nu} + \left[\theta_1 + \frac{\xi_1}{v_{PQ}}a(x)\right]\frac{g_1^2}{32\pi^2}B_{\mu\nu}\tilde{B}^{\mu\nu}$$

$$+ \frac{1}{2}(\partial_\mu a)^2 + \frac{1}{v_{PQ}}(\partial_\mu a)[j^{(PQ)\mu} - v_{PQ}(\partial_\mu a)] \tag{5.60}$$

where $j^{(PQ)\mu}$ is given in (5.55). Thus, the anomaly generates a potential for $a(x)$ and it is no longer true that all values of $\langle a \rangle$ are allowed in the vacuum nor that the axion field is massless. In fact, in the θ-vacuum, Peccei and Quinn showed that

$$\bar{\theta} + \frac{\xi_3}{v_{PQ}}\langle\bar{\theta}|a(x)|\bar{\theta}\rangle = 0 \tag{5.61}$$

so that the T-violating QCD θ-term is cancelled. We have already noted that the other θ-terms may be dropped, so what remains is an effective Lagrangian in which the physical axion field $\hat{a}(x)$, given by

$$\hat{a}(x) \equiv a(x) - \langle\bar{\theta}|a(x)|\bar{\theta}\rangle \tag{5.62}$$

has interactions with the gauge field strengths and the matter fermions:

$$\mathcal{L}_{\text{eff}} = \mathcal{L}_{sm} + \frac{\xi_3 g_3^2}{32\pi^2 v_{PQ}}\hat{a}(x)G_{\mu\nu}^a\tilde{G}^{a\mu\nu} + \frac{\xi_2 g_2^2}{32\pi^2 v_{PQ}}\hat{a}(x)W_{\mu\nu}^a\tilde{W}^{a\mu\nu}$$

$$+ \frac{\xi_1 g_1^2}{32\pi^2 v_{PQ}}\hat{a}(x)B_{\mu\nu}\tilde{B}^{\mu\nu} + \frac{1}{2}(\partial_\mu\hat{a})^2$$

$$+ \frac{1}{v_{PQ}}(\partial_\mu\hat{a})[j^{(PQ)\mu} - v_{PQ}(\partial_\mu\hat{a})]. \tag{5.63}$$

Effectively, the offending T-violation has been removed by replacing the θ-parameter by a dynamical (axion) field. So the next task is to determine the physical properties of the axion and their implications for experiment.

5.3.2 Visible and invisible axion models

The properties of the axion may be calculated using current algebra techniques [9–11, 14] or by an effective Lagrangian technique [9, 12, 13, 15]. The former give

$$m_a \simeq 0.62\,\text{eV}\left(\frac{10^7\,\text{GeV}}{f_a}\right) \tag{5.64}$$

where

$$f_a = \frac{v_{PQ}}{\xi_3} \tag{5.65}$$

fixes the strength of the axion coupling to the gluon field strength in (5.63). We can see roughly how this estimate arises by noting that this coupling provides an effective potential for the axion so that

$$m_a^2 = \left\langle \frac{\partial^2 V_{\text{eff}}}{\partial a^2} \right\rangle = -\frac{1}{f_a} \frac{g_3^2}{32\pi^2} \frac{\partial}{\partial a} \langle G_{\mu\nu}^a \tilde{G}^{a\mu\nu} \rangle$$

$$\sim \frac{\Lambda_{QCD}^4}{f_a^2}$$

where the last estimate is just on dimensional grounds [16]. Thus comparison with the current algebra result (5.64) would imply $\Lambda_{QCD} \sim 80$ MeV, which is a bit low but in the right ballpark. Although the axion is not massless, it is clear that any reasonable scale of symmetry breaking will give $f_a \gg \Lambda_{QCD}$, which implies a very light axion as the price for solving the strong CP problem. For example, $f_a \sim v \sim 250$ GeV and (5.64) gives $m_a \sim 24$ keV.

The effective Lagrangian (5.63) shows that the axion will decay to two photons with a Lagrangian of the form

$$\mathcal{L}_{a\gamma\gamma} = -g_\gamma \frac{\alpha_{em}}{\pi} \frac{a(x)}{f_a} E \cdot B. \tag{5.66}$$

(Without confusion, we may suppress the hat now.) This decay mode

$$a \rightarrow \gamma\gamma \tag{5.67}$$

will be dominant unless

$$m_a > 2m_e. \tag{5.68}$$

The strength g_γ may be derived from (5.63) and is given by

$$g_\gamma = \frac{\xi_1 + \xi_2}{2\xi_3} \tag{5.69}$$

so it is completely determined by the Peccei–Quinn (PQ) charge assignments.

Similarly, from (5.63) the axion coupling to the fermion f is

$$\mathcal{L}_f = -\frac{1}{v_{PQ}} (\partial_\mu a) \sum_{\chi=R,L} x_{f_\chi} \bar{f} \gamma_\mu a_\chi f \tag{5.70}$$

where

$$a_\chi = \tfrac{1}{2}(1 \pm \gamma_5) \qquad \text{for } \chi = R, L. \tag{5.71}$$

Equivalently, $\mathcal{L}_f$ can be written in the form

$$\mathcal{L}_f = i g_f \frac{m_f}{v_{PQ}} \bar{f} \gamma_5 f \tag{5.72}$$

with the strength g_f given in terms of the PQ charges of the right- and left-chiral components of f by

$$g_f = x_{f_R} - x_{f_L}. \tag{5.73}$$

There is by now overwhelming evidence that the original 'visible' axion, characterized by

$$f_a \sim v_{PQ} \sim v \sim 250 \, \text{GeV} \tag{5.74}$$

does not exist [15]. We mention briefly some of the laboratory-based experiments that lead to this conclusion. The coupling of the light quarks u, d to the axion may be expressed in terms of isoscalar and isovector combinations in an obvious way. The isovector part (λ_1) determines the mixing between the axion and the π^0 and, since the decay rate for pion beta-decay

$$\pi^+ \to \pi^0 e^+ \nu_e \tag{5.75}$$

is well known, the rate for the process

$$\pi^+ \to a e^+ \nu_e \tag{5.76}$$

can be reliably predicted in terms of the isovector amplitude (λ_1). Now, if the mass of the axion satisfies (5.68) it decays rapidly via the process

$$a \to e^+ e^- \tag{5.77}$$

and a bound on the branching ratio for this process can be inferred from the measured branching ratio [17] for the process

$$\pi^+ \to e^+ e^- e^+ \nu_e \tag{5.78}$$

This requires that the isovector amplitude is small,

$$|\lambda_1| \lesssim 2 \times 10^{-2} \tag{5.79}$$

and this is sufficient to exclude the 'short-lived visible axion' models satisfying (5.68), since λ_1 is predicted to be large in such models [15].

However, if the mass of the axion satisfies

$$m_a < 2m_e \tag{5.80}$$

it can only decay slowly, via the process (5.67). In this case, there are strong experimental bounds deriving from the failure to detect axion production in various beam dump experiments. In such experiments, many different processes may produce axions and while it is difficult to calculate individual processes reliably, they contribute incoherently and cannot all vanish. Thus, the production cross sections for the processes

$$pN \to aX$$
$$eN \to aX$$

and the interaction cross section for

$$aN \rightarrow X \tag{5.81}$$

can be confidently estimated and collectively they require [18]

$$m_a \lesssim 50 \text{ keV}. \tag{5.82}$$

Thus, if the solution of the strong CP problem is to be found using the PQ mechanism, the axion must be 'invisible' in these experiments. All models which achieve this use an $SU(3) \times SU(2) \times U(1)$ singlet scalar field σ having a non-zero $U(1)_{PQ}$ charge, which acquires a large VEV ($v_{PQ} \gg v$) so that the beam dump bound is satisfied. One way to achieve the invisibility is if the known quarks and leptons have zero $U(1)_{PQ}$ charge but there exist some new quarks (X), presumably very heavy, having non-zero PQ charge. Such a possibility was proposed by Kim [19] and by Shifman *et al* [20] and the axion is called the 'KSVZ' or 'hadronic' axion. The coupling to the scalar field σ is given by

$$\mathcal{L}^{KSVZ} = -h\overline{X}_L \sigma X_R + \text{h.c.} \tag{5.83}$$

and there is no (tree-level) coupling to the leptons. Another possibility, suggested by Dine *et al* [21] and by Zhitnitskii [22], is that the known quarks and leptons do carry PQ charge so, as in the original model, two Higgs doublets $H_{1,2}$ are required but they are coupled to the PQ field σ only via a term in the Higgs potential having the form.

$$V^{DFSZ} = \lambda H_1^t i \tau_2 H_2 \sigma + \text{h.c.} \tag{5.84}$$

This was discussed in [5]. The axion in this model is called the 'DFSZ' or 'GUT' axion. Although differing considerably in their physical input, the models make similar predictions for the coupling strength g_γ of the axion to two photons:

$$g_\gamma^{KSVZ} = -0.96$$
$$g_\gamma^{DFSZ} = 0.37.$$

5.3.3 Astrophysical constraints on axions

The experimental requirement discussed earlier that axions, if they exist, must be 'invisible' implies that their coupling to photons, leptons and hadrons is very weak. This is most naturally achieved by making $f_a \sim v_{PQ}$ very large which, from (5.64) in turn entails m_a being very small. For example, for a GUT axion, we might expect $f_a \sim v_{PQ} \sim v_{GUT} = O(10^{15} \text{ GeV})$ and then (5.64) gives $m_a \sim 10^{-8}$ eV. In principle, any weakly interacting particle having a mass smaller than typical stellar temperatures, i.e. in the keV–MeV range, can provide an additional mechanism for a star to cool, besides the standard neutrino emission. Of course, the interactions must be strong enough to ensure sufficiently copious production of the particle so that large amounts of energy can be carried away by the new coolant but weak enough for the coolant to stream away without undue hindrance from too many interactions. Since stellar evolution models are well developed and successful in accounting for the observed stellar lifetimes, the axion production cross sections and, hence, the strength of its various couplings are constrained by the error bars on the observational data [23–27].

For example, in globular cluster stars, axions may be produced by the Compton process

$$\gamma e \rightarrow a e \tag{5.85}$$

shown in figure 5.1, or by axion bremsstrahlung

$$e Z \rightarrow a e Z \tag{5.86}$$

shown in figure 5.2. The production cross section for both of these and, hence, the stellar cooling rate is proportional to g_{aee}^2 where, using (5.72),(5.73), (5.65) and (5.64),

$$g_{aee} = g_e \frac{m_e}{v_{PQ}} = \frac{(x_{eR} - x_{eL})m_a m_e}{\xi_3 (0.62 \times 10^{16} \, \text{eV}^2)}. \tag{5.87}$$

The observational data yield the constraint [23, 28]

$$|g_{aee}| \lesssim 0.5 \times 10^{-12} \tag{5.88}$$

so that

$$\left| \frac{(x_{eR} - x_{eL})}{\xi_3} \right| m_a \lesssim 0.62 \times 10^{-2} \, \text{eV} \tag{5.89}$$

which gives $m_a \lesssim 10^{-2}$ eV as the generic constraint on DFSZ models, taking the unknown PQ charges to be of order unity. Of course, the mass of the hadronic axion is unconstrained by these data.

The globular cluster data also constrain the axion–photon coupling, which enters via the Primakoff process

$$\gamma \leftrightarrow a \tag{5.90}$$

shown in figure 5.3, in which a photon is converted to an axion in the coherent electromagnetic field of a nucleus or an electron. The production cross section is proportional to $g_{a\gamma\gamma}^2$ where, from (5.66) and (5.64),

$$g_{a\gamma\gamma} = g_\gamma \frac{\alpha_{em}}{\pi f_a} = \frac{m_a g_\gamma \alpha_{em}}{\pi (0.62 \times 10^{16} \, \text{eV}^2)} \tag{5.91}$$

and the data yield the constraint [27]

$$|g_{a\gamma\gamma}| \lesssim 0.6 \times 10^{-10} \, \text{GeV}^{-1}. \tag{5.92}$$

Then

$$|g_\gamma| m_a \lesssim 0.16 \, \text{eV} \tag{5.93}$$

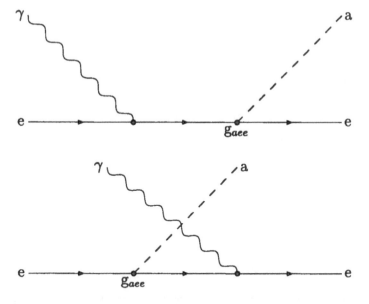

Figure 5.1. Axion production via the Compton process.

and $m_a \lesssim 0.4$ eV for both the DFSZ and KSVZ axions.

Similar arguments may be applied to the cooling of neutron stars. In the supernova SN 1987A, thermal neutrinos transported away the binding energy of the newly formed neutron star in about 10 s [29, 30], in accord with theoretical calculations. The possibility of other mechanisms for removing the energy is, therefore, constrained and, for axions, the axion–nucleon coupling is constrained, which, in turn, constrains the mass [31, 32] to satisfy

$$m_a \lesssim 0.01 \text{ eV}. \tag{5.94}$$

5.3.4 Axions and cosmology

If they exist, axions would be produced in the early universe and the relic axions have important implications for current and future observations. In principle, axions may be produced thermally or non-thermally and two distinct non-thermal mechanisms have been proposed.

The discussion of thermal production is straightforward. At high temperatures, axions are created (and destroyed) by photoproduction or gluoproduction on quarks:

$$\gamma q \leftrightarrow aq \tag{5.95}$$
$$gq \leftrightarrow aq.$$

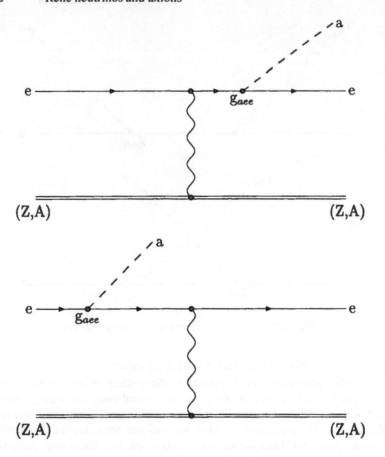

Figure 5.2. Axion bremsstrahlung.

When the temperature drops below that of the quark–hadron phase transition $T \lesssim \Lambda_{QCD} \sim 175$ MeV, by the pion–axion conversion process (5.90)

$$\pi N \leftrightarrow aN. \qquad (5.96)$$

Each of these processes has an associated absorption rate

$$\Gamma^{T}_{\text{abs}} = n_T \langle \sigma |v| \rangle_{\text{abs}} \qquad (5.97)$$

where n_T is the number density of the axion's target $T = q$ or N, σ is the scattering cross section, v is the relative velocity of the axion and the target T and $\langle \ldots \rangle$ denotes a thermal average. If the expansion rate of the universe is slow compared with the total absorption rate Γ_{abs}, then we expect that these processes will eventually achieve thermal equilibrium with the standard (relativistic) axion number density given by equation (5.2) with $g_a = 1$. However, if the axions

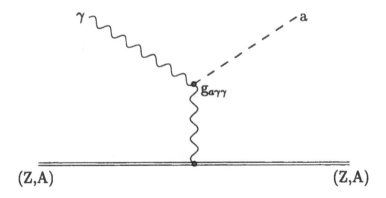

Figure 5.3. Photon-axion conversion: the Primakoff process.

interact too weakly, the total absorption rate is too slow for them ever to reach thermal equilibrium and their number density freezes out with a value below $n_{a,\text{eq}}$.

To quantify this, we use the abundance defined in (5.11) with the equilibrium value given in (5.12) with $g_{a,\text{eff}} = 1$. The Boltzmann equation determining the evolution of Y_a is

$$\frac{dY_a}{dt} = -\Gamma_{\text{abs}}(Y_a - Y_{a,\text{eq}}). \qquad (5.98)$$

Thus, $Y_a(t)$ always lies between its initial value and Y_a^{eq}:

$$Y_a(t) - Y_{a,\text{eq}} = (Y_a(0) - Y_{a,\text{eq})} \exp\left(-\int_0^t \Gamma_{\text{abs}}\, dt'\right). \qquad (5.99)$$

It is convenient to recast the integral in terms of the variable

$$x \equiv \frac{m_N}{T}. \qquad (5.100)$$

In the radiation era the scale factor $R(t) \propto t^{1/2}$, so that the Hubble rate

$$H = \frac{1}{2t} \propto T^2 \propto x^{-2} \qquad (5.101)$$

and the relic abundance may be written as

$$Y_a(x) = Y_{a,\text{eq}}\left[1 - \left(1 - \frac{Y_a(0)}{Y_{a,\text{eq}}}\right)\exp\left(-\int_0^x \frac{\Gamma_{\text{abs}}(x')}{x'H(x')}\, dx'\right)\right]. \qquad (5.102)$$

Below the quark–hadron phase transition, the nucleons are non-relativistic and have an equilibrium number density given by (5.1) in the limit $T \ll m_N$,

$$n_N = g_N \left(\frac{m_N^2}{2\pi x}\right)^{3/2} e^{-x}. \qquad (5.103)$$

The thermally-averaged cross section is

$$\langle \sigma |v| \rangle_{\text{abs}} = g_{aNN}^2 x^{-2} m_\pi^{-2} \tag{5.104}$$

with

$$g_{aNN} \approx \frac{m_N}{f_{PQ}}. \tag{5.105}$$

In the radiation-dominated era,

$$H = \left(\frac{8\pi\rho}{3m_P^2} \right)^{1/2} = \frac{2\pi}{3} \left(\frac{\pi g_*}{5} \right)^{1/2} \frac{m_N^2}{x^2 m_P}. \tag{5.106}$$

Putting all this together, we get

$$\frac{\Gamma_{\text{abs}}(x)}{H(x)} = \frac{3 g_N}{2\pi^3} \left(\frac{5}{8 g_*} \right)^{1/2} \left(\frac{m_N^{3/2} m_P^{1/2}}{f_{PQ} m_\pi} \right)^2 x^{-3/2} e^{-x}$$

$$\approx \left(\frac{10}{g_*} \right)^{1/2} \left(\frac{m_a}{1.2 \times 10^{-3} \text{ eV}} \right)^2 x^{-3/2} e^{-x} \tag{5.107}$$

using (5.64). Above the quark–hadron phase transition, the processes (5.95) dominate and Γ_{abs}/H scales as x, achieving its maximum value just after the transition. Thus, we can estimate the final relic abundance as

$$Y_a(\infty) = Y_{a,\text{eq}} \left\{ 1 - \left(1 - \frac{Y_a(0)}{Y_{a,\text{eq}}} \right) \right.$$

$$\times \exp \left[- \frac{3 g_N}{2\pi^3} \left(\frac{5}{8 g_*} \right)^{1/2} \left(\frac{m_N^{3/2} m_P^{1/2}}{f_{PQ} m_\pi} \right)^2 I(x_{qh}) \right] \right\}$$

$$= \frac{0.278}{g_{*,\text{dec}}} \left\{ 1 - (1 - 3.6 g_{*,\text{dec}} Y_a(0)) \right.$$

$$\times \exp \left[- \left(\frac{10}{g_*} \right)^{1/2} \left(\frac{m_a}{1.2 \times 10^{-3} \text{ eV}} \right)^2 I(x_{qh}) \right] \right\} \tag{5.108}$$

where

$$I(x_{qh}) \equiv \int_{x_{qh}}^{\infty} x'^{-5/2} e^{-x'} \, dx'$$

$$= -\frac{2}{3} [x_{qh}^{-3/2} e^{-x_{qh}} (2 x_{qh} - 1) + 2\sqrt{\pi} (\text{erf}(\sqrt{x_{qh}}) - 1)] \tag{5.109}$$

and $g_{*,\text{dec}}$ is the value of g_* at decoupling (freeze-out). The parameter $x_{qh} \sim 5$ is the value of x at the quark–hadron phase transition, so that $I \sim 10^{-4}$. Thus, if the mass of the axion were greater than about 0.1 eV, the relic abundance would be near to the equilibrium abundance. In fact, masses this large are already excluded by the data, so the relic abundance depends upon the initial value. For example, with m_a saturating the SN 1987A bound (5.94), we estimate that

$$\frac{Y_a(\infty)}{Y_{a,eq}} \sim 0.007 + 0.993\frac{Y_a(0)}{Y_{a,eq}}. \tag{5.110}$$

At high temperatures, it is plausible to assume that there are no axions, so that $Y_a(0) = 0$, in which case the relic abundance is very far from thermal. In any event, it is clear that thermal axions cannot provide anything like the measured matter density. As in (5.16), we define Ω_a^{th} to be the fraction of the closure energy density provided by thermal axions:

$$\Omega_a^{th} \equiv \frac{\rho_a^{th}}{\rho_c}. \tag{5.111}$$

Then, analogously to (5.35), we find that

$$m_a = \frac{g_{*,dec}}{10}\Omega_a^{th}h^2(130\,eV). \tag{5.112}$$

Saturating the measured value (5.22) of the current mass density would require $m_a \sim 18$ eV for closure, a value which is clearly excluded by the observational bounds already obtained.

In all of the foregoing discussion, it was tacitly assumed that the classical axion field had a constant value, in fact the value (5.61) needed to ensure that the strong $\bar{\theta}$-term vanishes. However, in the early universe, when the temperature $T \sim f_a \gg \Lambda_{QCD}$, the $U(1)_{PQ}$ symmetry is broken and massless axions are created. The potential which gives the axions a mass arises from non-perturbative instanton effects only when the temperature drops to $T \sim \Lambda_{QCD}$. Thus, at high temperatures, there is no reason why the axion should have the preferred value for which $\bar{\theta} = 0$. When instantons generate a potential for the axion field, it will roll towards the preferred value, so the foregoing assumption that the field is a constant is not true during this era. This 'misalignment' of the field with its ground-state value means that there is a non-zero axion field energy density which we shall now calculate.

We assume that the axion field is spatially homogeneous and depends only on time. Then, from (5.63), the effective axion action is

$$S = \int d^4x \sqrt{g}(\tfrac{1}{2}\dot{a}^2 - \tfrac{1}{2}m_a^2a^2 + \Gamma_a\dot{a})$$

$$= \int d^4x \, R^3(t)(\tfrac{1}{2}\dot{a}^2 - \tfrac{1}{2}m_a^2a^2 + \Gamma_a\dot{a})$$

where $R(t)$ is the cosmological scale factor and we have retained only the quadratic (mass) term in the axion potential. The equation of motion is

$$\frac{d}{dt}[R^3(\dot{a} + \Gamma_a)] + R^3m_a^2(T)a = 0. \tag{5.113}$$

The decay width Γ_a of the axion is tiny, so we may safely ignore it henceforth. Initially, at high temperatures $T \gg \Lambda_{QCD}$, the axion is massless and we assume that $\dot{a} = 0$. Then the field is constant:

$$a(t) = a_i \tag{5.114}$$

where a_i is the initial 'misaligned' value of the field. As the temperature falls $m_a^2(T)$ increases and the equation of motion is

$$\ddot{a} + 3H\dot{a} + m_a^2(T)a = 0 \tag{5.115}$$

where, as usual, $H \equiv \dot{R}/R$ is the Hubble parameter. Eventually, the temperature reaches T_i at which

$$m_a(T_i) = 3H(T_i) \tag{5.116}$$

and, thereafter, $a(t)$ oscillates with frequency $m_a(T)$. The energy density associated with the axion field is

$$\rho_a = \tfrac{1}{2}\dot{a}^2 + \tfrac{1}{2}m_a^2 a^2 \tag{5.117}$$

so, using (5.115),

$$\dot{\rho}_a = \dot{m}_a m_a a^2 - 3H\dot{a}. \tag{5.118}$$

Averaging over one oscillation

$$\langle \dot{a}^2 \rangle = m_a^2 \langle a^2 \rangle \tag{5.119}$$

and then (5.118) gives

$$\langle \dot{\rho}_a \rangle = \left(\frac{\dot{m}_a}{m_a} - 3H \right) \langle \rho_a \rangle \tag{5.120}$$

whose solution is

$$\langle \rho_a \rangle R^3(t) \propto m_a(T). \tag{5.121}$$

Thus, the axion number density $n_a = \langle \rho_a \rangle / m_a(T)$ scales as $R^{-3}(t)$, even though the axion mass is varying. The entropy density s also scales in this way, so assuming that there has been no entropy production since the axion field began to oscillate, their ratio is conserved. When the temperature $T = T_i$, given by (5.116),

$$\rho_a = \tfrac{1}{2}m_a^2(T_i)a_i^2 \tag{5.122}$$

since initially $\dot{a} = 0$, and

$$\left. \frac{n_a}{s} \right|_{T_i} = \frac{45 m_a(T_i)a_i^2}{4\pi^2 g_* T_i^3} = \frac{45 a_i^2}{2\sqrt{5\pi} g_* T_i m_P}. \tag{5.123}$$

The present (misaligned) axion energy density is given as a fraction of the closure energy density by

$$\Omega_a^{\text{mis}} = \frac{\rho_{a0}^{\text{mis}}}{\rho_c} = \left. \frac{n_a}{s} \right|_{T_i} m_a \frac{s_0}{\rho_c}. \tag{5.124}$$

Now, recall that the axion field satisfies (5.61), so a_i is given in terms of the initial value $\bar{\theta}_i$ of the angular field $\bar{\theta}$ by

$$a_i = f_a \bar{\theta}_i \qquad (5.125)$$

where f_a is given by (5.65), and $\bar{\theta}_i$ can be anywhere in the range $(-\pi, \pi)$. The temperature T_i found by solving (5.116) is about 1 GeV for an axion of mass $m_a = 10^{-5}$ eV and

$$\frac{T_i}{1 \text{ GeV}} \simeq \left(\frac{m_a}{10^{-5} \text{ eV}}\right)^{0.18}. \qquad (5.126)$$

Then

$$\Omega_a^{\text{mis}} \simeq 0.16 h^{-2} \left(\frac{10}{g_*}\right)^{1/2} \left(\frac{m_a}{10^{-5} \text{ eV}}\right)^{-1.18} \bar{\theta}_i^2. \qquad (5.127)$$

Note the negative power dependence on m_a. If we replace $\bar{\theta}_i$ by its root mean square (rms) value $\pi/\sqrt{3}$ in the range $(-\pi, \pi)$, we see that the axion energy density from the vacuum misalignment does not exceed the measured matter density (5.22), provided that the mass of the axion is *greater* than about 10^{-5} eV.

However, it is not clear that we are justified in replacing $\bar{\theta}_i$ by its rms value. In each causally connected domain, we expect $\bar{\theta}_i$ to have an independent value. If these values are randomly set, then it is reasonable to replace $\bar{\theta}_i$ by its rms value provided that the observable universe is composed of many such causally connected domains. We shall see later that there is good evidence that there was a period of inflation in the early universe and, if the reheating temperature after this is over is lower than that of the PQ symmetry breaking, then the observable universe is composed of only about *one* causal region and we have no *a priori* reason for selecting any particular value of $\bar{\theta}_i$ in our patch and, consequently, no way of estimating Ω_a^{mis}. In any case, (5.127) is only a rough estimate. There are theoretical uncertainties, which amount to a factor $\Delta \simeq \frac{1}{3}$–3, deriving from the PQ-model dependence and the nature of the QCD phase transition and also from anharmonic corrections which give a factor $f(\bar{\theta}_i)$ when the initial value $\bar{\theta}_i$ is in a region where other terms in the axion potential are important, besides just the quadratic terms which we have retained.

Further, the homogeneous oscillations of the axion field correspond to the creation of *zero momentum* axions and it has been argued that non-zero momentum axions, with a momentum spectrum $g(k)$, are created before the temperature drops to $T \sim \Lambda_{QCD}$ by other non-perturbative effects. In consequence, the axion density deriving from the 'misalignment' effect is an *under*estimate of the actual density, as we shall see. The $U(1)_{PQ}$ symmetry with which we are concerned is directly analogous to the global $U(1)$ symmetry relevant to a superfluid ^{4}He condensate at low temperatures. In this system, it is known that besides the ordinary bulk superfluidity, analogous to our homogeneous axion field, there are also vortex configurations in which the phase of the order parameter (or the pair wavefunction) varies spatially, although its magnitude remains constant (determined by the density of the superfluid condensate). Such topological configurations arise because the fundamental group of the manifold S^1 associated with symmetry group $U(1)$ of the ground state is non-trivial:

$$\pi_1(S^1) = Z. \tag{5.128}$$

As we traverse *any* closed path threaded by a vortex, the phase of the condensate varies continuously and changes by an integral multiple of 2π when we return to the starting point. By shrinking the size of the closed path, it is clear that there is a linear vortex on which the phase of the order parameter is undefined. In a real superfluid, there is a cylindrical core region, centred on this line, in which the magnitude of the order parameter varies, approaching zero on the line.

Similar considerations apply to our $U(1)_{PQ}$ symmetry. In the early universe when the PQ symmetry is broken, we expect the axion field to vary spatially, since it is uncorrelated beyond the horizon. These topological considerations (5.128) indicate that a random 'axion string' network will, therefore, be formed [33] just as vortex configurations are formed in superfluid ^{4}He. The thickness of the core region is $\delta \sim f_a^{-1}$. Roughly, there are two types of string: long strings, spanning the horizon, and small string loops. The loops oscillate and radiate axions, and this is the dominant energy-loss mechanism [11, 34]. The axions are massless when they are emitted and the emission continues until they acquire a mass via instanton effects when the temperature drops to $T \sim \Lambda_{QCD}$. A numerical simulation of a random network of (global) axion strings has been performed recently [35, 36]. This shows that, after a short initial period of relaxation, the network evolves to a 'scaling' regime, in which the large-scale behaviour of the network scales with the Hubble radius and the energy density is given by

$$\rho_a^{\text{string}} = \frac{\xi \mu}{t^2} \tag{5.129}$$

where ξ is a constant and μ is the string tension per unit length. Such behaviour was predicted theoretically by Albrecht and Turok [37]. The radiated axions have a momentum spectrum $g(k)$ which is peaked around wavelengths of order of the horizon scale $(\overline{k^{-1}} \sim (4\pi H)^{-1})$ and which decays exponentially for shorter wavelengths. The contribution Ω_a^{string} to the current fractional relic axion energy density is calculated as follows:

$$\Omega_a^{\text{string}} \simeq (0.39 \pm 0.26)h^{-2} \left(\frac{m_a}{10^{-5}\,\text{eV}}\right)^{-1.18} \tag{5.130}$$

which is somewhat larger than, but comparable with, the value obtained from the misalignment mechanism if we take the rms value for $\bar{\theta}_i$. So applying the measured matter density bound (5.22) requires the axion mass to be greater than about 10^{-5} eV, as before. The numerical simulation was performed on a 256^3 lattice but it has been noted [38] that this might not be sufficient to observe logarithmic corrections, proportional to $t^{-2} \ln t$, to the scaling behaviour (5.129). Such corrections would have the effect of enhancing axion production at later times, thereby reducing the lower bound on the axion mass needed to avoid overclosure. Of course, if the reheating temperature after inflation is lower than that of the PQ symmetry breaking, then axion strings and the radiated axions are washed out and essentially all of the relic axions come from the coherent oscillations of the zero mode discussed earlier.

Axions produced at a temperature around $T \sim \Lambda_{QCD}$ when they acquire a non-zero mass are non-relativistic and they are, therefore, candidates for cold dark matter (CDM). Studies of large-scale structure formation indicate that CDM is an important component and, if so, it gets trapped in the gravitational potential and contributes to the galactic halo. Such axions might be detected experimentally using a cavity permeated by a strong static magnetic field [39] via the Primakoff process (5.90). When the cavity frequency is tuned to the axion mass, the galactic halo axions convert resonantly into photons of that frequency. A cylindrical cavity of radius 1 m has a lowest TM mode of frequency $f = 115$ MHz, corresponding to an axion of mass $m_a = 0.475 \times 10^{-6}$ eV. Experiments using a tunable cavity of this size have produced exclusion zones in the axion mass *versus* axion–photon coupling $(m_a, g_{a\gamma\gamma})$ plane. All of these are normalized assuming that the local halo (CDM) density is entirely in the form of axions. The conversion power of the resonant cavity is proportional to $g_{a\gamma\gamma}^2$ and, until recently, the power sensitivity levels were too high to bound theoretically favoured models. However, a recent experiment at LLNL [40] has excluded KSVZ axions in the range

$$2.77 \times 10^{-6} \text{ eV} < m_a < 3.3 \times 10^{-6} \text{ eV} \qquad \text{(excluded)} \qquad (5.131)$$

and further experiments are underway at LLNL and Kyoto with sufficient sensitivity to detect DFSZ axions at even a fraction of the local halo density.

5.4 Exercises

1. Verify the form (5.63) giving the interactions of the axion with the gauge fields and fermions.
2. Show that axion decay into two photons is described by the Lagrangian (5.66) with g_γ given by (5.69).
3. Show that axion decay into an electron–positron pair is proportional to g_{aee}^2 where g_{aee} is given by (5.87).
4. Verify the expression (5.108) for the final axion relic abundance.

5.5 General references

The books and review articles that we have found most useful in preparing this chapter are:

- Peccei R D 1989 The Strong CP Problem in *CP Violation* ed C Jarlskog (Singapore: World Scientific)

- Kolb E R and Turner M S 1990 *The Early Universe* (Reading, MA: Addison-Wesley)
- Muryama H, Raffelt G, Hagmann C, van Bibber K and Rosenberg L J 2000 Axions and other very light bosons (*Review of Particle Properties* Groom D E *et al* Particle Data Group) *Eur. Phys. J.* C **15** 1

Bibliography

[1] Spergel D N *et al* 2003 *Astrophys. J. Suppl.* **148** 175, arXiv:astro-ph/0302209
[2] Gerstein G and Zel'dovich Ya B 1966 *Zh. Eksp. Teor. Fiz. Pis'ma Red.* **4** 174
[3] Cowsik R and McClelland J 1972 *Phys. Rev. Lett.* **29** 669
[4] Belavin A A, Polyakov A M, Schwartz A S and Tyupkin Yu S 1975 *Phys. Lett.* B **59** 85
[5] Bailin D and Love A 1993 *Introduction to Gauge Field Theory* revised edn (Bristol: IOP) p 212
[6] Adler S L 1969 *Phys. Rev.* **177** 2426
[7] Baluni V 1979 *Phys. Rev.* D **19** 2227
     Crewther R, di Vecchia P, Veneziano G and Witten E 1979 *Phys. Lett.* B **88** 123
[8] Peccei R D and Quinn H R 1977 *Phys. Rev. Lett.* **38** 1440
     Peccei R D and Quinn H R 1977 *Phys. Rev.* D **16** 1791
[9] Weinberg S 1978 *Phys. Rev. Lett.* **40** 229
[10] Ellis J and Gaillard M K 1979 *Nucl. Phys.* B **150** 141
[11] Donnelly T W *et al* 1978 *Phys. Rev.* D **18** 1607
[12] Kandaswamy J, Salmonsen P and Schechter J 1978 *Phys. Rev.* D **17** 3051
[13] Georgi H, Kaplan D B and Randall L 1986 *Phys. Lett.* B **169** 73
[14] Bardeen W A and Tye S-H 1981 *Phys. Lett.* B **74** 199
[15] Bardeen W A, Peccei R D and Yanagida T 1987 *Nucl. Phys.* B **279** 401
[16] Peccei R 1989 *CP Violation* ed C Jarlskog (Singapore: World Scientific) p 503
[17] Egli S *et al* 1989 *Phys. Lett.* B **222** 533
[18] Groom D E *et al* 2000 Particle Data Group *Euro. Phys. J.* C **15** 1
[19] Kim J 1970 *Phys. Lett.* B **43** 10
[20] Shifman M A, Vainshtein A I and Zakharov V I 1980 *Nucl. Phys.* B **166** 493
[21] Dine M, Fischler W and Srednicki M 1981 *Phys. Lett.* B **104** 199
[22] Zhitnitskii A P 1980 *Sov. J. Nucl. Phys.* **31** 260
[23] Dicus D A, Kolb E W, Teplitz V L and Wagoner R V 1978 *Phys. Rev.* D **18** 1829
[24] Sato K 1978 *Prog. Theoretical Phys.* **60** 1942
[25] Turner M S 1990 *Phys. Rep.* **197** 67
[26] Raffelt G G 1990 *Phys. Rep.* **198** 1
[27] Raffelt G G 1996 *Stars as Laboratories for Fundamental Physics* (Chicago, IL: University of Chicago Press)
[28] Raffelt G G and Weiss A 1995 *Phys. Rev.* D **51** 1495
[29] Hirata K *et al* 1987 *Phys. Rev. Lett.* **58** 1490
[30] Bionta R M *et al* 1987 *Phys. Rev. Lett.* **58** 1494
[31] Janka H-T, Keil W, Raffelt G G and Seckel D 1996 *Phys. Rev. Lett.* **76** 2621
[32] Keil W *et al* 1997 *Phys. Rev.* D **56** 2419
[33] Vilenkin A and Everett A E 1982 *Phys. Rev. Lett.* **48** 1867
[34] Vilenkin A and Vachaspati T 1987 *Phys. Rev.* D **35** 1138

[35] Yamaguchi M, Kawasaki M and Yokoyama J 1999 *Phys. Rev. Lett.* **82** 4578
[36] Yamaguchi M 1999 *Phys. Rev.* D **60** 103 511
[37] Albrecht A and Turok N 1989 *Phys. Rev.* D **40** 973
[38] Battye R A and Shellard E P S 1999 *Tegernsee 1999, Beyond the Desert* pp 565–72, arXiv:astro-ph/9909231
[39] Sikivie P 1983 *Phys. Rev. Lett.* **51** 1415
     Sikivie P 1985 *Phys. Rev.* D **32** 2988
[40] Hagmann C *et al* 1998 *Phys. Rev. Lett.* **80** 2043

Chapter 6

Supersymmetric dark matter

6.1 Introduction

The cosmological bounds on the masses of various known or hypothethical relic particles derive from the requirement that the total energy density of the relic particles X does not exceed the measured present total energy density ρ_0. In terms of dimensionless quantities, this gives

$$\Omega_{X,0} < \Omega_0 \tag{6.1}$$

where $\Omega_{X,0}$ and Ω_0 are defined in (5.16) and (1.41) with Ω_0 given by (5.19). Some of the relics are known. For example, the present photon energy density $\rho_\gamma = \frac{\pi^2}{15} T_0^4$, with $T_0 = 2.73$ K $= 2.35 \times 10^{-4}$ eV the present temperature of the cosmic microwave background, gives $\Omega_\gamma h^2 \simeq 2.471 \times 10^{-5}$. Thus,

$$\Omega_\gamma = 5.1 \times 10^{-5} \tag{6.2}$$

taking $h = 0.71^{+0.04}_{-0.03}$ as in (5.14). Similarly, and as noted previously, the measured primordial deuterium and helium abundances require that the baryon energy density gives $\Omega_b h^2 = 0.019 \pm 0.003$. Thus,

$$\Omega_b = 0.039 \pm 0.004 \tag{6.3}$$

and baryons constitute not more than a few percent of the total. Relic neutrinos also contribute and we may invert (5.35) to obtain a contribution of

$$\Omega_{\nu\bar{\nu}} = \frac{m_\nu}{47.4 \,\text{eV}} \tag{6.4}$$

for each relativistic species with $m_\nu \gg T_0$. The current experimental bound for the electronic species from the Mainz and Troitsk tritium beta-decay experiment [1] is far more restrictive:

$$m_{\nu_e} < 2.2 \,\text{eV} \qquad \text{at 95\% CL.} \tag{6.5}$$

Further, this bound effectively applies to *all* neutrino species, since the SuperK [2] and SNO [3] data show that the mass differences for atmospheric and solar neutrinos are tiny:

$$\Delta m^2_{\nu,\text{atmos}} \simeq 3 \times 10^{-3} \, \text{eV}^2 \tag{6.6}$$

$$\Delta m^2_{\nu,\text{solar}} \simeq 7 \times 10^{-5} \, \text{eV}^2. \tag{6.7}$$

Thus the sum of the neutrino masses is at most, 6.6 eV and likely to be much smaller. (When $m_\nu \ll T_0$, the relativistic fermion density gives $\Omega_\nu = 0.23\Omega_\gamma$.) In addition, simulations of structure formation in a neutrino-dominated universe are unable to reproduce the observed structure. The best fit to all of the cosmological data [4] gives

$$\Omega_\nu < 0.015 \tag{6.8}$$

which for three degenerate species implies

$$m_{\nu_e} < 0.23 \, \text{eV} \qquad \text{at 95\% CL.} \tag{6.9}$$

It is, therefore, clear that relic neutrinos too constitute only a small fraction of the total energy density. We might have anticipated that the dominant contribution to the present energy density would come from the matter comprising the galaxies and the contribution of the baryons *is* the largest. However, it is nowhere near large enough to account for the total energy density. Thus, we may be confident that the major contributions to Ω_0 are *not* from known sources. In any case, as we shall shortly see, there is strong evidence that there is a large amount of invisible 'dark matter' in the universe [5].

The possibility of dark matter was first suggested by Kapteyn [6] in 1922, who noted that its mass could be estimated from the velocity distribution of stars in our galaxy. The strongest evidence for its existence comes from measurements of rotation speeds of spiral galaxies. If we consider a star moving with speed $v(r)$ in a circular orbit of radius r outside of a spherically symmetric mass distribution with a total mass $M(r)$ interior to r, then

$$G_N M(r) = r v(r)^2. \tag{6.10}$$

The speed $v(r)$ can be determined for luminous objects such as stars or gas clouds by measuring the Doppler shifts in emission or absorption lines and the mass distribution $M(r)$ is then inferred from (6.10). The mass of a spiral galaxy can be determined by taking r to be the radius within which most of the light is emitted. In this way, the average galactic mass $\langle m_{\text{gal}} \rangle$ can be calculated. Combining this with the measured number density n_{gal} determines the average energy density

$$\langle \rho_{\text{lum}} \rangle = n_{\text{gal}} \langle m_{\text{gal}} \rangle. \tag{6.11}$$

These measurements show that the contribution of luminous matter is

$$\Omega_{\text{lum}} \lesssim 0.01. \tag{6.12}$$

In other words, luminous matter accounts for less than 1% of the mass of the universe. When these techniques are applied to the rare stars and neutral hydrogen (HI) clouds beyond the radius where light from the galaxy is emitted, it is found that $M(r)$ continues to increase, reaching a maximum in the range 150–300 km s^{-1} within a few kpc, and then remaining constant out to the largest radii at which HI clouds can be found. If there were no matter outside of the luminous region, then from (6.10) $v(r)$ should fall off as $r^{-1/2}$. Thus, the measurement of roughly constant values of $v(r)$ in over 1000 galaxies indicates that the galaxies have huge 'halos' of dark matter, with mass 3–10 times that of the luminous component. The rotation curve for our own Milky Way galaxy is difficult to measure, because the observer is inside the galaxy but there is little doubt that our galaxy too is immersed in a dark matter halo. Further, by studying the motion of galactic clusters a universal mass density corresponding to [7]

$$\Omega_m h^2 \simeq 0.1\text{--}0.3 \tag{6.13}$$

can be inferred. In fact, it was measurements of cluster galaxy dynamics that led to the discovery of dark matter by Zwicky [8] in 1933. Recent data on the acoustic peaks in the cosmic microwave background (CMB), combined with independent data from simulations of cluster formation, high-z supernovae, quasars, and the Lyman alpha forest, give the best-fit values [4]

$$\Omega_0 = 1.02 \pm 0.02 \qquad \Omega_m h^2 = 0.135^{+0.008}_{-0.009} \tag{6.14}$$

$$\Omega_\Lambda = 0.65 \pm 0.05 \qquad \Omega_b h^2 = 0.0224 \pm 0.0009 \tag{6.15}$$

where $\Omega_m \equiv \rho_m/\rho_c$ is the total matter contribution, distinguished from the cosmological constant contribution $\Omega_\Lambda \equiv \rho_{\text{vac}}/\rho_c = \Lambda/3H_0^2$, and $\rho_c \equiv 3M_P^2 H_0^2$ is the critical density defined in (1.37), and $h = 0.71^{+0.04}_{-0.03}$. Clearly, $\Omega_m \neq \Omega_b$. So there must be non-baryonic dark matter and the first problem is to identify its nature. There is also a second problem, which is to explain the discrepancy between the observed luminous matter density given in equation (6.12) and the calculated baryon density (6.3) required for the success of the primordial nucleosynthesis calculation. We shall have little to say about the latter problem, save to note that it seems at least possible that it can be solved by a combination of dark stars, intracluster gas and the Lyman alpha forest [9].

In this chapter, we first characterize the general properties that dark matter particles possess, whatever they are. Since there are no satifactory candidates within the standard $SU(3) \times SU(2) \times U(1)$ theory of strong and electroweak interactions, it is natural to look for suitable candidates in the (minimal) supersymmetric version of the standard model, the MSSM. One possibility, that gravitinos make up the dark matter, arises in any locally supersymmetric theory. It is studied in section 6.3. However, the most popular view is that dark matter is made of neutralinos. The parameters of the MSSM that control the mass and other properties of the neutralino are detailed in section 6.4. In the following section, we discuss the bounds that can be put on the neutralino mass using cosmological data and also the constraints on the parameters of the MSSM that arise from other data whose theoretical prediction depends upon these parameters. We discuss the prospects for the experimental detection of (neutralino) dark matter in section 6.6.

6.2 Weakly interacting massive particles or WIMPs

To have survived until the present epoch, any (non-baryonic) dark matter particles must either be stable or have a lifetime comparable with the present age of the universe. Further, if the dark matter particles have electromagnetic or strong interactions, they would bind to nucleons and form anomalous heavy isotopes. Such isotopes have been sought but not found [10]. Thus, the dark matter particles can, at best, participate in weak (and gravitational) interactions or, at worst, only in gravitational interactions. One obvious possibility satisfying the foregoing constraints is that the dark matter consists of neutrinos. However, we have already noted that the present data on neutrino masses (from tritium decay, atmospheric and solar neutrino experiments) show that although neutrinos might barely account for the inferred mass density (5.22) or (6.14), simulations of galaxy formation and cluster formation require *cold* dark matter. That is, the dark matter is made of weakly interacting massive particles (WIMPs). No such particles exist in the standard model but they do in its enlargement to the minimal supersymmetric standard model (MSSM).

We can estimate the relic density of WIMPs using the same techniques as those used for relic neutrinos in section 5.2. The difference is that the equilibrium abundance for a cold (i.e. non-relativistic) fermion species X is obtained from (5.1) by taking the limit $T \ll m_X$. The result is

$$n_{X,\text{eq}} = g_X \left(\frac{m_X T}{2\pi} \right)^{3/2} e^{-m_X/T}. \tag{6.16}$$

Roughly speaking, freeze-out of such relics occurs when their annihilation rate Γ_A becomes equal to the Hubble rate H. The annihilation rate is given by

$$\Gamma_A = n_{X,\text{eq}} \langle \sigma_A |v| \rangle \tag{6.17}$$

where σ_A is the annihilation cross section, v is the relative velocity of the annihilating WIMPs, and $\langle \ldots \rangle$ denotes an averaging over a thermal distribution of velocities of each particle at the decoupling (freeze-out) temperature T_{dec}. The Hubble rate is

$$H = \sqrt{\frac{8\pi G_N \rho}{3}} = 1.66\sqrt{g_{*,T}} \frac{T^2}{m_P}. \tag{6.18}$$

The abundance $Y_{X,T_{\text{dec}}}$ of X-particles at freeze-out is, therefore, given by

$$Y_{X,T_{\text{dec}}} \equiv \frac{n_{X,\text{eq},T_{\text{dec}}}}{s_{\text{dec}}} = \frac{H_{\text{dec}}}{s_{\text{dec}} \langle \sigma_A |v| \rangle} \tag{6.19}$$

where

$$s_{dec} = \frac{2\pi^2}{45} g_{*,S,T_{dec}} T_{dec}^3 \qquad (6.20)$$

is the entropy density at freeze-out, with $g_{*,S,T}$ defined in (5.10) As before, this gives the current abundance $Y_{X,0}$. Equating (6.17) and (6.18) then yields

$$\Omega_{X,0} \equiv \frac{\rho_{X,0}}{\rho_c} = Y_{X,0} \frac{s_0}{\rho_c} m_X$$

$$= \frac{m_X}{\langle \sigma_A |v| \rangle} \frac{s_0}{\rho_c} \frac{H_{dec}}{s_{dec}} \qquad (6.21)$$

$$= \frac{m_X}{T_{dec}} \frac{1}{\sqrt{g_{*,T_{dec}}}} \left(\frac{2 \times 10^{-27} \text{cm}^3\text{s}^{-1}}{\langle \sigma_A |v| \rangle} \right). \qquad (6.22)$$

The proportionality of $\Omega_{X,0}$ to the inverse of $\langle \sigma_A |v| \rangle$ means that the relic abundance is reduced as σ_A increases and this might have been anticipated: the more efficiently annihilation proceeds, the fewer relics remain. Taking $g_X = 2$, the freeze-out temperature satisfies

$$\left(\frac{m_X}{T_{dec}} \right)^{-1/2} e^{m_X/T_{dec}} = 0.076 \langle \sigma_A |v| \rangle \frac{m_X m_P}{\sqrt{g_{*,T_{dec}}}} \equiv K \qquad (6.23)$$

which may be solved iteratively

$$\frac{m_X}{T_{dec}} \simeq \ln K + \tfrac{1}{2} \ln \ln K. \qquad (6.24)$$

For a typical value $g_{*,T_{dec}} = 60$ and a typical weak cross section

$$\langle \sigma_A |v| \rangle = c \frac{\alpha_{em}^2}{8\pi m_X^2} = c \left(\frac{100 \text{ GeV}}{m_X} \right)^2 2.5 \times 10^{-27} \text{ cm}^3 \text{ s}^{-1} \qquad (6.25)$$

where c is of order unity, this gives

$$\frac{m_X}{T_{dec}} \simeq 22 + \ln c - \ln \left(\frac{m_X}{100 \text{ GeV}} \right). \qquad (6.26)$$

Thus,

$$\Omega_{X,0} \simeq \frac{6 \times 10^{-27} \text{cm}^3\text{s}^{-1}}{\langle \sigma_A |v| \rangle} \qquad (6.27)$$

and, using (6.22), a typical weak cross section (6.25) gives

$$\Omega_{X,0} = \frac{2.3}{c} \left(\frac{m_X}{100 \text{ GeV}} \right)^2 \qquad (6.28)$$

remarkably close to (6.14) for $m_X \simeq 100 \text{ GeV}$ since $h^2 \simeq \tfrac{1}{2}$.

The foregoing analysis shows that the dark matter might well be WIMPs but there remains the question of what the WIMPs actually are. Since there are no satisfactory candidates within the standard model, we must investigate plausible extensions of it. The most favoured, which has been studied in great detail in recent years, is the MSSM. (See, for example, [11]). We shall say more about the MSSM in section 6.4. First we discuss the possibility that dark matter is made of gravitinos, particles that occur in *any* locally supersymmetric theory.

6.3 The gravitino problem

In a supersymmetric theory, all particles have an associated superpartner, called a 'sparticle', whose spin differs by $\frac{1}{2}$ from that of the original particle. In a (locally supersymmetric) supergravity theory, the sparticle associated with the (spin-2) graviton has spin $\frac{3}{2}$ and is called the 'gravitino'. When supersymmetry is broken, the gravitino acquires a non-zero mass (see chapter 5 of [11], for example)

$$m_{3/2} = e^{G_0/2} m_P \tag{6.29}$$

where G_0 is the expectation value of G in the physical vacuum, with G defined in (2.144), and $m_P = G_N^{-1/2} = 1.22 \times 10^{19}$ GeV is the Planck mass. If gravitinos have survived until the present epoch, then their energy density $\rho_{3/2,0}$ could, in principle, dominate but not exceed [12] the present total energy density ρ_0 of the universe. Thus,

$$\Omega_{3/2} < \Omega_0 \tag{6.30}$$

where Ω_0 is defined in (1.41) and

$$\Omega_{3/2} \equiv \frac{\rho_{3/2,0}}{\rho_c}. \tag{6.31}$$

As for neutrinos, we can use this to bound $m_{3/2}$. Since gravitinos interact only gravitationally, their interaction rate

$$\Gamma_{3/2,\text{int}} \sim G_N^2 T^5 \sim \frac{T^5}{m_P^4} \tag{6.32}$$

just on dimensional grounds. They decouple when

$$\Gamma_{3/2,\text{int}} = H \sim \frac{T^2}{m_P} \tag{6.33}$$

which occurs when $T \sim m_P$ and while they are still relativistic. After decoupling, the number of gravitinos per comoving volume is constant so, as in (5.24),

$$Y_{3/2,\text{dec}} = Y_{3/2,0}. \tag{6.34}$$

Thus, as in (5.33),

$$\Omega_{3/2}H_0^2 = \frac{8\zeta(3)}{3\pi} \frac{g_{3/2\text{eff}}g_{*S,T_0}}{g_{*S,T_{\text{dec}}}} G_N T_0^3 m_{3/2} \tag{6.35}$$

where $g_{3/2,\text{eff}} = \frac{3}{4} \times 4$, since the gravitino has spin-$\frac{3}{2}$. Putting all of this together gives

$$m_{3/2} = \Omega_{3/2}h^2 g_{*S,T_{\text{dec}}}(4.4 \text{ eV}) \lesssim 1 \text{ keV} \tag{6.36}$$

taking $g_{*S,T_{\text{dec}}} \sim 200$ and using the value (1.42) for Ω_0. This is far smaller than the $\lesssim 10$ TeV scale needed to protect the hierarchy and to give TeV-scale masses to the sparticle spectrum, remembering that $m_{3/2}$ controls the size of the supersymmetry-breaking masses for matter.

A more likely scenario is that gravitinos are heavier but decayed before the present epoch. The fastest decay mode of the gravitino is into a standard model particle and its supersymmetric partner, with rate

$$\Gamma_{3/2,\text{max}} \simeq \frac{m_{3/2}^3}{m_P^2}. \tag{6.37}$$

When the temperature $T \sim m_{3/2}$, the expansion rate $H \simeq m_{3/2}^2/m_P$, which is faster by a factor $m_P/m_{3/2}$. This means that the equilibrium condition $\Gamma \geq H$ can only be reached at temperatures far below $m_{3/2}$ and, at these temperatures, collision processes are too weak to produce gravitinos. So the gravitino population can only be reduced by decays. Until they decay, the cosmic energy density is dominated by gravitinos with energy density

$$\rho_{3/2} = \frac{3\zeta(3)}{\pi^2} \frac{g(T)}{g(T_{\text{dec}})} m_{3/2} T^3 \tag{6.38}$$

where $g(T)$ is the effective number of massless degrees of freedom at temperature T. The expansion rate

$$H = \sqrt{\frac{8\pi}{3} \frac{\rho_{3/2}^{1/2}}{m_P}} \tag{6.39}$$

becomes equal to the decay rate $\Gamma_{3/2}$ when $T \simeq T_{3/2}$, where

$$T_{3/2} \simeq \left(\frac{\pi g(T_{\text{dec}})}{8\zeta(3)g(T_{3/2})} \right)^{1/3} \left(\frac{\Gamma_{3/2}^2 m_P^2}{m_{3/2}} \right)^{1/3}. \tag{6.40}$$

When they decay the energy is thermalized and reheats the universe to a temperature

$$T'_{3/2} \simeq \frac{(90\zeta(3))^{1/4}}{\pi} \left(\frac{m_{3/2}T_{3/2}^3}{g(T_{3/2})} \right)^{1/4}. \tag{6.41}$$

Using (6.40), this gives

$$T'_{3/2} \simeq \left(\frac{45g(T_{\text{dec}})}{4\pi^3 g^2(T_{3/2})} \right)^{1/4} \sqrt{\Gamma_{3/2} m_P}. \qquad (6.42)$$

If we use the fastest decay rate (6.37), we find that $T'_{3/2} < T_0 \sim 3$ K if $m_{3/2} \lesssim 10$ MeV. The increase in temperature after reheating

$$\frac{T'_{3/2}}{T_{3/2}} > 45^{1/4} \sqrt{2} \zeta(3)^{1/3} \pi^{-13/12} g(T_{3/2}) g^{-1/12}(T_{\text{dec}}) \left(\frac{m_P}{m_{3/2}} \right)^{1/6} \qquad (6.43)$$

results in a large increase in the entropy density, by a factor $(m_P/m_{3/2})^{1/2}$ since $s \propto T^3$. So if the gravitinos decay after nucleosynthesis, the baryon number to entropy ratio n_B/s during nucleosynthesis must have been much higher than now. This leads to too much helium and not enough deuterium [13]. To avoid these problems, we require that $T'_{3/2} > 0.4$ MeV, so that any previously formed helium nuclei are broken, the neutron-to-proton ratio is restored to its equilibrium value and nucleosynthesis restarts. If we use the fastest decay rate (6.37) in (6.42), this gives

$$m_{3/2} \gtrsim 10 \text{ TeV} \qquad (6.44)$$

which might, just, be consistent with solving the hierarchy problem. Whether or not the maximal decay rate (6.37), associated with the decay to a particle and sparticle, actually occurs depends on the details of the supergravity model used. For example, if the gravitino is the lightest supersymmetric particle (LSP), then if it decays it can only decay to non-supersymmetric particles, with a decay rate less the maximal. This decreases both $T_{3/2}$ and $T'_{3/2}$, thereby exacerbating the problem.

6.4 Minimal supersymmetric standard model (MSSM)

Although it is not required theoretically, it is customary to impose an 'R-parity' invariance on the supersymmetric standard model to ensure the absence of fast proton decay. The quantity R is defined as $R = (-1)^{3(B-L)+2S}$, where B is the baryon number, L is the lepton number and S is the spin. It is assumed to be multiplicatively conserved in all interactions. Then the most general MSSM has a total of 124 independent parameters [14]. These are comprised of three gauge coupling constants $(g_{1,2,3})$, three gaugino masses $(M_{1,2,3})$ and two gaugino phases, four Higgs/Higgsino sector mass parameters $(m_{H_u}^2, m_{H_d}^2, B, \mu)$ and one phase, nine fermion masses, 21 scalar squark and slepton masses, 39 mixing angles and 41 phases and $\bar{\theta}$ (the QCD θ-parameter discussed in section 5.3.1). This model is sometimes referred to as MSSM-124 [15]. In contrast, the standard model has 'only' 19 parameters: the three coupling constants, two Higgs parameters (m_H^2, v), nine fermion masses, three mixing angles and one phase, and $\bar{\theta}$. Since there are two Higgs doublet chiral superfields H_u and H_d in the MSSM, there are two VEVs, v_u and v_d. The combination $v_u^2 + v_d^2$ is fixed by the measured value of m_Z and is, therefore, not a free parameter but the ratio $\tan\beta \equiv v_u/v_d$ is a free parameter. The MSSM has five physical Higgs particles of which two ($H^{\pm}$) are charged, two (h^0, H^0) are neutral scalars and one (A^0) is a neutral pseudoscalar. The masses are all fixed in terms of the (known) gauge coupling strengths and three parameters, two of which ($\mu, \tan\beta$) have already been defined. Without loss of generality, the third may be taken to be m_A. Apart from $\bar{\theta}$ and the CKM mixing angles and phase, the unknown parameters of the MSSM therefore consist of 63 masses and mixing angles and 43 phases. Even if all of the phases are set to zero, it is still not feasible to explore the remaining parameter space ($M_{1,2,3}$, μ, $\tan\beta$, m_A, 21 scalar squark and slepton masses and 36 mixing angles).

The number of parameters is, therefore, drastically reduced by making further assumptions. In the low-energy approach [16], special phenomenolgically viable points in the parameter space are selected. For example, the five scalar (squark and slepton) symmetric 3×3 mass matrices and the three trilinear coupling A-matrices might be assumed to be generation independent or that they are flavour-diagonal in a basis where the quark and lepton mass matrices are diagonal. Neither of these has any strong theoretical motivation. Alternatively, in the high-energy approach that we shall follow, the parameters of the MSSM are treated as running parameters. In other words, the parameters 'run' or evolve with the renormalization scale in a way determined by the renormalization group equations. Then a structure is imposed on the parameters at some high energy scale. This would be the case if there is an underlying GUT symmetry, for example. In such a model, it is assumed that the gauginos all have a common mass $m_{1/2}$ at some (*a priori* unknown) unification scale m_X:

$$M_1(m_X) = M_2(m_X) = M_3(m_X) = m_{1/2}. \tag{6.45}$$

The gaugino masses at the electroweak scale are determined using renormalization group equations and at the electroweak scale their ratios are determined by the gauge coupling strengths:

$$\frac{M_3}{M_2} = \frac{\alpha_3}{\alpha_2} \qquad \frac{M_1}{M_2} = \frac{5\alpha_1}{3\alpha_2} \tag{6.46}$$

where $\alpha_3 \equiv g_3^2/4\pi$ etc. Similarly, it is also assumed that all scalars, except possibly the Higgs soft masses squared $m_{1,2}^2$, have a common squared-mass m_0^2 and that the trilinear cofficients have a common value A, at the unification scale.

$$m_{\tilde{Q}_L}^2(m_X) = m_{\tilde{u}_L^c}^2(m_X) = m_{\tilde{d}_L^c}^2(m_X) = m_0^2 I_3 \tag{6.47}$$

$$m_{\tilde{L}}^2(m_X) = m_{\tilde{e}_L^c}^2(m_X) = m_0^2 I_3 \tag{6.48}$$

$$A_u = A_d = A_e = A I_3. \tag{6.49}$$

This happens (in supergravity models) if the origin of the supersymmetry breaking is a 'hidden' sector which shares only gravitational interactions with the 'observable' sector that we inhabit. (The unification scale, as well as the value of the unified gauge coupling strength, is determined from the measured low-energy values of the coupling constants using the renormalization group equations.) An advantage of this approach is that one of the diagonal Higgs mass-squared parameters is typically driven negative by the renormalization group running, so electroweak symmetry breaking is thereby generated radiatively and the scale at which this happens is intimately connected to the low-energy supersymmetry breaking. The minimal supergravity (mSUGRA) model starts with seven parameters (if we allow for a non-minimal Kähler potential), namely $m_1^2, m_2^2, \mu(m_X), A, B, m_{1/2}, m_0^2$. Using the renormalization group equations, these determine m_Z, $\tan \beta$ (and m_A^2): it is customary to choose $\tan \beta$ as an input parameter and m_Z is, of course, fixed. Then $|\mu|$ and B are outputs and the remaining unknowns are $m_1^2, m_2^2, \epsilon(\mu(m_X)), \tan \beta, m_{1/2}, m_0$. When $m_1^2 = m_2^2 = m_0^2$, the model is called the 'constrained' MSSM or CMSSM.

Since $R = +1$ for all particles in the standard model and $R = -1$ for all of their supersymmetric partners (sparticles), it is easy to see that the lightest supersymmetric particle (LSP) must be stable. To be a WIMP candidate, the LSP must also be a colour-singlet and electrically neutral and there are relatively few sparticles with these properties. One posiibility is a sneutrino $\tilde{\nu}$. However, this possibility has been excluded. An accelerator-based limit from the 'invisible' width of the Z boson requires $m_{\tilde{\nu}} \gtrsim 44.7$ GeV [17] but, in this case, direct relic searches in low-background experiments require $m_{\tilde{\nu}} \gtrsim 20$ TeV [18]. Another possibility that arises in supergravity models is the gravitino, which is essentially undetectable. Also, as we saw in section 6.3, gravitino dark matter might raise theoretical problems that supersymmetry is supposed to have solved. However, in most supergravity models, the gravitino is not the LSP and is unstable. The most popular candidate by far is that the LSP is a *neutralino* [19].

6.5 Neutralino dark matter

There are four neutralinos $\chi_n^0 (n = 1, 2, 3, 4)$ in the MSSM, each of which is a linear combination of the four $R = -1$ Majorana fermions: the Wino $\tilde{W}^3$, the partner of the $SU(2)_L$ gauge boson; the Bino $\tilde{B}$, partner of the $U(1)_Y$ gauge boson; and the two neutral Higgsinos $\tilde{H}_u$ and $\tilde{H}_d$. Thus,

$$\chi_n^0 = N_{1n}\tilde{B} + N_{2n}\tilde{W}^3 + N_{3n}\tilde{H}_u + N_{4n}\tilde{H}_d \qquad (n = 1, 2, 3, 4) \qquad (6.50)$$

and the coefficients $N_{in}(i = 1, 2, 3, 4)$ are the normalized eigenvectors of the neutralino mass matrix

$$M_\chi = \begin{pmatrix} M_1 & 0 & -m_Z c_\beta s_W & m_Z s_\beta s_W \\ 0 & M_2 & m_Z c_\beta c_W & -m_Z s_\beta c_W \\ -m_Z c_\beta s_W & m_Z c_\beta c_W & 0 & -\mu \\ m_Z s_\beta s_W & -m_Z s_\beta c_W & -\mu & 0 \end{pmatrix} \tag{6.51}$$

where $c_\beta \equiv \cos\beta$, $s_\beta \equiv \sin\beta$, $c_W \equiv \cos\theta_W$ and $s_W \equiv \sin\theta_W$. The mass eigenvalues are conventionally labelled in ascending order, so χ_1^0 is the lightest neutralino and χ_4^0 the heaviest. Besides the (known) parameters m_Z and θ_W that appear in the standard model, the neutralino mass matrix involves four of the further 63 parameters that specify the (fairly) general MSSM discussed earlier. These are $M_{1,2}$, the (soft) masses of the $U(1)_Y$ and $SU(2)_L$ gauginos, and the Higgsino mixing parameter μ. In mSUGRA models, (6.46) holds and the neutralino masses and mixing angles are determined by only three parameters.

The question to be addressed then is whether for certain values of these limited parameter sets the MSSM has the neutralino χ_1^0 as the LSP and, if so, whether the predicted relic density is consistent with the observational data (6.14). To answer the latter question, the cross section for neutralino annihilation must be calculated in the MSSM and then used to calculate the relic density which is compared with the observational data on cold dark matter. Subtracting the baryonic contribution Ω_b from the total matter contribution Ω_m in (6.14) gives the 2σ range for the cold dark matter density satisfying

$$0.094 < \Omega_{CDM}h^2 < 0.129. \tag{6.52}$$

Before discussing these calculations, we should note that a precise determination of the relic density requires the solution of the Boltzmann equation governing the evolution of the number density n_χ. The estimate (6.27) is a fairly good estimate when $\sigma_A|v|$ is approximately constant, independent of v. Since the neutralinos are non-relativistic, we may generally expand the annihilation cross section as

$$\sigma_A|v| = a + bv^2 + \cdots \tag{6.53}$$

where the a term receives contributions only from s-wave scattering, the b-term from s- and p-waves, and so on. If $a \gg b$, then $\sigma_A|v|$ is indeed approximately constant. However, as we shall see, this is often a poor approximation because the dominant annihilation channel has the s-wave suppressed because of CP-invariance considerations. Thus, the p-wave is dominant and the estimate (6.27) of the relic abundance is a poor approximation. The true abundance can be computed by a numerical integration of the Boltzmann equation but an improved analytical approximation can also be found by solving in both the early- and late-time limits and then matching the two solutions near freeze-out. The result is [20]

$$\Omega_{\chi,0} = Y_{\chi,0}\frac{s_0}{\rho_c}m_\chi \tag{6.54}$$

with the present neutralino abundance $Y_{\chi,0}$ given by

$$Y_{\chi,0} = \frac{3.785}{\sqrt{g_{*,T_{dec}}} m_P m_\chi} \left(\frac{m_\chi}{T_{dec}}\right) \frac{1}{a + 3(b - a/4)T_{dec}/m_\chi} \tag{6.55}$$

and the freeze-out temperature satisfying

$$\frac{m_\chi}{T_{dec}} = \ln[0.0764 m_P (a + 6b T_{dec}/m_\chi) c (2+c) m_\chi (g_{*,T_{dec}} m_\chi/T_{dec})^{-1/2}] \tag{6.56}$$

which can be solved iteratively; c is a numerical constant of order unity determining when the early- and late-time solutions are matched: $c \simeq \frac{1}{2}$ typically gives a 5–10% precision. In special circumstances this estimate can be wrong by factors of two or more. These are (i) when the annihilation occurs near an s-channel pole; (ii) when the annihilation occurs near a mass threshold; and (iii) when there is 'co-annihilation', i.e. when there is another particle (χ') (e.g. a squark) with a mass that only slightly exceeds m_χ, and the χ can be converted to a χ' via scattering from standard model particles. If the annihilation cross section for the χ's is larger than that of the χs, then the abundance of both is controlled by the annihilation of the heavier and more strongly interacting particle. These special cases are important in practice and allow certain regions of the MSSM parameter space that would otherwise be forbidden.

The coefficients a and b in (6.53) are bounded above by partial-wave unitarity arguments, with the bounds being of order m_χ, essentially on dimensional grounds. This leads to a model-independent lower bound on the relic abundance [21]

$$\Omega_{\chi,0} \gtrsim \left(\frac{m_\chi}{200\text{TeV}}\right)^2. \tag{6.57}$$

Using the upper bound in (5.22) then gives $m_\chi \lesssim 100$ TeV. Of course, in the MSSM models with which we are concerned, the cross sections are proportional to α_{em}^2, so the largest cosmologically acceptable WIMP mass will be reduced by a factor of $\alpha_{em} \sim 10^{-2}$ from this most conservative bound. Thus, in supersymmetric models, we expect $m_\chi \lesssim 1$ TeV to be required by the cosmological constraint.

The calculation of the annihilation cross section in the MSSM is straightforward in principle but quite complicated in practice and we shall only comment on the salient features. The most important channels for neutralino annihilation are those that appear in lowest order (tree-level) perturbation theory, see figure 6.1. These are annihilation into a pair of fermions

$$\chi_1^0 \chi_1^0 \to f\bar{f} \qquad (f = q, l, \nu) \tag{6.58}$$

and into a pair of bosons

$$\chi_1^0 \chi_1^0 \to W^+W^-, \quad Z^0 Z^0, \quad W^\pm H^\mp, \quad Z^0 A^0, \quad Z^0 H^0, \quad Z^0 h^0, \quad H^+ H^- \tag{6.59}$$

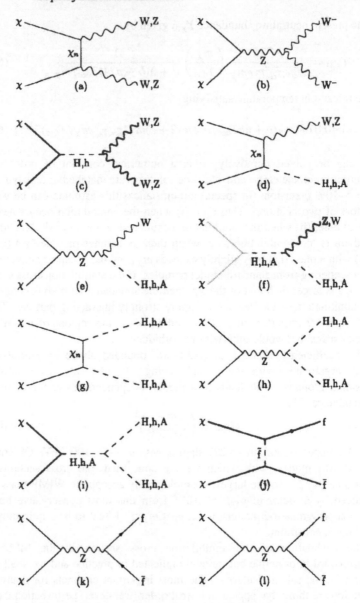

Figure 6.1. Feyman diagrams for neutralino annihilation processes: χ_n is any of the neutralino or (in the t-channel) chargino states; h is the lightest (neutral) Higgs scalar, H is the other neutral Higgs or charged Higgs; A is the pseudoscalar; f is a fermion and $\bar{f}$ is the corresponding sfermion. (Not all processes are allowed.)

and all three pairs from A^0, H^0, h^0.

Analytic expressions for the a terms for all of these processes are compact but the b terms are very involved [22]. Neutralinos are Majorana fermions and are, therefore, their own antiparticles. This means that, in an s-wave state, the two neutralinos must have their spins oppositely directed because of Fermi statistics. Therefore, if the neutralinos annihilate to a fermion (f)–antifermion pair ($\bar{f}$), then f and $\bar{f}$ must also have their spins antiparallel, and this implies that the amplitude acquires a factor m_f to account for the helicity flip. Another way to see this is to note that the initial s-wave $\chi_1^0 \chi_1^0$ state has $CP = -1$, so CP-invariance requires that the final state also has $CP = -1$. Annihilation into light fermion pairs will always be kinematically allowed but the previous argument shows that the s-wave contribution to, and therefore the a term in, the annihilation cross section is proportional to m_f^2/m_χ^2. Thus, annihilation into light quarks and leptons is negligible compared with annihilation into c, b and t-quark pairs: the latter occurs only when $m_\chi > m_t$ and dominates all other channels when it is open. CP-invariance also affects other amplitudes. For example, annihilation into Higgs bosons can be important when such channels are open. However, the s-wave amplitudes for the final states $h^0 h^0$, $H^0 H^0$, $H^0 h^0$, $A^0 A^0$, $H^+ H^-$ are identically zero because of CP-invariance: the same is true of the $Z^0 A^0$ final state.

The allowed parameter space is restricted by other data [23] besides the cosmological bounds (5.22). Specifically, the LEP bound on m_h, and $b \to s\gamma$ data, both force the parameter $m_{1/2}$ to larger values, while the BNL E821 measurement of the $g - 2$ factor of the muon [24] favours relatively low values of m_0 and $m_{1/2}$, at least for $\mu > 0$—the actual bounds are dependent on $\tan \beta$. There is also the requirement that the neutralino *is* the LSP and that the parameters allow radiatively driven electroweak symmetry breaking. The current position seems to be [25] that the MSSM *can* simultaneously satisfy all of these constraints. In the most constrained model, the CMSSM, there is a 'bulk' region in the $(m_{1/2}, m_0)$ plane with relatively low values of m_0 and $m_{1/2}$ in which both the cosmological and the non-cosmological constraints are satisfied, see figure 6.2 taken from [25]. In this region, supersymmetry is relatively easy to detect at colliders. The constraints deriving from the precision WMAP data have substantially reduced the size of this region. The bulk region is essentially defined by using the expression (6.54) and, in this region, the neutralino is essentially the Bino ($\tilde{B}$), i.e. $N_{10} \gg N_{20}, N_{30}, N_{40}$ in (6.50). In this case the annihilation proceeds mainly via t-channel sfermion exchange.

Extending from the bulk region to larger values of $m_{1/2}$ is a co-annihilation 'tail', where the neutralino LSP is almost degenerate with the next-to-lightest sparticle, usually the stau $\tilde{\tau}$. At larger values of m_0, close to the region where radiative electroweak symmetry breaking is no longer possible, there is a 'focus-point' region in which the neutralino has a larger Higgsino component, i.e. N_{30} or N_{40} in (6.50) are non-negligible. Lastly, when both m_0 and $m_{1/2}$ are large there may be a 'funnel' where rapid direct-channel annihilations via the A and H Higgs

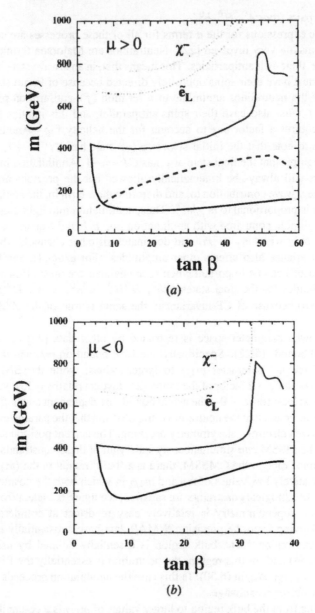

Figure 6.2. The ranges of m_χ allowed by cosmology and other constraints, for (a) $\mu > 0$ and (b) $\mu < 0$. Upper limits without (upper solid line) and with (dashed line) the $g_\mu - 2$ constraint are shown for $\mu > 0$: the lower limits are shown as lower solid lines. Note the sharp increases in the upper limits for $\tan \beta \gtrsim 50$, $\mu > 0$ and $\tan \beta \gtrsim 35$, $\mu < 0$ due to the rapid-annihilation funnels. Also shown as dotted lines are the $\tilde{e}_L$ and $\chi^\pm$ masses at the tips of the co-annihilation tails.

boson poles occurs. For $\mu > 0$, the neutralino mass is constrained to satisfy

$$108 \text{ GeV} \lesssim m_\chi \lesssim 370 \text{ GeV} \qquad (6.60)$$

with the minimum value occurring at $\tan \beta = 23$. For $\mu < 0$, there is no compatibility with the $g - 2$ data when the LEP data are used Thus, $\mu > 0$ is clearly favoured but if the $g - 2$ data are excluded the corresponding bounds are

$$160 \text{ GeV} \lesssim m_\chi \lesssim 430 \text{ GeV}. \qquad (6.61)$$

It is beyond our scope to discuss the status of the CMSSM in any further detail and the interested reader is referred to [25, 26]. The main point is that the MSSM and even the CMSSM are consistent with all current data and have a neutralino with mass in the range (6.60). In the absence of new theoretical motivation for particular values of the parameters, the most urgent need is for more experimental data. One way to obtain this is to detect neutralino dark matter and to ascertain its properties.

6.6 Detection of dark matter

The most direct signal for neutralino dark matter would be to observe its scattering from nuclei in a detector. By fitting both the luminous and dark matter to the measured rotation curve in our galaxy, the dark matter density at the position of the solar system is found to be of order 0.3–0.7 GeV cm^{-3}. If the halo of the Milky Way consists of WIMPs, then this means that hundreds to thousands of them pass through every square centimetre each second. For a typical neutralino with $m_\chi \sim 100$ GeV scattering from a xenon nucleus with $m_{Xe} \sim 130$ GeV, with a typical WIMP speed $\bar{v} \sim 270$ km s^{-1}, the nuclear recoil energy is below 100 keV (exercise 3). This energy is transferred to atomic electrons and produces detectable ionization. With a typical MSSM cross section, assuming coherent interaction with the xenon nucleus, this gives an event rate of less than 1 kg^{-1} day^{-1}. This is about 10^6 times lower than the ambient rate from background recoils due to gammas from the surrounding natural radioactivity. Nevertheless, it is feasible to distinguish between the two because the rate of energy loss with distance ($\mathrm{d}E/\mathrm{d}x$) is a factor of 10 lower for nuclear recoils. However, any background neutrinos, produced by cosmic-ray muons for example, produce nuclear recoils that are indistinguishable from WIMP recoils. Thus, the detector must be shielded from the muons by placing it deep underground. The velocity of the earth through the galactic halo varies during the year as the earth orbits the sun. This leads to an annual modulation of the dark matter event rate, with a maximum each year on 2nd June ± 1.3 days when the earth's motion is aligned with the sun's motion around the galactic centre and a minimum six months later. Due to the high inclination of the earth's orbital plane, this only amounts to a 5–7% change in the mean recoil rate. This annual modulation is the signature sought by all of the current detectors (DAMA, ZEPLIN-I, EDELWEISS and CDMS).

However, it is possible that the much larger low-energy background might be subject to other modulating effects. The direction of the WIMP 'wind' felt by the detectors is strongly peaked in the direction opposite to the solar motion, so the recoil directions will be strongly peaked in the same direction. If observed, this feature, combined with the annual modulation, would be the most convincing demonstration of the existence of WIMPs.

6.6.1   Neutralino–nucleon elastic scattering

If WIMPs solve the dark matter problem, they must have some small but finite coupling to ordinary matter—otherwise, they would not have annihilated in the early universe and would be overabundant today. By crossing symmetry, the amplitude for WIMP annihilation into a quark–antiquark pair is related to the elastic scattering of the WIMPs by quarks. Thus, we expect that WIMPs will have a small coupling to nuclei and may, therefore, be detectable by nuclear scattering. The calculations are considerably simplified because the WIMP velocity ($v_\chi/c \sim 10^{-3}$) is extremely non-relativistic. This feature also simplifies the conversion of the scattering cross sections from quarks into the scattering cross sections from the nuclei making the detector. In the non-relativistic limit, the axial vector current $\bar{\chi}_1^0 \gamma_\mu \gamma_5 \chi_1^0$ is just the WIMP spin and it is coupled to the nucleon spin. Since the neutralino is a Majorana fermion, it has no vector current $\bar{\chi}_1^0 \gamma_\mu \chi_1^0 = 0$. So the only other possible term in the effective interaction is the scalar $\bar{\chi}_1^0 \chi_1^0$ which couples to the mass of the nucleus—in the non-relativistic limit the 'tensor' current reduces to the scalar.

At tree level, the axial vector (spin) interaction receives contributions from t-channel Z boson and s-channel squark exchange, while the scalar interaction gets contributions from t-channel Higgses H, h as well as s-channel quark exchange. See figure 6.3. Since the lightest Higgs might be considerably lighter than the lightest squarks, its contribution to the scalar interaction could be significant if the neutralino state (6.50) has a substantial Higgsino component. Because it is proportional to the mass number A of the nucleus, the scalar amplitude will dominate for heavy nuclei. For a neutralino that is a pure Bino $\tilde{B}$, this occurs for $A \gtrsim 20$ in the large squark mass limit and this is confirmed by numerical surveys of the supersymmetric parameter space where scalar dominance for $A \gtrsim 30$ is almost always found [27].

It is of interest to examine the theoretical implications for the direct detection experiments of restricting the parameters of the CMSSM to the region allowed by the cosmological and other constraints. This programme has so far been undertaken [28] only in the case that $A = 0$. The LEP lower limit on m_h and the $b \to s\gamma$ data provide upper limits on the cross sections, while the $g - 2$ data provide lower limits, at least if $\mu > 0$. In that case, the overall conclusion is that the spin-independent cross section σ_{SI} satisfies

$$2 \times 10^{-10} \text{ pb} \lesssim \sigma_{SI} \lesssim 6 \times 10^{-8} \text{ pb} \tag{6.62}$$

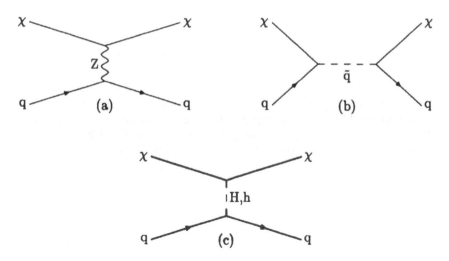

Figure 6.3. Lowest-order Feynman diagrams for neutralino–quark elastic scattering: (a) and (b) contribute to the spin-dependent amplitude; (b) and (c) contribute to the scalar amplitude.

and for the spin-dependent cross section, σ_{SD},

$$2 \times 10^{-7} \text{ pb} \lesssim \sigma_{SD} \lesssim 10^{-5} \text{ pb} \qquad (6.63)$$

If the $g - 2$ constraint is dropped and $\mu < 0$ is tolerated, then there is no lower limit on σ_{SI}. In general, it is found that σ_{SI} is relatively large in the bulk region but falls off in the co-annihilation tail. There is strong cancellation in σ_{SI} when $\mu < 0$.

6.6.2   WIMP annihilation in the sun or earth

If the galactic halo is composed of neutralino WIMPs, the WIMPs have a small probability of elastic scattering with the sun and/or the earth. The WIMPs that scatter to a velocity smaller than the escape velocity become gravitationally bound to that body[1]. Once captured, the WIMPs undergo further scattering from the elements of that body and settle to the core in a relatively short time period. Thus, the sun and earth are like (inefficient) cosmological vacuum cleaners, constantly sucking in WIMPs that are then stored in their cores. WIMPs that have accumulated in this way can annihilate, essentially at rest, to produce standard model particles most of whose decay products are absorbed without any observable consequences. However, some of the decay products include energetic muonic neutrinos (ν_μ, $\bar{\nu}_\mu$) that can pass through the sun and earth and

---

[1] For the sun, the escape velocity at the centre is $v_c = 1354$ km s^{-1} and at the surface $v_s = 795$ km s^{-1}; for the earth $v_c = 14.8$ km s^{-1}, $v_s = 11.2$ km s^{-1}.

be detected in astrophysical neutrino detectors on earth. The most promising technique for detection is via the observation of upward-moving muons produced by charged-current interactions of the muonic neutrinos with the rock below the detector. Neutralinos annihilate almost always to two-body final states, each carrying energy m_χ, and the decays of these can produce muonic neutrinos with energies between $m_\chi/3$ and $m_\chi/2$. The lower bound on the neutralino mass is conservatively 40 GeV, so the energy of the muonic neutrinos is several GeV. The only background is from atmospheric neutrinos produced by cosmic-ray spallation. This is a well-modelled and easily subtracted background.

The first step in calculating the WIMP-induced neutrino rate is to determine the annihilation rate Γ_A in the sun (or earth). This is given by

$$\Gamma_A = \tfrac{1}{2}C_A N^2 \tag{6.64}$$

where N is the number of WIMPs in the sun (or earth) and

$$C_A = \langle \sigma_A |v| \rangle \frac{V_2}{V_1^2}. \tag{6.65}$$

$\langle \sigma_A |v| \rangle$ is the annihilation cross section multiplied by the relative velocity in the limit of zero relative velocity, i.e. the a term of (6.53), and $V_{1,2}$ are effective volumes

$$V_j \equiv \left(\frac{3 m_p^2 T}{2 j m_\chi \rho} \right)^{3/2} \tag{6.66}$$

where T and ρ are the core temperature and density respectively. The time evolution of N is given by

$$\dot{N} = C - C_A N^2 \tag{6.67}$$

where C is the WIMP accretion rate and the second term arises because of the depletion caused by annihilations. The two processes equilibriate when $\dot{N} = 0$ and then

$$\Gamma_A = \tfrac{1}{2}C. \tag{6.68}$$

In other words, the annihilation rate is entirely determined by the accretion rate. One might wonder whether enough time has elapsed for equilibrium to become established but, in all cases of interest, it turns out that there has been [29].

Thus the next step is to calculate the capture rate C, which is, in turn, determined by the elastic scattering of the WIMPs with the nuclei of the sun (or earth). We have already noted that there are just two channels for elastic scattering: axial (spin-dependent) and scalar (spin-independent). The former contributes only to WIMP capture by the sun, since a negligible fraction of the earth's mass is in nuclei with spin. The latter contributes to capture by both the sun and the earth.

It is beyond our scope to give any details of these calculations. We merely note that it is relatively straightforward to estimate the axial contribution in terms of the elastic (axial) scattering cross section σ_0 for WIMPs on protons. The result is

$$C_{\text{ax}}^{\text{sun}} = 1.3 \times 10^{25} \text{ s}^{-1} \left(\frac{1 \text{ GeV}}{m_\chi}\right) \left(\frac{\sigma_0}{10^{-40} \text{ cm}^2}\right) S\left(\frac{m_\chi}{m_N}\right) \tag{6.69}$$

assuming a halo density of 0.3 GeV cm^{-3} and a dark matter velocity dispersion of $\bar{v} \sim 270$ km s^{-1}. S is a suppression factor with the properties

$$S(1) = 1 \qquad S(x) \sim \frac{3\langle v_{\text{esc}}^2 \rangle}{2\bar{v}^2 x} \qquad \text{as } x \to \infty. \tag{6.70}$$

The capture rate by the scalar interaction is much more complicated and requires the scattering cross section from several nuclei. We refer the interested reader to [30].

Once the capture rate and, hence the annihilation rate is known, the calculation of the flux of high-energy neutrinos is straightforward. It is given by

$$\left(\frac{d\phi}{dE}\right)_i = \frac{\Gamma_A}{4\pi R^2} \sum_F B_F \left(\frac{dN}{dE}\right)_{F,i} \qquad (i = \nu_\mu, \bar{\nu}_\mu) \tag{6.71}$$

where R is the sun–earth distance, or the radius of the earth, for neutrinos from the sun or earth respectively, B_F is the branching ratio for annihilation into channel F and $(dN/dE)_{F,i}$ is the differential energy spectrum for neutrinos of type i at the surface of the sun (or earth) expected in channel F at the core of the sun (or earth). The cross section for the production of a muon via a charged-current interaction is proportional to the neutrino energy, and the range of the muon in the rock is roughly proportional to the muon energy. Thus, the rate for the observation of neutrino-induced through-going muons is proportional to the second moment of the neutrino energy spectrum:

$$\int \left(\frac{dN}{dE}\right)_{F,i} E^2 \, dE. \tag{6.72}$$

For neutralinos giving a relic density in the range (5.22), this gives a rate for upward muons of

$$\Gamma_{\text{detector}}^{\text{ax}} = 1.65 \times 10^{-4} \text{ m}^{-2} \text{ yr}^{-1} \left(\frac{m_\chi}{1 \text{ GeV}}\right) S\left(\frac{m_\chi}{m_N}\right) \tag{6.73}$$

for WIMPs with only an axial coupling. S is the suppression factor occurring in (6.69). As already noted, this is relevant only for neutrinos from the sun. The results of the analogous calculation for WIMPs with only a scalar coupling cannot easily be summarized. Suffice it to say that fluxes as high as 10^{-2} m^{-2} yr^{-1}, the current experimental upper bound on the rate, and of at least 10^{-4} m^{-2} yr^{-1}, the expected sensitivity of the next generation of km^2 detectors, can be obtained for parameters giving 10 GeV $\lesssim m_\chi \lesssim$ 1 TeV. If $m_\chi \gtrsim 80$ GeV, the signals from the sun and earth are of comparable strength and the earth's signal is greater when $m_\chi \lesssim 80$ GeV, see figure 34 in [31].

6.6.3   WIMP annihilation in the halo

The foregoing proposals for observing WIMPs are the most promising techniques currently available. However, WIMPs have other potentially observable effects. In particular, their annihilation in the galactic halo can produce anomalous cosmic rays [32] which may be distinguishable from the familiar background cosmic rays.

These background cosmic rays occasionally include antiprotons produced by spallation of primary cosmic rays on interstellar hydrogen atoms. The flux of such antiprotons cuts off at energies below about 1 GeV, essentially for kinematic reasons because the primary cosmic-ray spectrum falls rapidly as the energy increases. WIMP annihilation, in contrast, can easily produce low-energy antiprotons as a result of hadronization of the decay products. Since the background production of antiprotons with energies in the range 100–1000 MeV is well understood, it is possible, in principle, to observe the anomalous antiprotons, provided that the WIMP mass is not too large.

Another signal could be the observation of 'line' source positrons arising from the direct annihilation of WIMPs into an electron–positron pair. Of course, there are other sources of positrons arising from the showering of other annihilation products but these will have a broad energy spectrum that is indistinguishable from the background. Although propagation through the galaxy would broaden the line, there are no other sources of such a peak in the energy range 10–1000 GeV. Observation of such a peak would give a direct measurement of the WIMP mass. Unfortunately (Majorana) neutralino annihilation into an e^+e^- pair is helicity suppressed, as previously noted. However, if the neutralino state (6.50) contains a significant Higgsino component, the annihilation process $\chi_1^0 \chi_1^0 \to W^+ W^-$ followed by $W^+ \to e^+ \nu_e$ will produce a positron with energy peaked around $m_\chi / 2$.

Similarly, WIMP annihilation in the halo into two photons would produce a monochromatic line at an energy of the WIMP mass. Of course, since they are electrically neutral, there is no direct coupling to photons but equally their (weak) interaction with other matter generates a small but non-zero cross section for annihilation into two photons via a loop diagram. Estimates of the cross section suggest that the signal would be barely observable with current detectors. However, cold dark matter predicts cusps in the density in the cores of galaxies. It is doubtful whether such cusps are compatible with observations but a (residual) peak in the density would assist the generation of a visible signal.

6.7   Exercises

1. Verify that the abundance $\Omega_{X,0}$ of cold dark matter X is given by (6.22) and, hence, check the estimate (6.28).
2. Show that the increase in temperature following reheating after the gravitinos decay is given by (6.43) and, hence, derive the bound (6.44) on $m_{3/2}$.

3. Show that a neutralino with mass $m_\chi \sim 100$ GeV scattering from a xenon nucleus with $m_{Xe} \sim 130$ GeV, with a typical WIMP speed $\bar{v} \sim 270$ km s^{-1}, produces a nuclear recoil energy which is below 100 keV.

6.8 General references

We have found the following article particularly useful in preparing this chapter.

- Jungman G, Kamionkowski M and Griest K 1996 *Phys. Rep.* **267** 195

Bibliography

[1] Weinheimer C 2002 *Proc of CLII, Course of Int. School of Physics 'Enrico Fermi'* Varenna, Italy 2002, arXiv:hep-ex/0210050
[2] Eguchi K *et al* 2003 *Phys. Rev. Lett.* **90** 021802, arXiv:hep-ex/0212021
[3] Ahmed Q R *et al* 2002 *Phys. Rev. Lett.* **89** 011301 arXiv:hep-ex/0204008
[4] Spergel D N *et al* 2003 *Astrophys. J. Suppl.* **148** 175, arXiv:astro-ph/0302209
[5] Trimble V 1987 *Ann. Rev. Astron. Astrophys.* **25** 425
    Sadoulet B 1999 *Rev. Mod. Phys.* **71** S197
    Turner M S and Tyson J a 1999 *Rev. Mod. Phys.* **71** S145
[6] Kapteyn J C 1922 *Astrophys. J.* **55** 302
[7] Bahcall J N *et al* 2999 *Science* **284** 1481
[8] Zwicky F 1933 *Helv. Phys. Acta* **6** 110
[9] Silk J 2003 Talk given at PASCOS'03, Mumbai, India
[10] Smith P F and Bennett J R J 1979 *Nucl. Phys.* B **179** 525
    Smith P F *et al* 1982 *Nucl. Phys.* B **206** 33
    Verkerk P *et al* 1992 *Phys. Rev. Lett.* **68** 1116
[11] Bailin D and Love A 1994 *Supersymmetric Field Theory and String Theory* (Bristol: IOP)
[12] Pagels H and Primack J 1982 *Phys. Rev. Lett.* **48** 223
[13] Weinberg S 1982 *Phys. Rev. Lett.* **48** 1303
[14] Dimopoulos S and Sutter D 1995 *Nucl. Phys.* B **452** 496, arXiv:hep-ph/9504415
    Sutter D 1997 *Stanford PhD Thesis* arXiv:hep-ph/9704390
[15] Haber H E 1998 *Nucl. Phys. B Proc. Suppl.* A–C **62** 469
[16] Particle Data Group 2002 *Phys. Rev.* D **66** 010001
[17] Ellis J R, Falk T, Olive K A and Schmitt M 1996 *Phys. Lett.* B **388** 97, arXiv:hep-ph/9607292
[18] Ahlen S *et al* 1987 *Phys. Lett.* B **195** 603
    Calwell D D *et al* 1988 *Phys. Rev. Lett.* B **61** 510
    Beck M *et al* 1994 *Phys. Lett.* B **336** 141
[19] Ellis J *et al* 1984 *Nucl. Phys.* B **238** 453
[20] Bernstein J, Brown L and Feinberg G 1985 *Phys. Rev.* D **32** 3261
    Scherrer R and Turner M S 1986 *Phys. Rev.* D **33** 1585
    Scherrer R and Turner M S 1986 *Phys. Rev.* D **34** 3263(E)
[21] Griest K and Kamionkowski M 1990 *Phys. Rev. Lett.* **64** 615
[22] See Jungman *et al op. cit.* section 6

[23] Lahanas A B, Nanopoulos D V and Spanos V C 2001 *Phys. Lett.* B **518** 94, arXiv:hep-th/0107151, hep-ph/0112134
      Ellis J R, Olive K A and Santoso Y 2002 *New J. Phys.* **4** 32, arXiv:hep-ph/0202110
      Chattopadhyay U, Corsetti A and Nath P 2002 *Phys. Rev.* D **66** 035003, arXiv:hep-ph/0201001
      Arnowitt R *et al* 2002 *Phys. Lett.* B **538** 121, arXiv:hep-ph/0203069
      Roszkowski L, Ruiz de Austri R and Nihei T 2001 *JHEP* **0108** 024, arXiv:hep-ph/0106334
[24] Bennett G W *et al* 2002 *Phys. Rev. Lett.* **89** 101804
[25] Ellis J *et al* 2003 *Phys. Lett.* B **565** 176, arXiv:hep-ph/0303043
[26] Olive K 2002 Invited talk at *10th Int. Conf. on Supersymmetry and Unification of Fundamental Interactions (SUSY02)* held at DESY, Hamburg, Germany, June 17-22, 2002, arXiv:hep-ph/0211064
[27] Drees M and Nojiri M 1993 *Phys. Rev.* D **48** 3483
[28] Ellis J R, Ferstl A and Olive K A 2002 *Phys. Lett.* B **532** 318, arXiv:hep-ph/0111064
[29] Griest K and Seckel D 1987 *Nucl. Phys.* B **1283** 681
[30] Kamionkowski M *et al* 1995 *Phys. Rev. Lett.* **74** 5174
[31] Jungman G, Kamionkowski M and Griest K 1996 *Phys. Rep.* **267** 195
[32] Kamionkowski M 1994 Particle astrophysics, atomic physics, and gravitation *Proc. Moriond Workshop, Villars sur Ollon, Switzerland, 22–29 January* ed J Van Tran Thanh, G Fontaine and E Hinds (Gif-sur-Yvette: Editions Frontières)

Chapter 7

Inflationary cosmology

7.1 Introduction

The inflationary universe scenario was devised by Guth [1] to provide a resolution to two major puzzles in the standard model of the universe, namely the horizon and flatness problems. As a by-product, inflation also solves problems associated with excessive abundances of particle relics. There are two basic versions of cosmological inflation which we shall refer to as 'old' and 'new' (or 'slow-roll') inflation. In both versions, the universe undergoes a period of very rapid expansion driven by a large cosmological constant in the false vacuum. This period of inflation ends when the universe has evolved to the true vacuum with zero cosmological constant. In the case of old inflation, the universe supercools in some high-temperature phase before undergoing a first-order phase transition to some low-temperature phase. In the case of new inflation, some scalar field rolls in a very flat region of a potential (where the vacuum energy is large and positive) and eventually rolls to a minimum of the potential with zero vacuum energy (cosmological constant).

In the first part of this chapter, we shall discuss old inflation and its successes and shortcomings. The second part of the chapter contains an exposition of new inflation (slow roll inflation.) We shall see that, as well as providing a solution to these cosmological problems, slow-roll inflation is capable of accounting for the size of the density perturbations in the cosmic microwave background radiation.

7.2 Horizon, flatness and unwanted relics problems

We discuss these three puzzles in the standard model of cosmology in turn.

7.2.1 The horizon problem

The cosmic microwave background radiation (CMBR) is very homogeneous. However, in the standard model, the present universe consists of many regions which were causally disconnected up to the time of recombination of electrons and photons after which the photons we now observe as the cosmic microwave background underwent no further scattering. The puzzle is how these causally disconnected regions could have ended up with the same microwave background temperature.

We can estimate how many causally disconnected regions there were at the time of recombination. From (3.117), the proper distance at time t from any point to the particle horizon is

$$d_H(t) = 2t \tag{7.1}$$

where we have taken $n = \frac{1}{2}$ for a radiation-dominated universe. In particular, at the recombination time t_r,

$$d_H(t_r) = 2t_r. \tag{7.2}$$

(We are assuming that for most of the time from $t = 0$ up until t_r the universe was radiation dominated. This is a reasonable approximation because the temperature at which the transition from radiation dominance to matter dominance occurs and the temperature at which recombination of electrons and protons occurs are well within an order of magnitude of each other, respectively 0.37 eV and 0.26 eV.) We need to know how many horizon volumes at time t_r have expanded to fill the presently observable region of the universe (the present horizon volume). Thus, we need to know the radius of the region at time t_r that has expanded to the radius of the presently observable universe. Let the volume of the observable universe at the present time t_0 be $V_0(t_0)$ and let the horizon volume at the recombination time be $V_r(t_r)$. Since RT is constant, because of conservation of entropy,

$$\frac{V_0(t_r)}{V_r(t_r)} = \frac{V_0(t_0)}{V_r(t_r)} \frac{R^3(t_r)}{R^3(t_0)} = \frac{V_0(t_0)}{V_r(t_r)} \left(\frac{T_0}{T_r}\right)^3 \tag{7.3}$$

In view of (3.117),

$$\frac{V_0(t_r)}{V_r(t_r)} = \left(\frac{t_0}{t_r}\right)^3 \left(\frac{T_0}{T_r}\right)^3 \tag{7.4}$$

With the universe matter dominated from approximately the time of recombination to the present time, so that

$$R(t) \propto t^{2/3} \tag{7.5}$$

we have

$$t \propto T^{-3/2}. \tag{7.6}$$

Thus,

$$\frac{V_0(t_r)}{V_r(t_r)} \simeq \left(\frac{T_r}{T_0}\right)^{3/2} \simeq 3.6 \times 10^4 \tag{7.7}$$

for $T_r \simeq 3.0 \times 10^3$ K and $T_0 \simeq 2.73$ K. This is the number of horizon volumes at the recombination time that expanded to fill the presently observable universe (the present horizon volume.) As advertised, this is a large number.

7.2.2 The flatness problem

The present value of Ω, the ratio of the density of the universe to the critical density, is not more than one order of magnitude different from 1. Since, as in section 1.3,

$$\Omega - 1 = \frac{k}{H^2 R^2} \qquad (7.8)$$

the departure of Ω from 1 is a measure of the extent to which the universe is curved (when $k \neq 0$). The value of Ω varies with time and we can estimate how close to 1 it would have had to have been at earlier times to be as close to 1 as it is today. The conclusion we shall come to shortly is that Ω would have been extraordinarily close to 1 in the early universe to be consistent with the present value of Ω.

First, let us study the way in which $\Omega - 1$ varies as the scale factor of the uuniverse $R(t)$ changes. Recalling that

$$\Omega = \frac{\rho}{\rho_c} = \frac{8\pi G_N}{3H^2} \rho \qquad (7.9)$$

and combining with (7.8), we recover the Friedmann equation

$$H^2 R^2 = \frac{8\pi}{3} G_N \rho R^2 - k. \qquad (7.10)$$

Assuming a radiation-dominated universe, ρ is proportional to T^4 and, for entropy conservation, RT is constant. Thus, we can write

$$\rho = a R^{-4}. \qquad (7.11)$$

Then, using (7.10),

$$\Omega - 1 = \frac{k}{\frac{8}{3}\pi G_N a R^{-2} - k}. \qquad (7.12)$$

It is clear, therefore, that $\Omega \to 1$ as $R \to 0$. However, as the universe expands, $\Omega \to 0$ as $R \to \infty$ if $k = -1$, and $\Omega \to \infty$ as $R \to R_{max} = (\frac{8}{3}\pi G_N a)^{1/2}$ if $k = 1$.

Next, let us estimate the value of $\Omega - 1$ in the early universe. Write

$$\rho = \xi T^4 \qquad (7.13)$$

where, from (2.22),

$$\xi = \frac{\pi^2}{30} \left(N_B + \frac{7}{8} N_F \right). \qquad (7.14)$$

Then, from (7.12), (7.11) and (7.13),

$$\Omega - 1 = \frac{k}{\frac{8}{3}\pi G_N \xi T^4 R^2 - k} \simeq \frac{\hat{k}}{\frac{8}{3}\pi G_N \xi T^2} \qquad (7.15)$$

for large values of T, where

$$\hat{k} = \frac{k}{R^2 T^2}. \tag{7.16}$$

In an adiabatically expanding universe, the constancy of RT implies that $\hat{k}$ is a (dimensionless) constant. We can estimate $\hat{k}$ from (7.15), (7.13) and (7.10) as

$$\hat{k} = \frac{8\pi G_N \rho}{3T^2} \simeq \frac{(\Omega_0 - 1)H_0^2}{T_0^2} \tag{7.17}$$

where H_0 and T_0 are the present values: $H_0 \simeq 1.54 \times 10^{-42}$ GeV and $T_0 \simeq 2.73$ K $\simeq 2.35 \times 10^{-13}$ GeV. For Ω_0 differing from 1 by no more than an order of magnitude,

$$|\hat{k}| \lesssim 2 \times 10^{-58}. \tag{7.18}$$

We are left with an unnaturally small number for $|\hat{k}|$ (unless k is strictly zero).

This bound on $|\hat{k}|$ can now be translated into a bound on $|\Omega - 1|$ at early times. By way of illustration, we take the value of ξ obtained in an $SU(5)$ supersymmetric GUT and estimate the value of $|\Omega - 1|$ at the grand unification scale and at the Planck scale. In this case, as in section 2.7, $N_B + \frac{7}{8}N_F = \frac{675}{4}$ and so

$$\xi = 55.5. \tag{7.19}$$

Recalling that $G_N = m_P^{-2}$, where the Planck mass $m_P = 1.22 \times 10^{19}$ GeV, we get the bound at the grand unification scale, $T_c = 2 \times 10^{16}$ GeV,

$$|\Omega - 1| \lesssim 1.66 \times 10^{-55} \tag{7.20}$$

and at the Planck scale

$$|\Omega - 1| \lesssim 1.66 \times 10^{-61} \tag{7.21}$$

Again, these are unnaturally small numbers (unless k is strictly zero). The problem is to find a way that conditions in the early universe could have produced such small numbers.

7.2.3 The unwanted relics problem

It is not infrequently the case that particles produced in the early universe are calculated to have unacceptably large relic densities in the present universe, either because they provide too large a contribution to the mass of the universe or for other reasons. For example, as discussed in section 3.10, unacceptably large monopole densities are produced in some GUTs. A mechanism is needed to dilute these densities to acceptable values.

7.3   Old inflation

A solution to all three problems discussed in the last section is for there to have been a period of very rapid expansion of the universe (cosmological inflation) during which the scale factor of the universe grew by a large amount. We shall discuss shortly how much expansion is sufficient to solve these problems. A simple mechanism to produce the required expansion is for the universe to have supercooled in a false vacuum prior to undergoing a first-order phase transition to the true vacuum in the way described in section 2.9. In that case, a large positive vacuum energy, constant until the phase transition is completed, can drive a period of expansion in a de Sitter universe. After some supercooling has occurred, the vacuum energy density in the Friedmann equation will dominate the radiation energy density and the curvature term, and the Friedmann equation simplifies to

$$H^2 = \frac{\dot{R}^2}{R^2} = \frac{8\pi G_N}{3} V \tag{7.22}$$

where V is the vacuum energy density in the false vacuum. This is equivalent to the de Sitter equation with cosmological constant

$$\Lambda = 8\pi G_N V = 8\pi m_p^{-2} V. \tag{7.23}$$

Moreover, the Hubble constant during the inflationary era has a constant value given by

$$H^2 = \tfrac{1}{3}\Lambda. \tag{7.24}$$

During cosmological inflation, the scale factor of the universe grows exponentially

$$R(t) \propto \exp\left(\sqrt{\frac{\Lambda}{3}}t\right) = e^{Ht}. \tag{7.25}$$

The exponential expansion means that by the time the transition to the true vacuum occurs, the scale factor of the universe may have increased by many orders of magnitude. The phase transition will be completed by the formation of bubbles of the true vacuum, as discussed in section 2.9. Once formed, the bubbles will tend to coalesce and the energy stored in the walls of the bubbles will be released resulting in the universe reheating. Thereafter, the universe will evolve as a (in the first instance) radiation-dominated Friedmann–Robertson–Walker (FRW) universe. However, the initial conditions for the evolution of the FRW universe will have been drastically modified by the period of inflation. If sufficient inflation has occurred, the various problems discussed in the previous section will be solved. We now estimate how much inflation is required for this purpose.

Consider first the horizon problem. This problem will be resolved if the presently observable universe lies in a *single* region which was causally connected at the time of decoupling of photons from matter (the recombination time), rather than containing of the order of 3.6×10^4 such regions as in section 7.2. The usual expression (3.117) for the distance to the particle horizon does not hold during the period of cosmological inflation. Instead,

$$d_H(t) = R(t) \int_0^t \frac{dt'}{R(t')} \qquad (7.26)$$

with $R(t)$ given by (7.25) if we neglect the period of growth of $d_H(t)$ during the period of radiation-dominated expansion which preceded cosmological inflation. Thus, during the inflationary period,

$$d_H(t) = H^{-1}(e^{Ht} - 1). \qquad (7.27)$$

The exponential growth of $d_H(t)$ during this period means that, once the universe has reheated to around the critical temperature after the phase transition has been completed, the horizon volume is exponentially greater than it was at this temperature prior to inflation. (Because the energy density in the false vacuum is of order T_c^4 and the radiation density in the FRW universe is given by (2.22), reheating to a temperature of order T_c occurs.) Thus, after the subsequent period of radiation-dominated expansion in the FRW universe, the size of the horizon volume at $t = t_r$ is exponentially greater than in the standard model of cosmology. It is then easy for one horizon volume at $t = t_r$ to contain many times over the volume which will expand to the presently seeable universe.

Consider next the flatness problem. Let $T = T_c$ be the temperature at which the low-temperature minimum of the effective potential (the true vacuum) becomes the absolute minimum. When the phase transition is first order, the universe will supercool to a temperature $T = T_s$ before the phase transition is completed by tunnelling out of the false vacuum, as described in section 2.9, and reheating of the universe to a temperature $T = T_R$ occurs, with $T_R \sim T_c$. The period when supercooling is occurring is a period of non-adiabatic expansion which modifies the discussion of the flatness problem given earlier. The flatness problem was cast in section 7.2 as the unnatural smallness of $|\Omega - 1| = |k/H^2 R^2|$ at early times, e.g. at the (supersymmetric) grand unification scale. During inflation, assumed to occur at that scale, H^2 is given by (7.24) and is constant. At the same time, R grows exponentially. In (7.20), $|\Omega - 1|$ was of order 10^{-55}. Thus, if R^2 grows by more than about 55 orders of magnitude during inflation, we end up with a 'natural' value of $|\Omega - 1|$ of order 1 at the supersymmetric grand unification scale. It is usual to measure inflation in terms of e-folds (one e-fold being growth of R by a factor of e). In terms of e-folds, what we require to overcome the flatness problem is around 64 e-folds of inflation[1]. This is sufficient to solve the horizon problem discussed above.

Finally, turning to the unwanted relics problem, let us consider, for definiteness, the magnetic monopole problem. In section 3.10, the excessive

---

[1] The WMAP data which suggests that $|\Omega_0 - 1|$ may be two orders of magnitude less than 1 indicates 66 e-folds may be nearer the mark. This makes little difference and we shall use 64 e-folds throughout.

magnetic monopole contribution to the predicted density of the universe today derived from the size of the ratio of the monopole number density to the entropy density at the time of the grand unified phase transition at which the monopoles were produced or, equivalently, from the size of the ratio of the number of monopoles to the entropy. The problem can be solved if a great deal of entropy is generated by inflation.

During supercooling, the entropy does not change. However, the non-adiabatic reheating results in increased entropy. The entropy density prior to supercooling is of order T_c^3. If the reheating temperature $T_R \sim T_c$, the entropy density is still of this order after reheating. However, because the volume of any region has been inflated by the inflation of R^3, the entropy in that region has increased by a factor of $e^{3H \Delta t}$, where Δt is the duration of the period of exponential expansion. If there is sufficient inflation to solve the flatness problem (about 64 e-folds of inflation), then the entropy increases by a factor of 2.4×10^{83}. In section 3.10, we found that for the case of a supersymmetric grand-unified phase transition, $\Omega_M h^2$ was 18–19 orders of magnitude greater than the predicted upper bound for Ωh^2 (and a few orders of magnitude more in the case of a first-order phase transition). Since the entropy generation resulting from inflation reduces $\Omega_M h^2$ by 83 orders of magnitude, the relic monopole density today is insignificant. A similar discussion applies to other particle relics.

For all its successes, old inflation has a fatal flaw. It is not possible to make a 'graceful exit' [2] from the period of inflationary expansion in a supercooling de Sitter universe to a reheated FRW universe. The problem arises because the phase transition is completed in the way described in section 2.9 by the formation of bubbles of the true vacuum inside the false vacuum. In the first instance, the vacuum energy of the de Sitter phase emerges as energy in the bubble walls. For the universe to thermalize, it is necessary for the bubble walls to undergo many collisions with other bubble walls. The trouble is that, on the one hand, sufficient inflation requires the nucleation rate for the true vacuum to be sufficiently low to allow a long period of supercooling. On the other hand, if bubbles of true vacuum are to form sufficiently rapidly for the bubbles to overlap and collide in an expanding universe, then this same nucleation rate needs to be sufficiently high. It turns out that these two requirements cannot be reconciled. More precisely, it is found that for nucleation rates low enough for sufficient inflation the universe always consists of clusters of bubbles of true vacuum with a few bubbles in each cluster surrounded by false vacuum.

7.4   New inflation

It is possible to retain the successes of old inflation while avoiding the graceful exit problem in an alternative formulation of cosmological inflation referred to as 'new' inflation [3–5] or 'slow-roll' inflation. The graceful exit problem derived from the slow rate of bubble formation at a first-order phase transition when

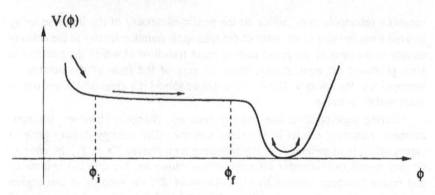

Figure 7.1. Slow-roll inflation. The slow-roll region is between ϕ_i and ϕ_f.

the nucleation rate was sufficiently low to allow the universe to remain in the false vacuum long enough for sufficient inflation to occur. In new inflation, the inflationary period begins with the scalar field (expectation value) ϕ, the 'inflaton', which, for the time being, we shall take to be real, in a region of the effective potential $V(\phi)$ which is very flat. The scalar field may have reached this region by tunnelling through a barrier between a false vacuum and the true vacuum or by the false vacuum having ceased to be a local minimum as the temperature dropped. The scalar field is then assumed to roll slowly down the the flat region of the potential. We shall discuss the scalar field dynamics involved shortly. While this process is occurring, the value of the potential is positive and can drive inflation. If $V(\phi)$ is sufficiently flat in the relevant region, the inflationary process can last long enough to solve the cosmological problems discussed earlier. Eventually, ϕ reaches a steeper region of the potential, descends more rapidly towards the absolute minimum of $V(\phi)$ with $V = 0$, overshoots and starts to oscillate about the absolute minimum. Quantum mechanical particle creation damps the oscillation and converts the vacuum energy into the energy of particles. Thermalization of the emitted particles creates a radiation-dominated FRW universe. The whole process is displayed in figure 7.1. We shall now discuss each stage of the process in more detail.

To study the slow-roll stage, we require the equation of motion for (the expectation value of) the scalar field ϕ. The Lagrangian density for a (real) scalar field with effective potential $V(\phi)$ is

$$\mathcal{L} = \tfrac{1}{2}\partial_\mu\phi\partial^\mu\phi - V(\phi). \qquad (7.28)$$

(Then the action is $S = \int d^4x \sqrt{-g}\mathcal{L}$.) One way of deriving the equation of motion is as the covariantized Euler–Lagrange equation for this field, namely

$$D_\mu(\partial^\mu\phi) = -V'(\phi) \qquad (7.29)$$

with the covariant derivative defined by

$$D_\lambda V^\mu = \partial_\lambda V^\mu + \Gamma^\mu_{\lambda\rho} V^\rho \tag{7.30}$$

for any 4-vector V^μ. Assuming a homogeneous field ϕ, so that the spatial gradients are zero,

$$\nabla\phi = 0, \tag{7.31}$$

the Euler–Lagrange equation reduces to

$$\ddot{\phi} + \Gamma^i_{i0}\dot{\phi} + V'(\phi) = 0. \tag{7.32}$$

With the coefficients of affine connection for the Robertson–Walker metric as in section 1.2, the explicit equation of motion is

$$\ddot{\phi} + 3H\dot{\phi} + V'(\phi) = 0 \tag{7.33}$$

where H is the Hubble 'constant' (exercise 1). If there is a range of values of ϕ for which slow roll occurs, then in that region the motion is dominated by the 'frictional' term $3H\dot{\phi}$ and the $\ddot{\phi}$ term is neglected. Then the equation of motion simplifies to

$$\dot{\phi} = -\frac{V'(\phi)}{3H}. \tag{7.34}$$

We derive the conditions that $V(\phi)$ must satisfy to obtain

$$\left|\frac{\ddot{\phi}}{3H\dot{\phi}}\right| \ll 1. \tag{7.35}$$

The double time derivative $\ddot{\phi}$ may be estimated from (7.34) as

$$\ddot{\phi} = -\frac{1}{3H}V'(\phi)\dot{\phi} + \frac{1}{3}H^{-2}\dot{H}V(\phi). \tag{7.36}$$

An estimate of $\dot{H}$ is now required. When the vacuum energy density ρ_V dominates over the radiation density and the curvature terms, the Friedmann equation gives

$$H^2 = \frac{8\pi G_N}{3}\rho_V = \frac{8\pi}{3}m_P^{-2}\rho_V. \tag{7.37}$$

Also, the energy-momentum tensor for the scalar field is given by

$$T_{\mu\nu} = \frac{\partial\mathcal{L}}{\partial(\partial^\mu\phi)}\frac{\partial\phi}{\partial x^\nu} - g_{\mu\nu}\mathcal{L} \tag{7.38}$$

$$= \partial_\mu\phi\partial_\nu\phi - \tfrac{1}{2}g_{\mu\nu}\partial_\lambda\phi\partial^\lambda\phi + g_{\mu\nu}V(\phi). \tag{7.39}$$

For a homogeneous field ϕ, the vacuum energy density ρ_V is given by

$$\rho_V = T_{00} = \tfrac{1}{2}\dot{\phi}^2 + V(\phi). \tag{7.40}$$

If we assume that the energy density is dominated by the potential energy, then

$$H^2 = \frac{8\pi}{3} m_P^{-2} V(\phi).$$ (7.41)

We shall see shortly that this is a consistent approximation when the effective potential $V(\phi)$ is flat enough to satisfy the conditions for slow roll. Returning to (7.36) with H estimated from (7.41), we see that

$$\frac{\ddot{\phi}}{3H\dot{\phi}} = -\frac{1}{9H^2} V'(\phi) + \frac{8\pi m_P^{-2}}{54} H^{-4} (V'(\phi))^2.$$ (7.42)

To satisfy (7.35), we require that

$$\frac{|V'(\phi)|}{9H^2} \ll 1$$ (7.43)

and

$$\frac{8\pi m_P^2}{54} (V'(\phi))^2 H^{-4} \ll 1.$$ (7.44)

With H^2 given by (7.41), these slow-roll conditions are

$$\frac{m_P^2 |V'(\phi)|}{V(\phi)} \ll 24\pi$$ (7.45)

and

$$m_P^2 \left(\frac{V'(\phi)}{V(\phi)} \right)^2 \ll 48\pi.$$ (7.46)

Slow roll occurs in the range of ϕ for which $V(\phi)$ is flat enough to satisfy these two conditions.

It can now be seen that, when the slow-roll conditions are satisfied, the vacuum energy density *is* dominated by the potential energy. Using (7.34), the kinetic term in (7.40) is

$$\frac{1}{2}\dot{\phi}^2 = \frac{(V'(\phi))^2}{9H^2}$$ (7.47)

so that, using (7.41) and (7.46),

$$\frac{\frac{1}{2}\dot{\phi}^2}{V(\phi)} = \frac{m_P^2}{96\pi} \left(\frac{V'(\phi)}{V(\phi)} \right)^2 \ll 1.$$ (7.48)

If the slow roll occurs between times t_i and t_f, and the value of the scalar field evolves from ϕ_i to ϕ_f during this time, then the amount of inflation, measured as the number of e-folds, is given by

$$N_e \equiv \ln \frac{R_f}{R_i} = \int_{t_i}^{t_f} \frac{\dot{R}}{R} \, dt = \int_{t_i}^{t_f} H \, dt.$$ (7.49)

Thus, when $\dot{\phi}$ is given by (7.34),

$$N_e = \ln \frac{R_f}{R_i} = \int_{\phi_i}^{\phi_f} H \frac{d\phi}{\dot{\phi}} = -\int_{\phi_i}^{\phi_f} 3H^2 \frac{d\phi}{V'(\phi)} \tag{7.50}$$

and when the vacuum energy density is dominated by the potential energy, as in (7.41), the number of e-folds of inflation is

$$N_e \equiv \ln \frac{R_f}{R_i} = -8\pi m_P^{-2} \int_{\phi_i}^{\phi_f} \frac{V(\phi)}{V'(\phi)} d\phi. \tag{7.51}$$

If we make the approximation

$$V'(\phi) \simeq V'(\phi_i) + V''(\phi_i)(\phi - \phi_i) \tag{7.52}$$

and take

$$V'(\phi_i) \simeq 0 \tag{7.53}$$

for a flat potential, then

$$V'(\phi) \simeq V''(\phi_i)(\phi - \phi_i). \tag{7.54}$$

Substituting this into (7.34), we find that

$$\phi - \phi_i \simeq \exp\left(-\frac{V''(\phi_i)}{3H}(t - t_i)\right). \tag{7.55}$$

Then, the motion is slow over a time period

$$\tau \sim \frac{3H}{|V''(\phi_i)|} \tag{7.56}$$

and, using (7.49),

$$N_e = \ln \frac{R_f}{R_i} \sim H\tau \sim \frac{3H^2}{|V''(\phi_i)|}. \tag{7.57}$$

With H^2 given by (7.41), this gives

$$\ln \frac{R_f}{R_i} \sim \frac{8\pi V(\phi)}{m_P^2 |V''(\phi_i)|}. \tag{7.58}$$

Thus, when the slow-roll condition (7.45) is satisfied, $\ln R_f/R_i$ is large and we get many e-folds of inflation. Equation (7.58) is useful as an initial test of whether sufficient inflation can occur. The estimate of the number of e-folds of inflation can be sharpened up by performing the integration in (7.51) over the region between ϕ_i and ϕ_f which are the boundaries of the region in which the slow-roll conditions are satisfied.

All of this discussion assumes that the motion of ϕ across the flat region is that of a classical field. If there are significant quantum fluctuations, ϕ may cross the flat region more rapidly and these conclusions may no longer be valid. We shall see later that, in the de Sitter space of an inflating universe, there *are* substantial quantum fluctuations and we need to check that this effect is not sufficient to invalidate these estimates of N_e. This will be discussed in section 7.6.

7.5   Reheating after inflation

Eventually, slow roll ends when the inflaton field ϕ reaches a steeper region of the potential. The inflaton then descends more rapidly towards the absolute minimum of the potential, overshoots it and starts to oscillate about the absolute minimum. Assuming that the inflaton possesses couplings to matter fields, the oscillation is damped by quantum mechanical particle creation as vacuum energy is converted into energy of particles [6–9]. Denote the decay rate of the inflaton by Γ_ϕ. We shall assume that $\Gamma_\phi \lesssim H_{\text{osc}}$, where H_{osc} is the value of the Hubble constant when the slow-roll period ends and the oscillating period begins.

A simple way of introducing the damping by particle emission into the dynamics of the inflaton is to modify (7.33) to

$$\ddot{\phi} + 3H\dot{\phi} + \Gamma_\phi \dot{\phi} + V'(\phi) = 0. \tag{7.59}$$

Multiplying by $\dot{\phi}/2$ and recalling (7.40) for the vacuum energy density ρ_V, we find that

$$\dot{\rho}_V + (3H + \Gamma_\phi)\dot{\phi}^2 = 0. \tag{7.60}$$

For simple harmonic oscillations, the average of the kinetic energy over an oscillation is equal to the average of the potential energy over an oscillation, so that

$$\tfrac{1}{2}\langle \dot{\phi}^2 \rangle = \langle V(\phi) \rangle = \tfrac{1}{2}\langle \rho_V \rangle. \tag{7.61}$$

Averaging (7.60) over oscillations, we write

$$\dot{\rho}_V + (3H + \Gamma_\phi)\rho_V = 0 \tag{7.62}$$

where ρ_V is now understood to refer to the time-averaged quantity. We shall discuss later in this section the circumstances in which (7.62) is valid.

While $t \lesssim \Gamma_\phi^{-1}$, neglecting the time between the big bang and the start of oscillations, the development of ρ_V is given, to a good approximation, by

$$\dot{\rho}_V + 3H\rho_V = 0 \tag{7.63}$$

which is identical to the energy conservation equation in a matter-dominated FRW universe. Thus, we may regard the vacuum energy density as equivalent to a gas of non-relativistic ϕ particles. During this period,

$$\rho_V \propto R^{-3} \qquad R \propto t^{2/3} \qquad \text{and} \qquad \rho_V \propto t^{-2}. \tag{7.64}$$

When $t \sim \Gamma_\phi^{-1}$, rapid decay of the vacuum energy to emitted particles occurs and the universe reheats to a temperature T_R given by

$$\frac{\pi^2}{30}\left(N_B + \frac{7}{8}N_F\right)T_R^4 = \rho_V(t = \Gamma_\phi^{-1}) \tag{7.65}$$

where $N_B + \frac{7}{8}N_F = \frac{915}{4}$ for the supersymmetric standard model and $\frac{427}{4}$ for the standard model. If we define a scale M by

$$M^4 \equiv \rho_V(t = t_i),\qquad(7.66)$$

where t_i is the time that slow roll begins, then we still have

$$\rho_V(t = t_{osc}) \simeq M^4\qquad(7.67)$$

because $\rho_V \simeq V(\phi)$ during slow roll and $V(\phi)$ does not change significantly over the flat region in which slow roll occurs. The initial value H_{osc} of the Hubble constant when the oscillatory period starts is then given by

$$H_{osc}^2 = \frac{8\pi}{3}m_P^{-2}M^4.\qquad(7.68)$$

The time t_{osc} when the oscillatory period starts is then of order

$$t_{osc} \sim H_{osc}^{-1} = \sqrt{\frac{3}{8\pi}}m_P M^{-2}\qquad(7.69)$$

(or one or two orders of magnitude greater). Then, from (7.64), we see that

$$\frac{\rho_V(t = \Gamma_\phi^{-1})}{\rho_V(t = t_{osc})} = (\Gamma_\phi t_{osc})^2\qquad(7.70)$$

and, using (7.69) and (7.67), we have

$$\rho_V(t = \Gamma_\phi^{-1}) = \frac{3}{8\pi}(\Gamma_\phi m_P)^2.\qquad(7.71)$$

Consequently, the reheating temperature T_R of (7.65) is

$$T_R = \left(\frac{45}{4\pi^3(N_B + \frac{7}{8}N_F)}\right)^{1/4}(\Gamma_\phi m_P)^{1/2}.\qquad(7.72)$$

Note that this is not, in general, of order M.

This discussion depends upon (7.63) for the time development of the (time-averaged) vacuum energy density. This is known to be correct for the oscillatory period if the ϕ particles decay only into fermions. However, when the ϕ particles decay into pairs of bosons $\phi \to \chi\chi$, then it is possible for very rapid decay via parametric resonance to occur [10], a process which generates very large numbers of χ particles. This is referred to as 'preheating'. However, the (radiation) energy density in light χ particles rapidly becomes small compared to the (matter) energy density remaining in the oscillatory ϕ vacuum. Thereafter, the reheating process occurs as before and the estimate of T_R is not much altered.

Since all pre-existing baryon asymmetry will be diluted exponentially by the inflationary period, the baryon asymmetry we observe now must be generated after reheating has occurred or during reheating. If the reheating temperature T_R is sufficiently high, then the baryon asymmetry may be produced in the usual way by the decay of leptoquark bosons in a GUT. Lower reheating temperatures will suffice if the sphaleron mechanism applies instead. A further possibility is that the baryon asymmetry is produced by the decay of the oscillating vacuum state which exists after slow roll has ceased, i.e. by the decay of particles associated with the inflaton field ϕ. This is the situation discussed in section 4.6 where all of the entropy of the universe is produced by the decay of particles whose decay is also producing the baryon asymmetry. Then the baryon asymmetry is

$$\frac{n_B}{s} \sim \frac{\epsilon T_R}{m_\phi} \tag{7.73}$$

where ϵ is the net baryon number produced by the decay of a scalar particle associated with ϕ. There is the weaker requirement that the reheating temperature should be high enough for nucleosynthesis to occur so that T_R should be at least a few MeV.

7.6   Inflaton field equations

As discussed in section 7.4, estimates of the amount of inflation occurring during slow roll require that quantum fluctuations in the inflaton field $\hat{\phi}$ do not cause the flat region of the potential to be crossed too rapidly. We now show that, in the inflationary universe, $\langle \phi^2 \rangle$ grows linearly with time [11–13].

The inflaton field operator may be expanded in terms of plane-wave modes as

$$\hat{\phi}(t, x) = \frac{1}{(2\pi)^{3/2}} \int d^3k \, (\psi_k(t) e^{ik \cdot x} a_k + \text{h.c.}) \tag{7.74}$$

where the creation and annihilation operators a_k and $a_k^\dagger$ obey

$$[a_k, a_{k'}^\dagger] = \delta(k - k'). \tag{7.75}$$

For a massless field in a flat FRW space, the field equation

$$D_\mu(\partial^\mu \phi) = 0 \tag{7.76}$$

leads to

$$\ddot{\psi}_k(t) + 3H\dot{\psi}_k(t) + R^{-2}(t)k^2\psi_k(t) = 0. \tag{7.77}$$

With

$$R(t) = R_0 e^{Ht} \tag{7.78}$$

during the inflationary expansion and using the variable

$$\eta \equiv -R_0^{-1} H^{-1} e^{-Ht} \tag{7.79}$$

the equation for the plane wave mode $\psi_k(t)$ becomes

$$\psi_k'' - 3\eta^{-1}\psi_k' + k^2\psi_k = 0 \tag{7.80}$$

where the primes denote differentiation with respect to η. The general solution is given in terms of Hankel functions

$$\psi_k(\eta) = \left(\frac{\pi}{4}\right)^{1/2} \eta^{3/2} H[c_1(k)H_{3/2}^{(1)}(k\eta) + c_2(k)H_{3/2}^{(2)}(k\eta)] \tag{7.81}$$

where $k = |k|$ and the condition

$$|c_2|^2 - |c_1|^2 = 1 \tag{7.82}$$

follows from the canonical commutation relations for $\hat{\phi}$ and its conjugate momentum. Retaining only the positive frequency part [13] of $\psi_k(\eta)$, for modes which go through many oscillations in an expansion time, we take

$$c_2(k) = 1, \qquad c_1(k) = 0 \tag{7.83}$$

for $k \gg R_0 H$. Then, for $k \gg R_0 H$,

$$\psi_k(\eta) = \left(\frac{\pi}{4}\right)^{1/2} \eta^{3/2} H H_{3/2}^{(2)}(k\eta) \tag{7.84}$$

$$\simeq -(2k)^{-1/2} H \eta[1 - i(k\eta)^{-1}]e^{-ik\eta}. \tag{7.85}$$

and

$$|\psi_k(\eta)|^2 = \frac{H^2}{2k^3}(1 + k^2 R_0^{-2} H^{-2} e^{-2Ht}) \tag{7.86}$$

The quantum fluctuation $\langle\phi^2\rangle$ may now be estimated as follows. Because the modes with wavelengths greater than the horizon (in the sense of the comoving Hubble length $H^{-1}/R(t)$) are expected to be responsible for the growth of $\langle\phi^2\rangle$ with time [13], an approximation to $\langle\phi^2\rangle$ is obtained by cutting off the k integration at $k = R_0 H e^{Ht}$. Then, from (7.74) and (7.75),

$$\langle\phi^2\rangle = \frac{1}{(2\pi)^3} \int d^3k \, |\psi_k(\eta)|^2 \tag{7.87}$$

and, using (7.86), this gives linear growth in time:

$$\langle\phi^2\rangle \simeq \frac{H^3}{4\pi^2}t + (\text{constant}). \tag{7.88}$$

It is important for a consistent model of inflation that quantum fluctuations do not result in $\langle\phi\rangle$ crossing the flat region of the potential faster than the time required for semi-classical slow roll across this region. In (7.57), the time to roll across the flat region (the period of slow roll) was

$$\tau \sim H^{-1} N_e. \tag{7.89}$$

Thus, we require a flat region of width $\Delta\phi$ with

$$(\Delta\phi)^2 > \frac{H^3}{4\pi^2}\tau = \frac{H^2 N_e}{4\pi^2} \tag{7.90}$$

so that we require

$$\Delta\phi > \frac{H N_e^{1/2}}{2\pi}. \tag{7.91}$$

We shall see in the example in section 7.8 that there is often a stronger constraint from the rquirement of obtaining density perturbations of the size found by COBE.

7.7 Density perturbations

The quantum fluctuations in the inflaton field discussed in the previous section result in density perturbations [14–16] in the post-inflationary universe, which may be responsible for galaxy formation. The density perturbations arise because the quantum fluctuations in $\hat{\phi}$ give ϕ, i.e. the expectation value of $\hat{\phi}$, slightly different values in different regions of space. This results in perturbations to the value of the vacuum energy density.

Central to the discussion of the formation of density perturbations is the fact that a given comoving wavelength (i.e. a wavelength in units of the scale factor $R(t)$ of the universe) can start inside the horizon before inflation begins, cross outside the horizon at some time during inflation, and then cross back inside the horizon after inflation has ended and a radiation-dominated universe has been established. (By 'horizon' we shall mean here not the particle horizon but the comoving Hubble length $H^{-1}/R(t)$. This is a measure of the distance light travels during an appreciable amount of expansion of the universe. Inside of a comoving Hubble length, causal processes do not feel the expansion of the universe.)) This behaviour is a consequence of the fact that when $R(t)$ is increasing as a power t^p of t with $p < 1$, the comoving Hubble length increases with time, whereas when $R(t)$ is increasing exponentially with time, the comoving Hubble length decreases with time, while the comoving wavelength is, by definition, constant. (See figure 7.2.) The (classical) inflaton field perturbation $\delta\phi(t, x)$ may be expressed in terms of perturbations $\delta\tilde{\phi}(t, k)$ of momentum k as

$$\delta\phi(t, x) = \int d^3 k \, e^{ik \cdot x} \delta\tilde{\phi}(t, k) \tag{7.92}$$

where we have written the complete inflaton field $\phi(t, x)$ as

$$\phi(t, x) = \phi_0(t) + \delta\phi(t, x) \tag{7.93}$$

with $\phi_0(t)$ the homogeneous classical field. Quantum fluctuations $\delta\tilde{\phi}(t, k)$ develop when the comoving scale $|k|^{-1}/R(t)$ is inside the horizon and become

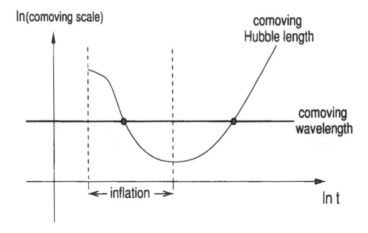

Figure 7.2. Behaviour of the comoving Hubble length during and after inflation.

'frozen in' when this comoving scale crosses outside the horizon, so that it is no longer subject to causal processes. When this comoving scale crosses back inside the horizon, the fluctuations reappear as classical density perturbations.

Calculation of $\delta\tilde{\phi}(t, k)$ requires a generalization of the equation of motion (7.33) of the homogeneous classical field $\phi_0(t)$ to include the spatial dependence of $\phi(t, x)$. Including this dependence (7.29) leads (exercise 2) to

$$\ddot{\phi} - R^{-2}\nabla^2\phi + 3H\dot{\phi} + V'(\phi) = 0 \qquad (7.94)$$

where a flat space has been assumed. (Note in passing that had we allowed ϕ to have spatial dependence in the discussion in section 7.4, this would have been damped out rapidly because of the exponential growth of $R(t)$ during the inflationary period.) Comparing (7.33) for $\phi_0(t)$ with (7.94) for $\phi(t, x)$, we see that $\delta\tilde{\phi}(t, k)$ obeys

$$(\ddot{\delta\tilde{\phi}}) + 3H(\dot{\delta\tilde{\phi}}) + R_0^{-2}e^{-2Ht}k^2\delta\tilde{\phi} + V''(\phi_0)\delta\tilde{\phi} = 0. \qquad (7.95)$$

For slow roll away from a maximum of the potential, we must have

$$V''(\phi_0) < 0. \qquad (7.96)$$

The perturbations start to grow when the fourth term in (7.95), which is the destabilizing influence, becomes larger than the third term. Thus, $\delta\tilde{\phi}(t, k)$ starts to grow at a time $t^*(k)$ (where $k \equiv |k|$) given by

$$R_0^{-2}e^{-2Ht^*}k^2 = -V''(\phi_0). \qquad (7.97)$$

For $t \gg t^*(k)$, the third term in (7.95) can be neglected and $\delta\tilde{\phi}$ obeys

$$\ddot{\delta\tilde{\phi}} + 3H\dot{\delta\tilde{\phi}} = -V''(\phi_0)\delta\tilde{\phi}. \qquad (7.98)$$

We also know that ϕ_0 obeys (7.33):

$$\ddot{\phi}_0 + 3H\dot{\phi}_0 = -V'(\phi_0). \tag{7.99}$$

Consequently,

$$\frac{\partial^2}{\partial t^2}(\dot{\phi}_0) + 3H\frac{\partial}{\partial t}(\dot{\phi}_0) = -V''(\phi_{0_i})(\dot{\phi}_0) \tag{7.100}$$

so that $\delta\tilde{\phi}$ obeys the same equation as $\dot{\phi}_0$ when H is nearly constant, as it will be during the period of slow roll. Thus, $\delta\tilde{\phi}(t, k)$ must be proportional to $\dot{\phi}_0$ with a k-dependent constant of proportionality which we write as $-\delta\tilde{\tau}(k)$:

$$\delta\tilde{\phi}(t, k) = -\delta\tilde{\tau}(k)\dot{\phi}_0(t). \tag{7.101}$$

Substituting into (7.93),

$$\phi(t, x) = \phi_0(t) - \delta\tau(x)\dot{\phi}_0(t) \tag{7.102}$$

where

$$\delta\tau(x) = \int d^3k\, e^{ik\cdot x}\delta\tilde{\tau}(k). \tag{7.103}$$

To first order in $\delta\tau$,

$$\phi(t, x) = \phi_0(t - \delta\tau(x)). \tag{7.104}$$

Thus, the scalar field fluctuations introduce a spatial dependence into the classical field $\phi_0(t)$ which is of the form of a spatially-dependent time lag. Also, for $t \ll t^*$, the $V''(\phi_0)$ term may be neglected and $\delta\tilde{\phi}$ is just the quantum fluctuation of a free massless scalar field in de Sitter space, which is known to be

$$\delta\tilde{\phi}(t, k) = \frac{H}{4\pi^{3/2}}(1 + R_0^{-2}k^2H^{-2}e^{-2Ht})^{1/2}. \tag{7.105}$$

In this limit, the equation obeyed by $\delta\tilde{\phi}(t, k)$ is identical to (7.77) and $\delta\tilde{\phi}(t, k)$ is the same as $|\psi_k|$ up to a normalization factor, as can be seen from (7.86). The normalization is determined by the requirement that $\delta\tilde{\phi}(t, k)$ is the rms fluctuation [14], so that

$$\delta\tilde{\phi}(t, k)^2 = \left(\frac{k}{2\pi}\right)^3 |\psi_k|^2. \tag{7.106}$$

Perturbations in a scalar field will produce density perturbations because the potential energy $V(\phi)$ is modified by perturbations in ϕ. Thus,

$$\delta\rho = \delta V = V'(\phi_0)\delta\tilde{\phi}. \tag{7.107}$$

A calculation of the evolution of the density perturbations using the formalism of Olson [14, 17] shows that when the comoving scale $|k|^{-1}/R(t)$ crosses back inside the horizon,

$$\frac{\delta\rho}{\rho} = 4H\delta\tilde{\tau}(k). \tag{7.108}$$

We may estimate $\delta\tilde{\tau}(k)$ by assuming that both the $t \gg t^*(k)$ and $t \ll t^*(k)$ expressions for $\delta\tilde{\phi}(t, k)$ are tolerable approximations for $t \simeq t^*(k)$. Equating (7.101) and (7.105) at $t = t^*(k)$ gives

$$\delta\tilde{\tau}(t^*, k) = \frac{-H}{4\pi^{3/2}\dot{\phi}_0(t^*)}(1 + R_0^{-2}k^2 H^{-2} e^{-2Ht^*})^{1/2}$$

$$= \frac{-H}{4\pi^{3/2}\dot{\phi}_0(t^*)}[1 + H^{-2}V''(\phi_0(t^*))]^{1/2}. \tag{7.109}$$

When the slow-roll conditions are satisfied,

$$V''(\phi_0) \ll H^2 \tag{7.110}$$

and we have

$$\delta\tilde{\tau}(t^*, k) \simeq \frac{-H}{4\pi^{3/2}\dot{\phi}_0(t^*)}. \tag{7.111}$$

Then

$$\frac{\delta\rho}{\rho} = \frac{-H^2(t^*)}{\pi^{3/2}\dot{\phi}_0(t^*)}. \tag{7.112}$$

In practice, $t^*(k)$ is a few Hubble times after the time when the comoving wavelength k^{-1} crossed outside the horizon. In evaluating (7.112), we shall always identify $t^*(k)$ with the horizon crossing time given by $k = R(t^*(k))H(t^*(k))$. Then (7.112) is consistent to within a factor of order one with other, more rigorous, treatments [15, 16].

The COBE observations require that

$$\frac{\delta\rho}{\rho} \sim 2 \times 10^{-5}. \tag{7.113}$$

This allows us to estimate the energy scale of the inflationary potential $V(\phi)$. Using (7.41) and (7.47), we have

$$\frac{\delta\rho}{\rho} \sim m_P^{-3}\frac{V(\phi)^{3/2}}{V'(\phi)}. \tag{7.114}$$

Then, using (7.113), it follows that

$$V(\phi)^{1/4} \sim \left(\frac{m_P V'(\phi)}{V(\phi)}\right)^{1/2}(10^{16}-10^{17}) \text{ GeV}. \tag{7.115}$$

Also, from the slow-roll condition (7.46),

$$\left(\frac{m_P V'(\phi)}{V(\phi)}\right)^{1/2} \ll 3.5. \tag{7.116}$$

Thus, we expect the energy scale of the inflationary potential is given by

$$V(\phi)^{1/4} \sim (10^{16}-10^{17}) \text{ GeV} \tag{7.117}$$

though it could be less if the inflationary potential is very flat.

7.8   A worked example

The discussion of previous sections may be illustrated by the following example of a suitable effective potential for inflation due to Steinhardt and Turner [5]. Consider the potential

$$V = V_0 - \beta\phi^3 + \lambda\phi^4 \tag{7.118}$$

where V_0, ϕ and λ are constants with $\beta, \lambda > 0$. It is assumed that the inflaton field ϕ starts rolling from $\phi = 0$, possibly because ϕ was zero in a high-temperature phase where some symmetry was restored. In that case, slow roll began at some lower temperature after the high-temperature phase ceased to be the stable vacuum. It is also assumed that V is zero at the absolute minimum to which ϕ rolls, corresponding to zero cosmological constant. At the absolute minimum,

$$\phi = \frac{3\beta}{4\lambda} \equiv \sigma \tag{7.119}$$

and $V(\sigma) = 0$ requires that

$$V_0 = \frac{27\beta^4}{256\lambda^3} \tag{7.120}$$

When $\phi \simeq \phi_0 = 0$, (7.41) then implies that

$$H^2 = H_0^2 = \frac{9\pi\beta^4}{32\lambda^3}m_P^{-2}. \tag{7.121}$$

The slow-roll condition (7.45) implies that slow roll occurs for ϕ in the range

$$0 \le \phi \lesssim \frac{27\pi\beta^4}{64\lambda^3}m_P^{-2} \equiv \phi_e \tag{7.122}$$

where it has been assumed that $\phi \ll \sigma$ in the slow-roll region. The second slow-roll condition (7.46) is automatically satisfied whenever

$$V''(\phi) \sim \frac{V'(\phi)}{\phi} \tag{7.123}$$

and $|\phi|$ is at least one order of magnitude less than m_P. This is true here because

$$V''(\phi) \simeq \frac{-2V'(\phi)}{\phi} \tag{7.124}$$

when $\phi \ll \sigma \lesssim m_P$. It is being assumed that σ is less than m_P so that we do not have to consider the effects of quantum gravity. Following (7.119), this is obviously arranged for

$$\beta \lesssim \tfrac{4}{3}\lambda m_P. \tag{7.125}$$

To avoid de Sitter fluctuations driving ϕ across the flat region faster than it would roll semi-classically, we need a width $\Delta\phi$ for the flat region with the property (7.91). Combining this with (7.122) leads to

$$\beta \gtrsim \frac{8}{9\sqrt{2}\pi^{3/2}} N_e^{1/2} \lambda^{3/2} m_P. \tag{7.126}$$

The criterion for sufficient inflation (7.57) yields

$$\beta \gtrsim 128 \left(\frac{32}{9\pi}\right)^{1/2} \lambda^{3/2} m_P \tag{7.127}$$

where $|V''(\phi_i)|$ has been evaluated at $\phi_i \sim H_0$ instead of $\phi_i \sim 0$ to allow for quantum fluctuations and N_e has been taken to be about 64. The condition (7.126) is clearly satisfied when (7.127) is.

For density fluctuations of the order required by COBE, (7.112) and (7.113) imply that

$$\frac{-H^2(t^*)}{\pi^{3/2}\dot{\phi}_0(t^*)} \sim 10^{-5}. \tag{7.128}$$

With the aid of (7.34), this leads to

$$\beta \sim \frac{10^5 H^3(t_e)}{\pi^{3/2}\phi^2(t_e)} \tag{7.129}$$

if $t^* \simeq t_e$, the time at which slow roll ends. With ϕ_e given by (7.122) and

$$H_e^2 \simeq \frac{8\pi}{3} m_P^{-2} V_0 \tag{7.130}$$

using (7.41), and assuming that $\beta/\lambda m_P \lesssim 1$ (which will turn out to be the case), we then find that

$$\beta \sim 8.5 \times 10^3 \lambda^{3/2} m_P. \tag{7.131}$$

In practice, this is not a particularly good approximation because $\dot{\phi}$ is not constant and $\dot{\phi}_0(t^*) < \dot{\phi}_0(t_e)$. A more careful calculation [5] gives a value of β several orders of magnitude larger. Note that even for the smaller value of β given by (7.131), the constraint for sufficient inflation (7.127) is satisfied with two orders of magnitude in hand and the de Sitter fluctuation constraint (7.126) is satisfied with four orders of magnitude in hand. Combining (7.131) with the condition (7.125) for σ to be less than m_P, we find that

$$\lambda < 2.5 \times 10^{-8}. \tag{7.132}$$

Then (7.131) implies that

$$\frac{\beta}{\lambda} m_P^{-1} \lesssim 1.3 \tag{7.133}$$

so that $\beta/\lambda m_P \lesssim 1$, as assumed earlier.

7.9 Complex inflaton field

In previous sections, we have been assuming that the inflaton field ϕ is a real scalar field. A simple extension is to take ϕ to be a complex scalar field [18] with the Lagrangian density

$$\mathcal{L} = \partial_\mu \phi \partial^\mu \phi^* - V(\phi, \phi^*). \tag{7.134}$$

For a homogeneous field, the Euler–Lagrange equations are then

$$\ddot{\phi} + 3H\dot{\phi} + \frac{\partial V}{\partial \phi^*} = 0 \tag{7.135}$$

with the Hubble constant given by

$$H^2 = \frac{8\pi}{3} m_P^{-2} V(\phi, \phi^*). \tag{7.136}$$

When the energy is dominated by the potential energy,

$$\dot{\phi}_1 = -\frac{1}{6H} \frac{\partial V}{\partial \phi_1} \tag{7.137}$$

$$\dot{\phi}_2 = -\frac{1}{6H} \frac{\partial V}{\partial \phi_2} \tag{7.138}$$

where we have separated ϕ into its real and imaginary parts

$$\phi = \phi_1 + i\phi_2. \tag{7.139}$$

For the slow-roll approximation to be valid,

$$\left| \frac{\ddot{\phi}_1}{3H\dot{\phi}_1} \right| \ll 1 \tag{7.140}$$

$$\left| \frac{\ddot{\phi}_2}{3H\dot{\phi}_2} \right| \ll 1. \tag{7.141}$$

These may be cast as the sufficient conditions (exercise 4)

$$m_P^2 \left| \frac{V_{11} V_1 + V_2 V_{12}}{V V_1} \right| \ll 48\pi \tag{7.142}$$

$$m_P^2 \left| \frac{V_{22} V_2 + V_1 V_{12}}{V V_2} \right| \ll 48\pi \tag{7.143}$$

$$m_P^2 \left| \frac{V_1^2 + V_2^2}{V^2} \right| \ll 96\pi \tag{7.144}$$

where $V_\alpha \equiv \partial V / \partial \phi_\alpha (\alpha = 1, 2)$ etc. The last condition ensures that the kinetic term may be neglected compared with the potential term in the vacuum energy density.

If slow roll occurs from $(\phi_1, \phi_2) = (\phi_{1_0}, \phi_{2_0})$ close to a saddle point or minimum of the effective potential, then we may write

$$V_1 \simeq V_{1_0} + V_{11_0}(\phi_1 - \phi_{1_0}) + V_{12_0}(\phi_2 - \phi_{2_0}) \qquad (7.145)$$
$$V_2 \simeq V_{2_0} + V_{12_0}(\phi_1 - \phi_{1_0}) + V_{22_0}(\phi_2 - \phi_{2_0}). \qquad (7.146)$$

In terms of the displacements,

$$x_\alpha \equiv \phi_\alpha - \phi_{\alpha_0} \qquad (\alpha = 1, 2) \qquad (7.147)$$

and, correct to linear order in x_α, the slow-roll equations may be written as

$$\dot{X} = A - MX \qquad (7.148)$$

with

$$X = \begin{pmatrix} x_1 \\ x_2 \end{pmatrix} \qquad A = \frac{1}{6H} \begin{pmatrix} V_{1_0} \\ V_{2_0} \end{pmatrix} \qquad (7.149)$$

and

$$M = \frac{1}{6H} \begin{pmatrix} V_{11_0} & V_{12_0} \\ V_{12_0} & V_{22_0} \end{pmatrix}. \qquad (7.150)$$

After diagonalizing M, we find that the displacements x_α are superpositions of eigensolutions with time dependence $e^{-\lambda_i t}$ where

$$\lambda_i = \frac{\mu_i}{6H} \qquad (i = 1, 2) \qquad (7.151)$$

with

$$2\mu_{1,2} = V_{11_0} + V_{22_0} \pm \sqrt{(V_{11_0} - V_{22_0})^2 + 4V_{12_0}^2}. \qquad (7.152)$$

The number of e-folds of inflation may then be written as

$$N_e = 2V_0 \min(-\mu_1^{-1}, -\mu_2^{-1}) \qquad (7.153)$$

if μ_1 and μ_2 are both negative. Otherwise, N_e is controlled by the negative μ_i. It is now necessary to have the potential sufficiently flat in all directions that there is slow roll no matter what direction of roll occurs off the maximum (or saddle point).

7.10 Chaotic inflation

Up to this point, it has been assumed that the initial conditions for slow-roll inflation are thermal. By this we mean that the field ϕ was at the minimum of the effective potential for a high-temperature phase until this minimum ceased to be the absolute minimum. Thereafter, ϕ appeared in the flat region of the potential, either by quantum mechanical or thermal tunnelling out of the metastable minimum or after the metastable minimum had ceased to exist. If the flat region is in the vicinity of a maximum or turning point of the effective potential, some fine tuning of the initial conditions may be required if ϕ is to start out in the flat region.

An alternative possibility [19] is that the initial conditions are provided by a chaotic quantum state which existed for times $t \lesssim t_P$, where

$$t_P \equiv m_P^{-1} \tag{7.154}$$

is the Planck time. A simple chaotic inflation model can be developed using the Lagrangian density for the inflaton field ϕ,

$$\mathcal{L} = \tfrac{1}{2}\partial_\mu \phi \partial^\mu \phi - V(\phi) \tag{7.155}$$

with

$$V(\phi) = \frac{1}{4}\lambda \phi^4 \tag{7.156}$$

and $\lambda \ll 1$, so that the potential is flat. At the Planck time, the uncertainty principle implies that $V(\phi)$ can only be measured with an accuracy of m_P^4. Thus, instead of ϕ being fixed at the minimum of $V(\phi)$ at $\phi = 0$ in all regions of space, we should expect ϕ to take values in the range

$$-\frac{m_P}{\lambda^{1/4}} \lesssim \phi \lesssim \frac{m_P}{\lambda^{1/4}} \tag{7.157}$$

in various regions of space (domains). These are the initial conditions for the domains. The evolution of ϕ for $t > t_P$ will permit a classical description, provided

$$V(\phi) \lesssim m_P^4 \qquad \partial_\mu \phi \partial^\mu \phi \lesssim m_P^4 \tag{7.158}$$

in all domains. Since $V(\phi)$ is, in general, non-zero, the various domains will undergo varying amounts of exponential expansion (inflation).

Consider one such domain with an initial homogeneous field $\phi(t_P)$. (As observed after (7.94), spatial dependence of ϕ is in any case damped out rapidly by the exponential growth of $R(t)$.) For the potential (7.156), the Hubble constant of (7.41) is

$$H = \left(\frac{2}{3}\pi\lambda\right)^{1/2} \phi^2 m_P^{-1}. \tag{7.159}$$

Neglecting the $\ddot{\phi}$ term in (7.33) for slow roll

$$\dot{\phi} = -\left(\frac{\lambda}{6\pi}\right)^{1/2} m_P \phi \tag{7.160}$$

so that

$$\phi = \phi(t_P) \exp\left[-\left(\frac{\lambda}{6\pi}\right)^{1/2} m_P(t - t_P)\right]. \tag{7.161}$$

The inflationary growth of the scale factor $R(t)$ is now given by (7.49) as

$$\ln \frac{R(t)}{R(t_P)} = \int_{t_P}^{t} H \, dt. \tag{7.162}$$

Then, substituting the solution (7.161) for ϕ into the expression (7.159) leads to

$$\ln \frac{R(t)}{R(t_P)} = \frac{\pi}{m_P^2} \phi^2(t_P) \left(1 - \exp\left[-\left(\frac{2\lambda}{3\pi} \right)^{1/2} m_P(t - t_P) \right] \right). \tag{7.163}$$

For $t \sim t_P$, this may be approximated by

$$R(t) \simeq R(t_P) \exp\left[\left(\frac{2\pi\lambda}{3m_P^2} \right)^{1/2} \phi^2(t_P)(t - t_P) \right]. \tag{7.164}$$

From (7.161), we see that the motion is slow roll over a period

$$\tau \sim \left(\frac{\lambda}{6\pi} \right)^{-1/2} m_P^{-1} \tag{7.165}$$

during which time we see from (7.164) that there are

$$N_e = 2\pi \left(\frac{\phi(t_P)}{m_P} \right)^2 \tag{7.166}$$

e-folds of inflation. Thus, there are at least 64 e-folds of inflation provided

$$\phi(t_P) \gtrsim 3.2 m_P. \tag{7.167}$$

This value is in the rquired range (7.157) provided

$$\lambda \lesssim 10^{-2}. \tag{7.168}$$

A region of the universe with such a value of $\phi(t_P)$ could, therefore, develop into a universe in which the horizon and flatness problems are solved, as required for the universe we occupy.

The observed value of $\delta\rho/\rho$ puts a more stringent constraint on the size of λ. We estimate $\delta\rho/\rho$ from (7.112) with $\dot{\phi}$ given by (7.160), H given by (7.159) and $\phi(t^*)$ given by (7.207) with $p = 4$. Then

$$\frac{\delta\rho}{\rho} = \frac{2\sqrt{6\lambda}}{3\pi^2} [N_e(\phi(t^*))]^{3/2} \tag{7.169}$$

where, as in section 7.12, $N_e(\phi(t^*))$ is the number of e-folds of inflation occurring after cosmologically interesting scales leave the horizon. Thus,

$$\lambda = \left(\frac{\delta\rho}{\rho} \right)^2 \frac{3\pi^4}{8} [N_e(\phi(t^*))]^{-3}. \tag{7.170}$$

For $|\delta\rho/\rho| \sim 10^{-5}$ and $N_e(\phi(t^*)) = 50$, we have

$$\lambda \simeq 3 \times 10^{-14}. \tag{7.171}$$

The spectral index discussed in section 7.12 has the value given by

$$n(k) - 1 = -0.06 \tag{7.172}$$

7.11   Hybrid inflation

In the slow-roll inflation models discussed so far, the inflaton ϕ which is undergoing slow roll is also responsible for the vacuum energy density that drives the inflationary expansion of the universe. In 'hybrid' inflation [20], two (real) scalar fields ϕ and ψ are involved. The field ϕ undergoes slow roll but most of the vacuum energy is due to the presence of ψ. When the value of ϕ drops below some critical value ϕ_c, the field ψ is destabilized and it rolls from a vacuum with positive energy density to the vacuum with zero energy density, so that inflation ends.

The simple original model, due to Linde [20] has a potential of the form

$$V = \frac{1}{2}m^2\phi^2 + \frac{\lambda_1}{4}(\psi^2 - M^2)^2 + \frac{\lambda_2}{4}\psi^2\phi^2 \tag{7.173}$$

so that

$$V = V_0 + \frac{1}{2}m^2\phi^2 - \frac{1}{2}m_\psi^2\psi^2 + \frac{\lambda_2}{4}\psi^2\phi^2 + \frac{\lambda_1}{4}\psi^4 \tag{7.174}$$

where

$$V_0 = \frac{\lambda_1}{4}M^4 \quad \text{and} \quad m_\psi^2 = \lambda_1 M^2. \tag{7.175}$$

The field ψ has an effective mass-squared

$$m_{\text{eff}}^2 = \lambda_2\phi^2 - m_\psi^2. \tag{7.176}$$

For $\phi^2 > \phi_c^2 \equiv m_\psi^2/\lambda_2$, the effective mass-squared is positive and the only minimum of the effective potential in the ψ direction is at $\psi = 0$. The parameters may be chosen [20] such that the curvature of the effective potential is much greater in the ψ-direction than in the ϕ-direction, so that, initially, ψ rolls to $\psi = 0$ while ϕ^2 remains larger than ϕ_c^2. After a period of slow roll, ϕ^2 eventually drops below ϕ_c^2. At that point, ψ starts to roll towards a true minimum of the effective potential which is at

$$\phi = 0 \quad \psi = \pm M. \tag{7.177}$$

This marks the end of the inflationary period. In the slow-roll region,

$$V(\phi) \simeq V_0 + \frac{1}{2}m^2\phi^2. \tag{7.178}$$

Thus, using (7.58),

$$N_e \simeq \frac{8\pi V_0}{m_P^2 m^2} \qquad (7.179)$$

if V_0 is the dominant term. N_e can be adjusted to be greater than 64. For example, for $V_0 \simeq (10^{16} \text{ GeV})^4$, we have $N_e \gtrsim 64$ for

$$m \lesssim 10^{13} \text{ GeV}. \qquad (7.180)$$

The parameter $\phi_c^2 = m_\psi^2 / \lambda_2$ gives enough freedom to obtain the observed value for $\delta\rho/\rho$. (In the previous example $\phi_c \simeq 3 \times 10^{17}$ GeV.) We refer the reader to the paper by Linde [20] and the book of Liddle and Lyth in the general references for this chapter, for more detail.

7.12   The spectral index

The dependence of the density perturbation $\delta\rho/\rho$ on the scale k will eventually allow different inflationary models to be distinguished by observations. Let us define

$$P(k) \equiv \left(\frac{\delta\rho}{\rho}\right)^2 = \pi^{-3} \left(\frac{H^2(t^*(k))}{\dot\phi_0(t^*(k))}\right)^2 \qquad (7.181)$$

where we have used (7.112). ($P(k)$ is proportional to the 'power spectrum'.) The spectral index $n(k)$ is defined by

$$n(k) - 1 \equiv \frac{d \ln P(k)}{d \ln k}. \qquad (7.182)$$

If $n(k)$ is a constant, this reduces to

$$P(k) \propto k^{n-1} \qquad (7.183)$$

so that $n = 1$ corresponds to a scale-independent spectrum.

The spectral index may be evaluated using the slow-roll conditions. Slow-roll parameters $\epsilon(\phi)$ and $\eta(\phi)$ may be defined by

$$\epsilon(\phi) \equiv \frac{1}{2} M_P^2 \left(\frac{V'(\phi)}{V(\phi)}\right)^2 \qquad (7.184)$$

and

$$\eta(\phi) \equiv M_P^2 \frac{V''(\phi)}{V(\phi)} \qquad (7.185)$$

where

$$M_P^2 \equiv \frac{m_P^2}{8\pi} = \frac{1}{8\pi G_N}. \qquad (7.186)$$

The derivatives $d\epsilon/d\ln k$ and $d\eta/d\ln k$ are required for the discussion of $n(k)$. From (7.34), during slow roll,

$$dt = -\frac{3H}{V'(\phi)}\,d\phi. \tag{7.187}$$

Moreover, the right-hand side of (7.181) is to be evaluated, as discussed after (7.112), when

$$k = R(t^*(k))H(t^*(k)). \tag{7.188}$$

During slow-roll inflation, the rate of change of H is small compared with the rate of change of R and so

$$d\ln k = H\,dt. \tag{7.189}$$

Combining this with (7.187) gives

$$\frac{d}{d\ln k} = -\frac{1}{3H^2}V'(\phi)\frac{d}{d\phi}. \tag{7.190}$$

Then, using (7.41) for H^2,

$$\frac{d}{d\ln k} = -M_P^2\frac{V'(\phi)}{V(\phi)}\frac{d}{d\phi}. \tag{7.191}$$

With the aid of this, we may now evaluate the required derivatives of the slow-roll parameters (7.184) and (7.185), with the result

$$\frac{d\epsilon}{d\ln k} = -2(\epsilon\eta - 2\epsilon^2) \tag{7.192}$$

and

$$\frac{d\eta}{d\ln k} = 2\epsilon\eta - \xi^2 \tag{7.193}$$

where

$$\xi^2 \equiv M_P^4\frac{V'(\phi)V''(\phi)}{V(\phi)^2}. \tag{7.194}$$

Returning to (7.181) and (7.182), we may first simplify $P(k)$ using (7.34) and (7.41) to obtain

$$P(k) = \frac{1}{6\pi^3\epsilon}\frac{V(\phi)}{M_P^4}. \tag{7.195}$$

Then differentiating with the aid of (7.191) gives [21]

$$n(k) - 1 = -6\epsilon + 2\eta. \tag{7.196}$$

Being slow-roll parameters, $\epsilon(\phi)$ and $\eta(\phi)$ are small. Consequently,

$$|n(k) - 1| \ll 1 \tag{7.197}$$

and $n(k)$ differs little from 1. The deviation of $n(k)$ from 1 at values of k of interest (k^{-1} between the present Hubble radius of 3000 Mpc and the smallest scale for large-scale structure observations of 1 Mpc) is determined by the slow-roll parameters $\epsilon(\phi)$ and $\eta(\phi)$ for the given model. The variation of $n(k)$ with k is also easily calculated using (7.192) and (7.193) to be [22]

$$\frac{d \ln n(k)}{d \ln k} = 16\epsilon\eta - 24\epsilon^2 - 2\xi^2. \tag{7.198}$$

The spectral index distinguishes models of inflation. Consider, for example, an inflationary potential of the form

$$V(\phi) = \lambda\phi^p \qquad (p > 0). \tag{7.199}$$

(The chaotic-inflation potential (7.156) is a special case with $p = 4$.) Then

$$\epsilon(\phi) = \tfrac{1}{2}M_P^2 p^2\phi^{-2} \qquad \text{and} \qquad \eta(\phi) = M_P^2 p(p-1)\phi^{-2} \tag{7.200}$$

so that

$$n(k) - 1 = -M_P^2 p(p+2)\phi^{-2} \tag{7.201}$$

with ϕ to be evaluated at $t = t^*(k)$ for scales k of interest. The key thing is the number of e-folds of inflation that occurs after cosmologically interesting scales leave the horizon. From (7.51), what we require is

$$N_e(\phi(t^*)) = -M_P^{-2}\int_{\phi(t^*)}^{\phi_e}\frac{V(\phi)}{V'(\phi)}\,d\phi \tag{7.202}$$

where ϕ_e corresponds to the end of slow roll. In the present model,

$$\frac{V(\phi)}{V'(\phi)} = p^{-1}\phi \tag{7.203}$$

leading to

$$\phi^2(t^*) - \phi_e^2 = 2N_e(\phi(t^*))pM_P^2. \tag{7.204}$$

Slow roll ends when $\epsilon(\phi) \sim 1$ and, from (7.200), we see that this happens when

$$\phi = \phi_e \sim pM_P. \tag{7.205}$$

Thus,

$$\phi^2(t^*) \simeq p^2 M_P^2 + 2N_e(\phi(t^*))pM_P^2. \tag{7.206}$$

For modest values of p and large values of $N_e(\phi(t^*))$,

$$\phi(t^*) \simeq \sqrt{2N_e(\phi(t^*))p}M_P \tag{7.207}$$

so that from (7.201)

$$n(k) - 1 \simeq -\frac{2+p}{2N_e(\phi(t^*))}. \tag{7.208}$$

For example [23], if $N_e(\phi(t^*)) = 50$, we have

$$n(k) - 1 \simeq -\frac{2+p}{100} \tag{7.209}$$

and $n(k)$ at observable scales differs from 1 by a few percent.

7.13 Exercises

1. Obtain an alternative derivation of the equation of motion (7.33) for ϕ from the energy–momentum conservation equation.
2. Generalize the equation of motion for the inflation field $\phi(t, x)$ to include spatial variation (7.94).
3. Redo the calculations of section 7.8 for the potential

$$V = V_0 - \alpha\phi^2 + \lambda\phi^4 \qquad \text{with } \alpha, \lambda > 0$$

4. Derive the suffiicient conditions (7.142) to (7.144) for the slow roll for a complex inflation field.

7.14 General references

The books and review articles that we have found most useful in preparing this chapter are:

- Kolb E W and Turner M S 1990 *The Early Universe* (Reading, MA: Addison-Wesley)
- Liddle A R and Lyth D H 2000 *Cosmological Inflation and Large-Scale Structure* (Cambridge: Cambridge University Press)
- Olive K A 1990 Inflation *Phys. Rep.* **190** 307

Bibliography

[1] Guth A H 1981 *Phys. Rev.* D **23** 347
[2] Guth A H and Weinberg E 1983 *Nucl. Phys.* B **212** 321
[3] Linde A 1982 *Phys. Lett.* B **108** 389
[4] Albrecht A and Steinhardt P J 1982 *Phys. Rev. Lett.* **48** 1220
[5] Steinhardt P J and Turner M S 1984 *Phys. Rev.* D **29** 2162
[6] Albrecht A, Steinhardt P J, Turner M S and Wilczek F 1982 *Phys. Rev. Lett.* **48** 1437
[7] Dolgov A D and Linde A D 1982 *Phys. Lett.* B **116** 329
[8] Abbott L, Farhi E and Wise M B 1982 *Phys. Lett.* B **117** 29
[9] Turner M S 1983 *Phys. Rev.* D **28** 1243
[10] Kofman L, Linde A and Starobinsky A A 1997 *Phys. Rev.* D **56** 3258
[11] Linde A D 1982 *Phys. Lett.* B **116** 335
[12] Vilenkin A and Ford L 1982 *Phys. Rev.* D **26** 1231
[13] Vilenkin A 1983 *Nucl. Phys.* B **226** 527
[14] Guth A H and Pi S Y 1982 *Phys. Rev. Lett.* **49** 1110
[15] Bardeen J A, Steinhardt P J and Turner M S 1983 *Phys. Rev.* D **28** 679
[16] Lyth D H 1985 *Phys. Rev.* D **31** 1792
[17] Olson D W 1976 *Phys. Rev.* D **14** 327
[18] Bailin D, Kraniotis V G and Love A 1998 *Phys. Lett.* B **443** 111
[19] Linde A D 1983 *Phys. Lett.* B **129** 177
[20] Linde A D 1991 *Phys. Lett.* B **259** 38

[21] Liddle A R and Lyth D H 1992 *Phys. Lett.* B **291** 391

[22] Kosowsky A and Turner M S 1995 *Phys. Rev.* D **52** 1739

[23] Liddle A R and Lyth D H 2000 *Cosmological Inflation and Large-Scale Structure* section 3.4.3 (Cambridge: Cambridge University Press)

Chapter 8

Inflation in supergravity

8.1 Introduction

In order for new inflation (slow-roll inflation) to produce sufficient e-folds of inflation, it is necessary for the effective potential for the inflaton to be very flat near the point where slow roll begins. In general, even if the tree-level potential has this property, this flatness may be lost once radiative corrections are included. For this reason, it is advantageous in constructing models of inflation to employ a supersymmetric theory where cancellations of radiative corrections are enforced by supersymmetry. If we want a supersymmetric theory that contains gravity, the natural approach, as discussed in section 2.8, is to construct a locally supersymmetric theory (supergravity). Once a superpotential and a Kähler potential have been chosen for the inflaton field, the Lagrangian terms and the effective potential for the inflaton follow.

As we shall see, viable inflationary theories can be constructed with simple choices of superpotentials and Kähler potentials. The positive value of the potential required for inflation may arise either from the first term in (2.156), the F-term, or from the second term in (2.156), the D-term, in the case that the inflaton is coupled to fields charged under a $U(1)$ gauge symmetry. Theories of chaotic inflation and hybrid inflation may also be constructed.

There are two potential problems arising from supergravity models of inflation which will be addressed in section 8.5 and section 8.6. The first is that gravitons (whose density was rendered negligible by inflation) may be produced by reheating after inflation. It is necessary to arrange that the reheating temperature is such that the gravitons do not have a serious effect on the abundances of deuterium and of ^{3}He relative to ^{4}He abundance.

The second problem (the so-called 'Polonyi' problem) results from the presence in supergravity theories of scalar fields with only gravitational strength interactions, which release the energy stored in their expectation values at very late times. This can lead to negligibly low helium and deuterium abundances at

226

temperatures too low for the necessary abundances to be recreated. It can also lead to too low a baryon number density.

8.2   Models of supergravity inflation

The key thing we require to study inflation is the effective potential for the inflaton field, which we assume to be a gauge-singlet scalar field ϕ. If we also assume minimal kinetic terms for ϕ arising from (2.151), then the effective potential is of the form

$$V = e^{\phi^*\phi} \left(\left| \frac{\partial W}{\partial \phi} + \phi^* W \right|^2 - 3|W|^2 \right) \tag{8.1}$$

as in (2.152), in units where M_P of (2.145) is one. In general [1], we may consider a superpotential which is a power series in ϕ,

$$W(\phi) = \mu^2 \sum_{n=0}^{\infty} \lambda_n \phi^n \tag{8.2}$$

where, as usual, we are not distinguishing notationally between the chiral superfield and the scalar field in that supermultiplet. When the expectation value of ϕ is real, the explicit effective potential corresponding to the superpotential (8.2) is (exercise 1)

$$\begin{aligned}
V = \mu^4 e^{\phi^2} [\lambda_1^2 &- 3\lambda_0 + 4\lambda_1(\lambda_2 - \lambda_0)\phi \\
&+ (\lambda_0^2 - \lambda_1^2 + 4\lambda_2^2 - 2\lambda_2\lambda_0 + 6\lambda_1\lambda_3)\phi^2 + 2(\lambda_1\lambda_0 + 6\lambda_2\lambda_3 + 4\lambda_1\lambda_4)\phi^3 \\
&+ (\lambda_1^2 + \lambda_2^2 + 9\lambda_3^2 + 2\lambda_2\lambda_0 + 2\lambda_1\lambda_3 + 2\lambda_0\lambda_4 + 16\lambda_2\lambda_4 + 10\lambda_1\lambda_5)\phi^4 \\
&+ \cdots].
\end{aligned} \tag{8.3}$$

A particularly simple case [2] is to take $W(\phi)$ quadratic in ϕ:

$$W(\phi) = \mu^2(\lambda_0 + \lambda_1\phi + \lambda_2\phi^2). \tag{8.4}$$

The form of $W(\phi)$ is further restricted by the requirement of the existence of a supersymmetry-preserving minimum of $V(\phi)$ with $V = 0$ for the following reasons. We have seen in section 7.7 that the energy scale of the inflationary potential is expected to be of order 10^{16}–10^{17} GeV. After inflation has occurred, ϕ rolls to a minimum of this potential. This minimum should have $V = 0$ because otherwise there would be a vacuum energy on the 10^{16}–10^{17} GeV scale which could not be cancelled by later supersymmetry breaking on the electroweak scale. There would then be a large cosmological constant. This minimum should be a supersymmetry-preserving minimum because otherwise there would be supersymmetry breaking on a scale too large for the hierarchy problem to be solved. Let this minimum be at $\phi = \sigma$. From (2.158), supersymmetry conservation requires

$$\frac{\partial W}{\partial \phi} + \phi W = 0. \tag{8.5}$$

From (8.1), $V = 0$ requires, in addition, that $W = 0$. Thus, we require that

$$\frac{\partial W}{\partial \phi} = W = 0 \qquad \text{at } \phi = \sigma. \tag{8.6}$$

In general, we require

$$\sum_{n=0}^{\infty} \lambda_n \sigma^n = 0 \qquad \text{and} \qquad \sum_{n=0}^{\infty} n \lambda_n \sigma^{n-1} = 0. \tag{8.7}$$

In the special case (8.4), these lead to

$$\sigma = -\frac{\lambda_1}{2\lambda_2} \qquad \text{and} \qquad 4\lambda_0 \lambda_2 = \lambda_1^2. \tag{8.8}$$

Then,

$$W(\phi) = \mu^2 \lambda_2 (\phi - \sigma)^2. \tag{8.9}$$

Then, from (8.1) the effective potential corresponding to (8.9) is (exercise 2)

$$V = e^{\phi^2} \mu^4 \lambda_2^2 [\phi^6 - 4\sigma\phi^5 + (6\sigma^2 + 1)\phi^4 - 4\sigma^3\phi^3 +$$
$$+ (\sigma^4 - 6\sigma^2 + 4)\phi^2 + 8\sigma(\sigma^2 - 1)\phi + \sigma^2(4 - 3\sigma^2)]. \tag{8.10}$$

Assuming that slow roll occurs from close to the origin, we need

$$V'(0) = 0. \tag{8.11}$$

Then

$$\sigma(\sigma^2 - 1) = 0. \tag{8.12}$$

For the minimum *not* to be at the origin (since ϕ must roll from a maximum), we do *not* want $\sigma = 0$. Thus, σ must be ± 1 and we take

$$\sigma = 1 \tag{8.13}$$

without loss of generality. Then (8.9) becomes

$$W(\phi) = \mu^3 \lambda_2 (\phi - 1)^2 \tag{8.14}$$

(still in units where $M_P = 1$) and (8.10) is

$$V(\phi) = e^{\phi^2} \mu^4 \lambda_2^2 (1 - \phi^2 - 4\phi^3 + 7\phi^4 - 4\phi^5 + \phi^6). \tag{8.15}$$

Expanding e^{ϕ^2} in powers of ϕ^2 for the purpose of studying the slow-roll region close to the origin,

$$V(\phi) = \mu^4 \lambda_2^2 (1 - 4\phi^3 + \tfrac{13}{2}\phi^4 - 8\phi^5 + \tfrac{23}{3}\phi^6 + \cdots). \qquad (8.16)$$

It follows that

$$V_0 \equiv V(0) = \mu^4 \lambda_2^2. \qquad (8.17)$$

Then, the Hubble parameter H relevant to slow roll is given by (7.41) and, in units with $M_P = 1$,

$$H^2 \simeq H_0^2 = \tfrac{1}{3} V_0 = \tfrac{1}{3} \mu^4 \lambda_2^2 \qquad (8.18)$$

recalling that $m_P^2 = 8\pi M_P^2$. Also, in units with $M_P = 1$, the slow-roll condition (7.45) is

$$\frac{|V''(\phi)|}{V(\phi)} \ll 3 \qquad (8.19)$$

and the slow-roll condition (7.46) is

$$\left(\frac{V'(\phi)}{V(\phi)} \right)^2 \ll 6. \qquad (8.20)$$

With $V(\phi)$ given by (8.16) and with ϕ close to zero, (8.19) requires that ϕ is in the range

$$0 \lesssim \phi \lesssim \frac{M_P}{8} \equiv \phi_e \qquad (8.21)$$

(8.20) is automatically satisfied whenever $V''(\phi) \sim V'(\phi)/\phi$ and $|\phi|$ is at least an order of magnitude less than m_P. We certainly satisfy the latter requirement because $\phi_e = \tfrac{1}{8} M_P$ and, for small ϕ, the former condition is also satisfied.

To avoid de Sitter fluctuations driving ϕ across the flat region too rapidly (faster than it would roll semi-classically) we need a width $\Delta\phi = \phi_e$ for the flat region with the property (7.91). With the Hubble constant given by (8.18), then

$$\mu^4 \lambda_2^2 < \frac{3\pi^2}{16N_e}. \qquad (8.22)$$

If we are able to arrange that $N_e \sim 64$, then this requires that

$$\mu^2 |\lambda_2| < 0.17 \qquad (8.23)$$

or, equivalently,

$$|H_0| < 0.098. \qquad (8.24)$$

Turning next to the number of e-folds of inflation, (7.57) requires that

$$N_e \sim \frac{3H^2}{|V''(\phi_i)|}. \qquad (8.25)$$

To allow for quantum fluctuations $V''(\phi_i)$ should be evaluated at $\phi_i \sim H_0$ rather than $\phi_i \sim 0$. (See, for example, [3].) A rough estimate may be made by keeping only the term linear in ϕ in $V'(\phi)$. Then,

$$N_e \sim \frac{|H_0|}{8\mu^4\lambda_2^2} \sim \frac{1}{8\sqrt{3}\mu^2|\lambda_2|}. \tag{8.26}$$

$N_e \sim 64$ is achieved for

$$\mu^2|\lambda_2| \simeq 1.13 \times 10^{-3}. \tag{8.27}$$

This value of $\mu^2|\lambda_2|$ satisfies with ease the bound (8.23) to avoid de Sitter fluctuations driving ϕ across the flat region too rapidly.

Using (7.112) and (7.113), density fluctuations of the order required by the COBE observations imply that

$$-\frac{H^2(t^*)}{\pi^{3/2}\dot{\phi}_0(t^*)} \sim 2 \times 10^{-5} \tag{8.28}$$

(see [3]). We certainly have $t^* \lesssim t_e$, the time at which slow roll ends. From (7.34),

$$\dot{\phi} = -\frac{V'(\phi)}{3H}. \tag{8.29}$$

If we make a rough estimate of $V'(\phi)$ from the term quadratic in ϕ, then

$$V'(\phi_e) = -12\mu^4\lambda_2^2\phi_e^2. \tag{8.30}$$

If we also approximate $H(t_e)$ by H_0 of (8.18),

$$\dot{\phi}(t_e) \simeq 4\sqrt{3}\mu^2|\lambda_2|\phi_e^2. \tag{8.31}$$

With these approximations,

$$-\frac{H^2(t_e)}{\pi^{3/2}\dot{\phi}(t_e)} \simeq -\frac{\mu^2|\lambda_2|}{12\sqrt{3}\mu^2|\lambda_2|\phi_e^2}. \tag{8.32}$$

With $t \sim t_e$ and $\phi_e \sim 1/8$, the value of $\mu^2|\lambda_2|$ consistent with the density fluctuations (8.28) is

$$\mu^2|\lambda_2| \simeq 3.6 \times 10^{-5}. \tag{8.33}$$

This value of $\mu^2|\lambda_2|$ is sufficiently small that we get from (8.26) more than 64 e-folds of inflation with ease. The condition (8.22) to avoid de Sitter fluctuations driving ϕ too rapidly across the flat region is also satisfied with ease. For smaller values of t^*, and so of $\phi(t^*)$, $\mu^2|\lambda_2|$ is even smaller.

To decide whether sufficient inflation occurs in practice, it is also necessary to consider the initial conditions, because sufficient inflation depends on ϕ starting rolling across the flat region between $\phi = 0$ and $\phi = \phi_e$ from a value of ϕ sufficiently close to $\phi = 0$. However, if initially the universe was in thermal equilibrium, then thermal effects may put ϕ well into the region $\phi > 0$ and prevent enough inflation. This is referred to as the 'thermal constraint' [4]. In general, for minimal kinetic terms the zero-temperature effective potential V derived from (2.147) is

$$V = e^G(G_i G^i - 3) \tag{8.34}$$

and from (2.162), the finite-temperature correction $\bar{V}_1^T$ to the effective potential is given by

$$\bar{V}_1^T = \text{constant} + \frac{N}{12} T^2 e^G (G_i G^i - 2) \tag{8.35}$$

in the limit of a large number N of chiral fields, which is a reasonable approximation in practice. If the finite-temperature effects are not to destroy the flatness of the potential, we must require that

$$\frac{\partial \bar{V}_1^T}{\partial \phi} = 0 = \frac{\partial V}{\partial \phi} \qquad \text{at } \phi = 0. \tag{8.36}$$

For a single gauge-singlet real scalar field ϕ with minimal kinetic terms, so that $G_i = G^i = G'(\phi)$, these require that

$$G'(0) = 0 \tag{8.37}$$

so that

$$V(0) < 0. \tag{8.38}$$

It is then impossible for ϕ to roll to a (supersymmetry-preserving) minimum with $V = 0$. Thus, the thermal constraint is a very powerful constraint. It may sometimes be evaded if the inflaton has non-minimal kinetic terms.

However, if a (weakly-coupled) inflaton field ϕ is out of thermal equilibrium for temperatures below the Planck scale, then the initial value of ϕ will, in general, have a broad distribution [5]. Consequently, it is unlikely that a randomly chosen horizon volume will possess a (smoothed-out) value of ϕ close enough to $\phi = 0$ for much inflation to occur. However, any horizon volume which *does* have a value of ϕ close to $\phi = 0$ will undergo inflation and, after inflation has occurred, most of space will be occupied by such regions. As a result, we are very likely to find ourselves in a region of space which derived from such a horizon volume at early times. For this reason, we shall not consider ourselves bound by the thermal constraint.

We consider next reheating in the context of this simple supergravity model. For a gauge-singlet inflaton field with only gravitational strength couplings, we expect a decay rate

$$\Gamma_\phi \sim \frac{m_\phi^3}{M_P^2}. \tag{8.39}$$

Also, from the double derivative of the effective potential, a mass-squared m_ϕ^2 of the inflaton of order

$$m_\phi^2 \sim \frac{\mu^4 \lambda_2^2}{M_P^5} \tag{8.40}$$

is to be expected, allowing for $\langle \phi \rangle \sim M_P$ owing to the effects of quantum gravity and restoring factors of M_P. Thus,

$$\Gamma_\phi \sim \frac{\mu^6 \lambda_2^3}{M_P^5}. \tag{8.41}$$

The reheating temperature of (7.72) is

$$T_R \sim (\Gamma_\phi M_P)^{1/2} \sim \frac{\mu^3 \lambda_2^{3/2}}{M_P^2} \tag{8.42}$$

ignoring the difference between m_P and M_P. Using the estimate (8.33), this gives

$$T_R \sim 10^{11} \text{ GeV}. \tag{8.43}$$

8.3   *D*-term supergravity inflation

To arrange for sufficient inflation in the model of the previous section, it was necessary to take $\mu^2 |\lambda_2| \sim 10^{-3}$, which is an unnatural fine-tuning of the superpotential. This is a generic feature of supergravity models where the positive value of the effective potential during inflation is due to a non-zero F-term in the sense that $|\frac{\partial W}{\partial \phi_i} + \phi^{i*} W|^2 \neq 0$ in (2.156). (The terminology is because $\partial W / \partial \phi_i + \phi^{i*} W$ is the generalization to the supergravity context of the auxiliary field usually denoted by F^i in the construction of the globally-supersymmetric Lagrangian.) When all relevant fields are gauge singlet and assuming minimal kinetic terms, the effective potential for the inflaton takes the form (8.1). There is a term quadratic in ϕ, namely $V_0 \phi^* \phi$, where $V_0 \equiv V(0)$. Keeping only this term and assuming a real inflaton,

$$V''(\phi) \sim V_0 \tag{8.44}$$

and then, from (8.25), the number of e-folds of inflation is

$$N_e \sim \frac{3H^2}{|V_0|}. \tag{8.45}$$

But when $\phi \simeq \phi_0 \simeq 0$,

$$H^2 \simeq \tfrac{1}{3} V_0. \tag{8.46}$$

Thus,

$$N_e \sim 1. \tag{8.47}$$

To obtain sufficient e-folds of inflation generally requires some fine-tuning of the parameters of the superpotential, so that other quadratic terms can cancel the one displayed. In the model just discussed in section 8.2, the choice of superpotential is such that a quadratic term in $V(\phi)$ does not occur. Nevertheless, there is a similar problem because sufficient inflation required the parameter $\mu^2|\lambda_2|$ in the superpotential to be $\lesssim 10^{-3}$.

This generic difficulty for F-term inflation can be avoided if the positive value of V required for inflation originates from the second term of (2.156) (referred to as the D-term because it generalizes the contribution of the auxiliary field denoted by D in the context of global supersymmetry). Because there is no factor of $e^{\phi^*\phi}$ in the D-term, the previous argument does not apply and D-term inflation [6] does not suffer from the generic problem discussed above.

A simple example [6] is provided by the superpotential

$$W = \lambda\phi\phi_+\phi_- \tag{8.48}$$

where ϕ is the inflaton field and $\phi_\pm$ are two other scalar fields with charges ± 1 under a $U(1)$ gauge symmetry. Let the Kähler potential K be chosen to have the minimal form, as in (2.151), and let the gauge kinetic function be minimal, as in (2.154). Then

$$G = \phi^*\phi + \phi_+^*\phi_+ + \phi_-^*\phi_- + \ln|W|^2 \tag{8.49}$$

and (exercise 3), using (2.156), the effective potential is

$$V = e^{|\phi|^2+|\phi_+|^2+|\phi_-|^2}\lambda^2[|\phi_+\phi_-|^2(1+|\phi|^2) + |\phi\phi_-|^2(1+|\phi_+|^2) \\ + |\phi\phi_+|^2(1+|\phi_-|^2)] + \tfrac{1}{8}g^2(|\phi_+|^2 - |\phi_-|^2 - \xi)^2. \tag{8.50}$$

A constant ξ (the Fayet–Iliopoulos) term has been included, which can be present for a $U(1)$ gauge symmetry. We assume that $\xi > 0$. It may be checked (exercise 4) that $(\phi_+, \phi_-) = (0, 0)$ is a minimum in the (ϕ_+, ϕ_-) space when the inflaton field ϕ satisfies

$$|\phi| > \frac{g\sqrt{\xi}}{\lambda} \equiv \phi_c. \tag{8.51}$$

For these minima, we see that

$$V = \tfrac{1}{2}g^2\xi^2. \tag{8.52}$$

Thus, the potential is then flat so far as the inflaton is concerned (and has a large positive curvature in the ϕ_+ and ϕ_- directions).

In a chaotic inflation scenario, we may assume the initial condition $|\phi| \gtrsim \phi_c$. Then inflation will occur. The amount of inflation will be very large because the only lack of flatness in the effective potential, so far as the inflaton is concerned, is due to radiative corrections. Note that the positive value of V driving inflation is essentially due to the D-term.

8.4 Hybrid inflation in supergravity

The hybrid inflation idea discussed in section 7.11 may be extended to the context of supergravity. A simple superpotential which allows hybrid inflation to be implemented [7] is

$$W = \phi(\lambda \psi_1 \psi_2 - \mu^2) \tag{8.53}$$

where ψ_1 and ψ_2 are a pair of (chiral) superfields in non-trivial conjugate representations of some non-Abelian gauge group and ϕ is a superfield neutral under any gauge group. (We use the same notation for superfields and the scalar field associated with them.) Assuming minimal kinetic terms, as in (2.151) and (2.156), the corresponding effective potential apart from the last (D-) terms in (2.156) deriving from the non-trivial gauge properties of ψ_1 and ψ_2, is as follows

$$V = e^{|\phi|^2}(|F_\phi|^2 + |F_{\psi_1}|^2 + |F_{\psi_2}|^2 - 3|W|^2) \tag{8.54}$$

where

$$F_\phi \equiv \frac{\partial W}{\partial \phi} + \phi^* W = (1 + |\phi|^2)(\lambda \psi_1 \psi_2 - \mu^2) \tag{8.55}$$

$$F_{\psi_1} \equiv \frac{\partial W}{\partial \psi_1} + \psi_1^* W = \lambda \phi(1 + |\psi_1|^2)\psi_2 - \mu^2 \psi_1^* \phi \tag{8.56}$$

$$F_{\psi_2} \equiv \frac{\partial W}{\partial \psi_2} + \psi_2^* W = \lambda \phi(1 + |\psi_2|^2)\psi_1 - \mu^2 \psi_2^* \phi. \tag{8.57}$$

If ψ_1 and ψ_2 roll rapidly to zero, then the effective potential for ϕ is

$$V = e^{\phi^2}\mu^4(1 - \phi^2 + \phi^4) \tag{8.58}$$

since ϕ is a real scalar field being neutral under any gauge group. Expanding in powers of ϕ^2,

$$V \simeq \mu^4(1 + \tfrac{1}{2}\phi^4). \tag{8.59}$$

The cancellation of the quadratic term in ϕ evades the generic problem with F-term inflation discussed in the previous section.

It may be seen by returning to the globally supersymmetric theory that it is indeed reasonable to take ψ_1 and ψ_2 fixed to zero. The globally supersymmetric effective potential is (following (2.120))

$$V = \left|\frac{\partial W}{\partial \phi}\right|^2 + \left|\frac{\partial W}{\partial \psi_1}\right|^2 + \left|\frac{\partial W}{\partial \psi_2}\right|^2 \tag{8.60}$$

$$= |\lambda \psi_1 \psi_2 - \mu^2|^2 + \lambda^2 \phi^2(|\psi_1|^2 + |\psi_2|^2) \tag{8.61}$$

(apart from the D-terms for the gauge non-singlets ψ_1 and ψ_2 and remembering that ϕ is neutral.) The absolute minimum of the effective potential is at

$$\phi = 0, \qquad \psi_1 = \psi_2 = \frac{\mu}{\sqrt{\lambda}}. \tag{8.62}$$

However, if

$$\phi > \frac{\mu}{\sqrt{\lambda}} \equiv \phi_c \qquad (8.63)$$

the fields ψ_1 and ψ_2 have positive effective squared masses and are confined to $\psi_1 = \psi_2 = 0$. The effective masses of ψ_1 and ψ_2 are contained in the terms

$$-\mu^2\lambda(\psi_1\psi_2 + \psi_1^*\psi_2^*) + \lambda^2\phi^2(\psi_1\psi_1^* + \psi_2\psi_2^*). \qquad (8.64)$$

Writing

$$\psi_1 \equiv \frac{1}{\sqrt{2}}(A_1 + iB_1) \qquad \text{and} \qquad \psi_2 \equiv \frac{1}{\sqrt{2}}(A_2 + iB_2) \qquad (8.65)$$

the effective mass terms are

$$\tfrac{1}{2}\lambda^2\phi^2(A_1^2 + B_1^2 + A_2^2 + B_2^2) - \mu^2\lambda(A_1 A_2 - B_1 B_2) \qquad (8.66)$$

and the mass-squared eigenvalues are $(\lambda^2\phi^2 \pm \mu^2\lambda)/2$, both of which are positive when $\phi > \phi_c$.

Now the model for inflation reduces to one with a single real scalar inflaton with potential (8.59). The slow-roll conditions (7.45) and (7.46), in units where the reduced Planck mass M_P is 1, are satisfied when

$$\frac{|V''(\phi)|}{V(\phi)} \ll 3 \qquad \text{and} \qquad \left(\frac{V'(\phi)}{V(\phi)}\right)^2 \ll 6. \qquad (8.67)$$

With the above potential, these give $\phi^2 \ll 1/2$ and $\phi^2 \ll 1.15$ respectively. Thus, the slow-roll region is

$$\phi^2 \ll \tfrac{1}{2}. \qquad (8.68)$$

To calculate the number of e-folds of inflation, it is necessary to consider the time dependence of ϕ during slow roll. In units where $M_P = 1$, (7.41) is $H^2 = V/3$ and in the slow-roll region $V(\phi) \simeq \mu^4$, so that

$$H \simeq \frac{\mu^2}{\sqrt{3}}. \qquad (8.69)$$

Then (7.34) is

$$\dot{\phi} = -\frac{2}{\sqrt{3}}\mu^2\phi^3 \qquad (8.70)$$

which shows that ϕ is decreasing with t for $\phi > 0$. In $M_P = 1$ units, (7.51) gives

$$N_e = -\int_{\phi_i}^{\phi_f} \frac{V(\phi)}{V'(\phi)} \, d\phi \qquad (8.71)$$

$$\simeq \tfrac{1}{4}(\phi_f^{-2} - \phi_i^{-2}) \qquad (8.72)$$

so, for $\phi_f \ll \phi_i$, we have

$$N_e \simeq \tfrac{1}{4}\phi_f^{-2}. \tag{8.73}$$

Thus, what sets the limit on the number of e-folds of inflation is how small ϕ_f can get. There is no limit set by slow roll because we have slow roll whenever $\phi^2 \ll 1/2$. However, there are radiative corrections to the effective potential that we should now take into account. At one-loop order, they are of the form

$$V_{\text{1-loop}} = \frac{\lambda^2 \mu^4}{8\pi^2} \ln \frac{\phi}{\phi_c} \tag{8.74}$$

so that

$$V'_{\text{1-loop}} = \frac{\lambda^2 \mu^4}{8\pi^2} \phi^{-1}. \tag{8.75}$$

This is of the same order as $V'(\phi)$ from the supergravity potential (8.59) (without radiative corrections) when

$$\phi \simeq \sqrt{\frac{\lambda}{4\pi}}. \tag{8.76}$$

For smaller values of ϕ, the contribution to $V'(\phi)$ in the slow-roll equation due to radiative corrections is larger than $V'(\phi)$ from supergravity at tree level. Thus, we must truncate the contribution to N_e from rolling in the uncorrected supergravity potential at

$$\phi_f \simeq \sqrt{\frac{\lambda}{4\pi}}. \tag{8.77}$$

Then the contribution to N_e from the period of slow roll before radiative corrections become important is

$$N_e \simeq \pi\lambda^{-1}. \tag{8.78}$$

This gives at least 64 e-folds of inflation (even without including any further slow rolling when radiative corrections have become important) for

$$\lambda \lesssim 0.05. \tag{8.79}$$

Recalling that $M_P^2 = m_P^2/8\pi$, (7.114) and (8.59) lead to

$$\frac{\delta\rho}{\rho} \simeq \frac{\mu^2}{2(8\pi)^{3/2}\phi^3(t^*)} \tag{8.80}$$

in units where $M_P = 1$. If we estimate t^* as the time at which we can no longer neglect radiative corrections, then from (8.76)

$$\phi(t^*) \simeq \sqrt{\frac{\lambda}{4\pi}} \tag{8.81}$$

and

$$\frac{\delta\rho}{\rho} \simeq \frac{\mu^2}{2^{5/2}(4\pi)^3\lambda^{3/2}}. \tag{8.82}$$

The requirement that $\delta\rho/\rho \simeq 2 \times 10^{-5}$ fixes μ in units of M_P once λ has been chosen. For λ chosen as in (8.79), we have

$$\mu \lesssim 0.05 M_P. \tag{8.83}$$

8.5 Thermal production of gravitinos by reheating

The thermal production of gravitinos in the early universe can cause problems, as discussed in section 6.3. At first sight, a possible solution to these problems is for the gravitino density to be diluted by inflation. However, a gravitino density can be produced by reheating after inflation and it is necessary for this density to be low enough that the problem is not recreated.

Gravitinos produced by reheating after inflation can have a serious effect on the abundances of deuterium (D) and ^{3}He relative to the ^{4}He abundance. The problem is that D and ^{3}He can be produced by photofission from ^{4}He by radiation from gravitino decay. These relative abundances are known to be very small and so we must avoid gravitino densities sufficiently large to violate these bounds.

The gravitino density $n_{3/2}$ produced during reheating by $2 \rightarrow 2$ scattering processes involving gauge bosons and gauginos has been estimated [8] to be given by

$$\frac{n_{3/2}}{n_\gamma} \simeq 2 \times 10^{-13} \left(\frac{T_R}{10^9 \text{ GeV}}\right) \tag{8.84}$$

where T_R is the reheating temperature. However, an estimate of the amount of D and ^{3}He produced by photofission from ^{4}He requires that

$$m_{3/2}\frac{n_{3/2}}{n_\gamma} \lesssim 3 \times 10^{-12} \text{ GeV}. \tag{8.85}$$

Thus, there is a bound on the reheating temperature:

$$T_R \lesssim \frac{1.5 \times 10^{10} \text{ (GeV)}^2}{m_{3/2}}. \tag{8.86}$$

For example, for $m_{3/2} = 100$ GeV, $T_R \lesssim 1.5 \times 10^8$ GeV. Subsequent calculations [9] have shown that the bound is less stringent than this formula suggests for larger values of $m_{3/2}$, e.g. for $m_{3/2} = 1$ TeV, $T_R \lesssim 2 \times 10^9$ GeV. There is also the danger of excessive gravitino production by decay of the inflaton. This is a very model-dependent matter but sufficient suppression can occur in particular models [10].

8.6   The Polonyi problem

This generic problem results from the presence in a theory of a light scalar field which has only gravitational strength interactions, with the consequence that it is decoupled during most of the history of the universe and eventually releases energy stored in its expectation value at a very late time [11]. This release of energy increases the entropy of the universe at a low temperature thereby producing a negligible baryon abundance and, worse still, negligible helium and deuterium abundances. The temperature may then be too low for the required abundances to be recreated.

A simple example of this problem, from which it derives its name, occurs in the context of the Polonyi model for supersymmetry breaking in supergravity. We describe first this mechanism for supersymmetry breaking. The Polonyi model is an example of a model in which supersymmetry breaking occurs in a 'hidden sector', by which is meant a sector of the theory which couples to the 'observable sector' of quarks, leptons, gauge fields, Higgs scalars, and their supersymmetric partners only through gravitational interactions. The hidden sector of the Polonyi model employs a single gauge-singlet scalar field $\tilde{\phi}$ (*not* the inflaton) and its supersymmetric fermionic partner with superpotential

$$\tilde{W}(\tilde{\phi}) = \tilde{\mu}^2(\tilde{\phi} + \beta). \tag{8.87}$$

(We are using the same notation for the chiral superfield and its scalar field component.) Minimal kinetic terms are chosen so that G of (2.144) has the form

$$G = \tilde{\phi}^*\tilde{\phi} + \ln|\tilde{W}|^2. \tag{8.88}$$

In (8.87), $\tilde{\mu}$ is a real parameter with dimensions of mass and

$$\beta = 2 - \sqrt{3} \tag{8.89}$$

in units where $M_P = 1$. The parameter β has been fixed to this value so that the effective potential of (2.147)

$$V = \tilde{\mu}^4 e^{\tilde{\phi}^*\tilde{\phi}}(|1 + \tilde{\phi}^*(\tilde{\phi} + \beta)|^2 - 3|\tilde{\phi} + \beta|^2) \tag{8.90}$$

has its absolute minimum at

$$\tilde{\phi} = \sqrt{3} - 1 \tag{8.91}$$

with $V = 0$ and, so, the desirable feature of a vanishing cosmological constant in the physical vacuum. At this minimum,

$$\frac{\partial \tilde{W}}{\partial \tilde{\phi}} + \tilde{\phi}^*\tilde{W} = \sqrt{3}\tilde{\mu}^2 \neq 0. \tag{8.92}$$

Consequently, supersymmetry is broken, as discussed after (2.152).

If the field $\tilde{\phi}$ starts (because of quantum fluctuations) at some value of $\tilde{\phi}$ which differs from the minimum, then the energy density stored in the expectation value of $\tilde{\phi}$ is of order $\tilde{\mu}^4$. The size of $\tilde{\mu}$ is related to the size of supersymmetry breaking effects. In supergravity theories with supersymmetry breaking in a hidden sector, the size of supersymmetry breaking effects transmitted gravitationally to the observable sector is on the scale of the gravitino mass $m_{3/2}$ and, for supersymmetry to solve the hierarchy problem, $m_{3/2}$ should be about 100 GeV to 10 TeV. In the Polonyi model, the gravitino mass is

$$m_{3/2} = \frac{\tilde{\mu}^2}{M_P} e^{(\sqrt{3}-1)^2/2}. \tag{8.93}$$

Thus, for $m_{3/2}$ in the range 100 GeV to 10 TeV, we have

$$10^{10} \text{ GeV} \le \tilde{\mu} \le 10^{11} \text{ GeV}. \tag{8.94}$$

(For a discussion of the gravitino mass in supergravity theories with hidden-sector supersymmetry breaking see, for example, [12].)

For the discussion of the entropy increase of the universe when the Polonyi field vacuum energy decays, we shall need to know the expectation value of the Polonyi field. In the context of cosmological inflation, we should determine this expectation value by minimizing the total effective potential of the Polonyi field and the inflaton. (There may also be effects of quantum fluctuations.) The superpotential of the Polonyi field $\tilde{\phi}$ is as in (8.87) and, for definiteness, we may take the superpotential for the inflaton field ϕ to be

$$W(\phi) = \mu^2 \lambda_2 (\phi - \sigma)^2 \tag{8.95}$$

as in (8.9). Thus, the total superpotential is

$$W_{\text{tot}}(\phi, \tilde{\phi}) = \tilde{W}(\tilde{\phi}) + W(\phi). \tag{8.96}$$

We also assume minimal kinetic terms so that

$$G = \tilde{\phi}^* \tilde{\phi} + \phi^* \phi + \ln |W_{\text{tot}}|^2. \tag{8.97}$$

Then the effective potential can be calculated from (2.147) in units where the reduced Planck mass $M_P = 1$. Taking $\tilde{\phi}$ to be real, working to quadratic order in $\tilde{\phi}$ (when $\tilde{\phi} \ll 1$ in the same units), and remembering that $\phi \simeq 0$ during slow roll, the minimum of the effective potential may be estimated to be (exercise 5)

$$\tilde{\phi} \simeq \frac{2\mu^2 \tilde{\mu} \lambda_2 \sigma^2}{\mu^4 \lambda_2^2 - 4\beta \mu^2 \tilde{\mu} \lambda_2 \sigma^2}. \tag{8.98}$$

The values of the parameters of the Polonyi model are given by (8.89) and (8.94), and the parameters of the superstring model of inflation that we are employing by (8.13) and (8.33). Using these values, $\tilde{\phi}$ may be estimated to be

$$\tilde{\phi} \simeq 3(10^{-3}-10^{-4}). \tag{8.99}$$

Table 8.1. Event sequence when the inflaton vacuum energy decays before the Polonyi field oscillates.

Time (t)	$\rho_\phi(t)$	$\rho_{\tilde\phi}(t)$	$\rho_{\rm rad}(t)$
$t_f < t < t_D \sim \Gamma_\phi^{-1}$	$\sim t^{-2}$	$\simeq \rho_{\tilde\phi}(t_f)$	0
$t = t_D \sim \Gamma_\phi^{-1}$			$\rho_\phi(t_D) \to \rho_{\rm rad}(t_D)$ $T_R \sim m_\phi^{3/2} M_P^{-1/2}$
$t_D < t < t_{\tilde\phi} \sim m_{\tilde\phi}^{-1}$	0	$\simeq \rho_{\tilde\phi}(t_f)$	$\sim T^4$
$t_{\tilde\phi} < t < \tilde t_D \sim \Gamma_{\tilde\phi}^{-1}$		$\sim T^3$	$\sim T^4$
$t = \tilde t_D \sim \Gamma_{\tilde\phi}^{-1}$		$\rho_{\tilde\phi}(\tilde t_D) \to$ radiation $\tilde T_R \sim m_{\tilde\phi}^{3/2} M_P^{-1/2}$	

We shall assume in what follows that, in reduced Planck-scale units,

$$\tilde\phi = f \tag{8.100}$$

where f is not many orders of magnitude less than unity. Considering the effects of quantum fluctuations will also lead to this conclusion.

To proceed further, we need to specify the temperature at which the inflaton vacuum energy decays. The crucial thing is whether this is before or after the Polonyi field starts to oscillate [13, 14].

8.6.1 Inflaton decays before Polonyi field oscillation

Let us consider first the case when the inflaton has already decayed and reheated the universe before the Polonyi field starts to oscillate. Then, the sequence of events is summarized in table 8.1. Inflation ends at $t = t_f$ and, at that time in the present model, the Polonyi field vacuum energy density is

$$\rho_{\tilde\phi}(t_f) \simeq \tfrac{1}{2} m_{\tilde\phi}^2 \tilde\phi^2 \tag{8.101}$$

provided that $\tilde\phi$ is significantly less than 1 in reduced Planck-scale units. Also, the inflaton vacuum energy density is

$$\rho_\phi(t_f) \simeq \mu^4 \lambda_2^2 \tag{8.102}$$

from (8.17), because ϕ does not roll much during inflation. The value of the Polonyi mass-squared is determined by

$$m_{\tilde\phi}^2 = \frac{\partial^2 V_{\rm eff}}{\partial \tilde\phi^2} = \beta^2 \tilde\mu^2 \tag{8.103}$$

where, from (8.93),

$$\tilde{\mu} \sim m_{3/2}^{1/2} \tag{8.104}$$

in the present units and β is given by (8.89).

For $t_{\tilde{\phi}} < t < \tilde{t}_D$, where the Polonyi field starts to oscillate at $t = t_{\tilde{\phi}}$ when the temperature is $T_{\tilde{\phi}}$, and its vacuum energy density decays at $\tilde{t}_D$,

$$\frac{\rho_{\tilde{\phi}}(t)}{\rho_{\text{rad}}(t)} = \frac{\rho_{\tilde{\phi}}(t_{\tilde{\phi}})}{\rho_{\text{rad}}(t_{\tilde{\phi}})} \frac{T_{\tilde{\phi}}}{T} = \frac{\rho_{\tilde{\phi}}(t_f)}{\rho_{\text{rad}}(t_{\tilde{\phi}})} \frac{T_{\tilde{\phi}}}{T} \tag{8.105}$$

is the ratio of the Polonyi field vacuum energy density to the radiation energy density ρ_{rad}. Since

$$\rho_{\text{rad}}(t_{\tilde{\phi}}) \sim T_{\tilde{\phi}}^4 \tag{8.106}$$

with the approximation (8.101), we have

$$\frac{\rho_{\tilde{\phi}}(t)}{\rho_{\text{rad}}(t)} \sim \frac{m_{\phi}^2 \tilde{\phi}^2}{2T_{\tilde{\phi}}^3 T}. \tag{8.107}$$

Further progress requires a calculation of $T_{\tilde{\phi}}$.

Between $t = t_D$, the time at which the inflaton vacuum energy density decays, when the temperature is T_D, and $t = t_{\tilde{\phi}}$, the universe is radiation dominated. Thus, in reduced Planck-scale units,

$$\left(\frac{\dot{T}}{T}\right)^2 = \left(\frac{\dot{R}}{R}\right)^2 = \frac{1}{3}\rho_{\text{rad}}(T) = \frac{1}{3}\rho_{\text{rad}}(T_D)\left(\frac{T}{T_D}\right)^4 \tag{8.108}$$

where we have used the fact that RT is constant whenever the number of particle species is constant, for conservation of entropy. This equation has the solution

$$t = \frac{\rho_{\text{rad}}(T_D)}{6T_D^4 T^2}. \tag{8.109}$$

At $t = t_{\tilde{\phi}}$, $\rho_{\text{rad}}(T_D)$ is one or two orders of magnitude larger than T_D^4, using (2.22) with $N_B + \frac{7}{8}N_F = \frac{915}{4}$ for the supersymmetric standard model or $\frac{427}{4}$ for the standard model respectively. Also, since $t_{\tilde{\phi}} \sim m_{\tilde{\phi}}^{-1}$, we have

$$T_{\tilde{\phi}} \sim m_{\tilde{\phi}}^{1/2}. \tag{8.110}$$

Returning to (8.107), for $t_{\tilde{\phi}} < t < \tilde{t}_D$,

$$\frac{\rho_{\tilde{\phi}}(t)}{\rho_{\text{rad}}(t)} \sim \frac{m_{\tilde{\phi}}^{1/2} \tilde{\phi}^2}{T}. \tag{8.111}$$

If the Polonyi field were not to decay, then at $T = 2.7 \text{ K} \simeq 10^{-31} M_P$, we would have

$$\frac{\rho_{\tilde{\phi}}(t)}{\rho_{\text{rad}}(t)} \sim 10^{31} m_{\tilde{\phi}}^{1/2} \tilde{\phi}^2. \tag{8.112}$$

If we want to avoid $\rho_{\tilde{\phi}}(t)/\rho_{\text{rad}}(t) > 1$, so that the $\tilde{\phi}$ energy density does not dominate the energy density of the universe and produce too large an expansion rate, we need

$$\tilde{\phi} \lesssim (10^{-15} - 10^{-16}) m_{\tilde{\phi}}^{-1/4}. \tag{8.113}$$

With $m_{\tilde{\phi}}$ given by (8.103) and (8.93) and for

$$m_{3/2} = 10^{-15} - 10^{-16} \tag{8.114}$$

in reduced Planck-scale units, we then require

$$\tilde{\phi} \lesssim 10^{-13} - 10^{-14}. \tag{8.115}$$

Comparing this with (8.100), this is an unnaturally small value by many orders of magnitude. Making (8.106) more precise only increases this by an order of magnitude.

In practice, the Polonyi vacuum energy decays but the previous discussion suggests that there may be a problem with entropy production when it does decay. The decay occurs at temperature $\tilde{T}_D$ with

$$t \sim \tilde{t}_D \sim \Gamma_{\tilde{\phi}}^{-1} \tag{8.116}$$

with

$$\Gamma_{\tilde{\phi}}^{-1} \sim m_{\tilde{\phi}}^3 \tag{8.117}$$

and then, from (8.111),

$$\frac{\rho_{\tilde{\phi}}(\tilde{t}_D)}{\rho_{\text{rad}}(\tilde{t}_D)} \sim \frac{m_{\tilde{\phi}}^{1/2} \tilde{\phi}^2}{\tilde{T}_D}. \tag{8.118}$$

We now require an estimate of the Polonyi field decay temperature $\tilde{T}_D$. Remembering that, between $t = t_{\tilde{\phi}}$ and $t = \tilde{t}_D$, the vacuum energy density for $\tilde{\phi}$ behaves like a gas of free non-relativistic particles behaving as in (7.64), $\rho_{\tilde{\phi}}(t)$ is growing relative to $\rho_{\text{rad}}(t)$. Consequently, we may expect $\rho_{\tilde{\phi}}$ to dominate the energy density of the universe by $t = \tilde{t}_D$. We therefore approximate the time dependence of the temperature by a universe dominated by the Polonyi field energy density $\rho_{\tilde{\phi}}$. (Recall that the inflaton vacuum energy density has already decayed and been converted to radiation.) Then we have to solve

$$\left(\frac{\dot{T}}{T}\right)^2 = \left(\frac{\dot{R}}{R}\right)^2 = \frac{1}{3} \rho_{\tilde{\phi}}(t_{\tilde{\phi}}) \left(\frac{T}{T_{\tilde{\phi}}}\right)^3 \tag{8.119}$$

with the result that

$$t \sim \left(\frac{\rho_{\tilde{\phi}}(t_{\tilde{\phi}})}{T_{\tilde{\phi}}}\right)^{-1/2} T^{-3/2}. \tag{8.120}$$

At $t = \tilde{t}_D$, and with $\tilde{t}_D$ given by (8.116), we get

$$\tilde{T}_D \sim m_{\tilde{\phi}}^2 \rho_{\tilde{\phi}}^{-1/3}(t_{\tilde{\phi}}).T_{\tilde{\phi}} \tag{8.121}$$

Using (8.111), and with $\rho_{\tilde{\phi}}(t_{\tilde{\phi}}) = \rho_{\tilde{\phi}}(t_f)$ given by (8.101), we have

$$\tilde{T}_D \sim m_{\tilde{\phi}}^{11/6} \tilde{\phi}^{-2/3}. \tag{8.122}$$

Substituting this into (8.118) gives

$$\frac{\rho_{\tilde{\phi}}(\tilde{t}_D)}{\rho_{\text{rad}}(\tilde{t}_D)} \sim m_{\tilde{\phi}}^{-4/3} \tilde{\phi}^{8/3}. \tag{8.123}$$

Consequently, the $\tilde{\phi}$ vacuum energy density dominates the density of the universe at the moment of decay provided

$$\tilde{\phi} > m_{\tilde{\phi}}^{1/2}. \tag{8.124}$$

With $m_{\tilde{\phi}}$ given by (8.103), this condition is

$$\tilde{\phi} > 10^{-8} \tag{8.125}$$

in reduced Planck-scale units. With the estimate (8.100) of $\tilde{\phi}$, (8.125) is satisfied with ease.

Provided that $\rho_{\tilde{\phi}}$ does dominate the energy density of the universe at the moment of decay, there is a further reheating of the universe (in addition to the reheating that occurred when the inflaton decayed) to a temperature

$$\tilde{T}_R \sim m_{\tilde{\phi}}^{3/2} \tag{8.126}$$

where $\Gamma_{\tilde{\phi}}$ is given by (8.117). For successful nucleosynthesis, we must have

$$\tilde{T}_R > 1 \text{ MeV} = 10^{-21} M_P. \tag{8.127}$$

This requires

$$m_{\tilde{\phi}} > 10^{-14} \tag{8.128}$$

in reduced Planck-scale units. The value of $m_{\tilde{\phi}}$ used here ($\sim 10^{-8}$) satisfies this bound with ease. There is then an entropy increase of

$$\Delta = \left(\frac{\tilde{T}_R}{\tilde{T}_D}\right)^3 \sim m_{\tilde{\phi}}^{-1} \tilde{\phi}^2 \sim 10^8 \tilde{\phi}^2. \tag{8.129}$$

Table 8.2. Event sequence when oscillations of the Polonyi field begin before the inflaton vacuum energy density has decayed.

Time (t)	$\rho_\phi(t)$	$\rho_{\tilde\phi}(t)$	$\rho_{\rm rad}(t)$
$t_f < t < t_{\tilde\phi} \sim m_{\tilde\phi}^{-1}$	$\sim t^{-2}$	$\simeq \rho_{\tilde\phi}(t_f)$	0
$t_{\tilde\phi} < t < t_D \sim \Gamma_\phi^{-1}$	$\sim t^{-2}$	$\sim t^{-2}$	0
$t = t_D \sim \Gamma_\phi^{-1}$			$\to \rho_{\rm rad}(t_D)$ $T_R \sim m_\phi^{3/2} M_P^{-1/2}$
$t_D < t < \tilde t_D \sim \Gamma_{\tilde\phi}^{-1}$	0	$\sim T^3$	$\sim T^4$
$t = \tilde t_D \sim \Gamma_{\tilde\phi}^{-1}$		$\rho_{\tilde\phi}(\tilde t_D) \to$ radiation $\tilde T_R \sim m_{\tilde\phi}^{3/2} M_P^{-1/2}$	

Whenever $\tilde\phi$ is not very much greater than 1 in reduced Planck-scale units, this is large and may dilute the baryon number density of the universe unacceptably. In that case, a reheat temperature $\tilde T_R$ large enough to recreate the required baryon number density is needed. For low-temperature baryogenesis,

$$\tilde T_R \gtrsim 100 \text{ GeV} \simeq 10^{-16} M_P \qquad (8.130)$$

is required and, using (8.126), this imposes the bound

$$m_{\tilde\phi} \gtrsim 10^{-10}\text{--}10^{-11}. \qquad (8.131)$$

In the model being studied here, $m_{\tilde\phi} \sim 10^{-8}$ using (8.103), (8.104) and (8.89) and so it should be possible to regenerate the baryon number by a low-temperature mechanism. However, for other types of light fields with only gravitational strength interactions, their masses may be too small for the entropy generation problem to be solved in this way (and the entropy generation may also be larger).

8.6.2 Inflaton decays after Polonyi field oscillation

We consider next the alternative possibility [13, 14] that the inflaton vacuum density does not decay until after the Polonyi field has already started to oscillate. This might result in too low a reheating temperature to regenerate the baryon number of the universe except with a low-temperature baryogenesis mechanism. The sequence of events is summarized in table 8.2. Between the end of inflation at $t = t_f$ and the start of Polonyi field oscillations at $t = t_{\tilde\phi}$, the inflaton vacuum energy density $\rho_\phi(t)$ decreases as t^{-2} and the Polonyi field vacuum energy density $\rho_{\tilde\phi}$ is essentially constant. Between $t = t_{\tilde\phi}$ and $t = t_D$, the time at which the inflaton vacuum energy density decays, both $t = \rho_\phi$ and $t = \rho_{\tilde\phi}$ decrease as t^{-2}.

At $t = t_{\tilde{\phi}}$, the inflaton vacuum energy has not yet decayed and this vacuum energy density is an effective matter density controlling the expansion of the universe, as discussed after (7.63). At this time, we expect

$$H \sim t_{\tilde{\phi}}^{-1} \tag{8.132}$$

or one or two orders of magnitude greater, and we also have

$$\rho_\phi = 3H^2 \tag{8.133}$$

in reduced Planck-scale unit, from (7.41). Thus, at $t = t_{\tilde{\phi}}$,

$$\rho_\phi \sim 3m_{\tilde{\phi}}^2. \tag{8.134}$$

Also, at $t = t_{\tilde{\phi}}$,

$$\rho_{\tilde{\phi}} \simeq \tfrac{1}{2} m_{\tilde{\phi}}^2 \tilde{\phi}^2 \tag{8.135}$$

where we can approximate $\tilde{\phi}$ by its value at $t = t_f$ because, to a good approximation, $\tilde{\phi}$ does not start to oscillate until $t = t_{\tilde{\phi}}$. Consequently, because ρ_ϕ and $\rho_{\tilde{\phi}}$ have the same time dependence between $t = t_{\tilde{\phi}}$ and $t = t_D$,

$$\frac{\rho_{\tilde{\phi}}(t_D)}{\rho_\phi(t_D)} = \frac{\rho_{\tilde{\phi}}(t_{\tilde{\phi}})}{\rho_\phi(t_{\tilde{\phi}})} \sim \frac{1}{6} \tilde{\phi}^2. \tag{8.136}$$

At $t = t_D$, the inflaton vacuum energy density decays and so immediately afterwards

$$\frac{\rho_{\tilde{\phi}}(t_D)}{\rho_{\text{rad}}(t_D)} \sim \frac{1}{6} \tilde{\phi}^2. \tag{8.137}$$

For $t < t_D$ but greater than $\tilde{t}_D$, the time at which the Polonyi field energy decays, $\rho_{\tilde{\phi}}$ decreases as T^3, whereas the radiation density decreases as T^4. Thus, for $\tilde{t}_D < t < t_D$,

$$\frac{\rho_{\tilde{\phi}}(t)}{\rho_{\text{rad}}(t)} = \frac{\rho_{\tilde{\phi}}(t_{\tilde{\phi}}) T_R}{\rho_{\text{rad}}(t_{\tilde{\phi}}) T} \sim \frac{\tilde{\phi}^2 T_R}{6T} \tag{8.138}$$

where T_R is the temperature to which the universe reheats when the inflaton vacuum energy decays.

First, consider what would happen if the Polonyi field $\tilde{\phi}$ were not to decay. For nucleosynthesis to be able to recreate the ^{4}He and deuterium densities diluted by the increase in entropy due to inflaton decay, we must have T_R larger than 1 MeV $\simeq 10^{-21} M_P$. Since

$$T_R \sim m_\phi^{3/2} M_P^{-1/2} \tag{8.139}$$

as a consequence of (7.72) with $\Gamma_\phi \sim m_\phi^3 M_P^{-2}$, it follows that

$$m_\phi \gtrsim 10^{-14} M_P \tag{8.140}$$

at $T = 2.7$ K $\sim 10^{-31} M_P$, and then, from (8.138), that

$$\frac{\rho_{\tilde{\phi}}(t_0)}{\rho_{\text{rad}}(t_0)} \gtrsim \frac{\tilde{\phi}^2}{6} \times 10^{10}. \tag{8.141}$$

To avoid $\rho_{\tilde{\phi}}$ dominating the energy density of the universe and producing an excessive expansion rate, we need

$$\tilde{\phi} \lesssim 10^{-5} \tag{8.142}$$

in reduced Planck-scale units. Thus, there will be a problem if $\tilde{\phi}$ is not very small on the reduced Planck scale.

This suggests that there might be a problem with entropy generation in the realistic case where $\tilde{\phi}$ decays before 2.7 K is reached. We must decide first whether or not the Polonyi vacuum energy density will dominate the energy density of the universe at the moment of decay. At $t = \tilde{t}_D$, when the Polonyi field vacuum energy decays, from (8.138)

$$\frac{\rho_{\tilde{\phi}}(\tilde{t}_D)}{\rho_{\text{rad}}(\tilde{t}_D)} = \frac{\tilde{\phi}^2 T_R}{6\tilde{T}_D} \tag{8.143}$$

Thus, we must next estimate the temperature $\tilde{T}_D$ at which this decay occurs. The Polonyi field energy density $\rho_{\tilde{\phi}}$ grows relative to the radiation energy density as the temperature drops for $t_D < t < \tilde{t}_D$. We therefore approximate the time dependence of the temperature by taking the energy density to be dominated by $\rho_{\tilde{\phi}}$. Then,

$$\left(\frac{\dot{T}}{T}\right)^2 = \left(\frac{\dot{R}}{R}\right)^2 = \frac{1}{3}\rho_{\tilde{\phi}}(t) = \frac{1}{3}\rho_{\tilde{\phi}}(t_D)\left(\frac{T}{T_{\tilde{\phi}}}\right)^3 \tag{8.144}$$

with solution

$$T = \left(\frac{4T_{\tilde{\phi}}^3}{3\rho_{\tilde{\phi}}(t_D)}\right)^{1/3} t^{-2/3}. \tag{8.145}$$

Using (8.116), (8.139) and (8.101), we see that

$$\rho_{\tilde{\phi}}(t_D) = \left(\frac{t_{\tilde{\phi}}}{t_D}\right)^2 \rho_{\tilde{\phi}}(t_{\tilde{\phi}}) \sim \frac{m_{\phi}^6}{m_{\tilde{\phi}}^2}\rho_{\tilde{\phi}}(t_f) \sim \frac{1}{2}m_{\phi}^6\tilde{\phi}^2 \tag{8.146}$$

where we have also used $t_{\tilde{\phi}} \sim m_{\tilde{\phi}}^{-1}$ and $t_D \sim \Gamma_{\phi}^{-1} \sim m_{\phi}^{-3}$. Now, from (8.145) and (8.146),

$$\tilde{T}_D \sim m_{\phi}^{-1/2}m_{\tilde{\phi}}^2\tilde{\phi}^{-2/3}. \tag{8.147}$$

Returning to (8.118),

$$\frac{\rho_{\tilde{\phi}}(\tilde{t}_D)}{\rho_{\text{rad}}(\tilde{t}_D)} \sim \frac{1}{4}\tilde{\phi}^{8/3}m_{\phi}^2 m_{\tilde{\phi}}^{-2} \sim \frac{1}{4}\tilde{\phi}^{8/3}T_R^{4/3}m_{\tilde{\phi}}^{-2}. \tag{8.148}$$

The Polonyi vacuum energy density dominates the energy density of the universe when this is greater than one. The condition for this is that

$$\tilde{\phi} \gtrsim m_\phi^{-3/4} m_{\tilde{\phi}}^{3/4} \sim T_R^{-1/2} m_{\tilde{\phi}}^{3/4}. \qquad (8.149)$$

Assuming that baryon number regeneration has to occur at reheating after the inflaton vacuum energy has decayed to radiation, we require

$$T_R \gtrsim 100 \text{ GeV} \qquad (8.150)$$

or, from (8.139),

$$m_\phi \gtrsim 10^{-11} M_P \simeq 10^7 \text{ GeV} \qquad (8.151)$$

for electroweak baryogenesis. Also, to regenerate the ^{4}He and deuterium densities after destruction of these nuclei by the decay products of the Polonyi field $\tilde{\phi}$, we require

$$m_{\tilde{\phi}} \gtrsim 10 \text{ TeV} \simeq 10^{-14} M_P \qquad (8.152)$$

much as in (8.140) for the dilaton.

If, for example, we take $m_\phi = 10$ GeV and $m_{\tilde{\phi}} = 10$ TeV, then for $\tilde{\phi} \sim 1$ in Planck-scale units, from (8.143),

$$\frac{\rho_{\tilde{\phi}}(\tilde{t}_D)}{\rho_{\text{rad}}(\tilde{t}_D)} \sim \frac{1}{4} \times 10^6 \qquad (8.153)$$

so that the $\tilde{\phi}$ vacuum energy density dominates the energy density of the universe. The increase in entropy of the universe in reheating after the $\tilde{\phi}$ vacuum energy density decays is

$$\Delta \sim \left(\frac{\tilde{T}_R}{\tilde{T}_D}\right)^3 \sim \tilde{\phi}^2 m_\phi^{3/2} m_{\tilde{\phi}}^{-3/2} \qquad (8.154)$$

where we have used (8.147) and $\tilde{T}_R \sim m_{\tilde{\phi}}^{3/2}$. For the same choices of m_ϕ and $m_{\tilde{\phi}}$, and $\tilde{\phi} \sim 1$, (8.154) gives

$$\Delta \sim 10^5 \qquad (8.155)$$

which, though quite large, may not be inconsistent with a sufficiently large baryon number density surviving.

If, however, we take $m_{\tilde{\phi}} \sim m_{3/2}^{1/2}$, as in the model being employed here, then with $\tilde{\phi} \sim 1$, $\rho_{\tilde{\phi}}(\tilde{t}_D) \gtrsim \rho_{\text{rad}}(\tilde{t}_D)$ when $T_R \gtrsim 10^6$ GeV, corresponding to $m_\phi \gtrsim 10^{-8} M_P$, and the Polonyi vacuum energy density then dominates the energy density of the universe at the moment of decay. (For lower values of T_R the Polonyi vacuum energy density does not dominate.) In that case, Δ is only of order 1 for $m_\phi \sim 10^{-8} M_P$ and no dangerous entropy generation need occur. Thus, the entropy generation problem is not present for moderate values of the parameters when the inflaton vacuum energy decays after the Polonyi field has started to oscillate.

8.7   Exercises

1. Derive the effective potential of (8.3) from (8.1) using the superpotential (8.2).
2. Derive the effective potential (8.10) for the superpotential (8.9) with minimal kinetic terms.
3. Derive the effective potential (8.50) for a model of D-term inflation.
4. Check that the D-term inflation model effective potential (8.50) has a minimum at $\phi_+ = \phi_- = 0$ when the inequality (8.51) is satisfied.
5. Estimate the minimum of the effective potential for the Polonyi field in the presence of the inflaton field.

8.8   General references

The books and review articles that we have found most useful in preparing this chapter are:

- Bailin D and Love A 1994 *Supersymmetric Gauge Field Theory and String Theory* (Bristol: IOP)

Bibliography

[1] Nanopoulos D V, Olive K A, Srednicki M and Tamvakis K 1983 *Phys. Lett.* B **123** 41
[2] Holmam H, Ramond P and Ross G G 1984 *Phys. Lett.* B **137** 343
[3] Steinhardt P J and Turner M S 1984 *Phys. Rev.* D **29** 2162
[4] Ovrut B A and Steinhardt P J 1983 *Phys. Lett.* B **133** 161
[5] Golwirth D S and Piran T 1992 *Phys. Rep.* **214** 223 and references therein
[6] Binetruy P and Dvali G 1996 *Phys. Lett.* B **388** 241, arXiv:hep-ph/9606342
[7] Linde A and Riotto A 1997 *Phys. Rev.* D **56** 1841, arXiv:hep-ph/9703209
[8] Ellis J, Kim J E and Nanopoulos D V 1984 *Phys. Lett.* B **145** 181
[9] Ellis J, Gelmini G B and Sarkar S 1992 *Nucl. Phys.* B **373** 399
[10] Ross G G and Sarkar S 1996 *Nucl. Phys.* B **461** 597, arXiv:hep-ph/9506283
[11] Coughlan G D *et al* 1983 *Phys. Lett.* B **131** 59
[12] Bailin D and Love A 1994 *Supersymmetric Gauge Field Theory and String Theory* (Bristol: IOP)
[13] Coughlan G D, Holman R, Ramond P and Ross G G 1984 *Phys. Lett.* B **140** 44
[14] de Carlos B, Casas J A, Quevedo F and Roulet E 1993 *Phys. Lett.* B **318** 447

Chapter 9

Superstring cosmology

9.1 Introduction

For energies small compared to the string scale (which is of the order of the Planck scale in weakly coupled heterotic string theories), heterotic string theory in ten dimensions reduces to four-dimensional supergravity theory once compactification of extra dimensions has taken place. Thus, the discussion in chapter 8 also applies to heterotic string theory with special choices of the superpotential and Kähler potential derived from specific string theories. However, there are other aspects of string theory cosmology that go beyond supergravity cosmology.

Superstring theories contain massless (in the first instance) fields, referred to as the 'dilaton' and, more generally, 'moduli', whose effective potential is flat. These are a field theory manifestation of degeneracies of the string vacuum. In particular, if the compactification of the theory from ten dimensions to four dimensions occurs on a six-dimensional torus, certain of these moduli, the 'T-moduli', correspond to the freedom (before non-perturbative effects are considered) to vary continuously the radii of the torus along the associated axes. Similarly, the (expectation value of the) dilaton S is associated with the freedom to redefine the strength of the gravitational coupling. The dilaton and moduli fields are expected to obtain masses on the electroweak scale when supersymmetry breaking occurs and a non-trivial effective potential is generated. The supersymmetric partners of the dilaton and moduli (the 'dilatino' and 'modulinos') have a cosmology similar to the gravitino and can produce unwelcome densities in the universe. The dilaton and moduli fields themselves have a cosmology similar to the Polonyi field discussed in section 8.6 and can produce excessive entropy by late decay. These problems will be discussed in section 9.2 together with a possible solution through the thermal inflation mechanism.

Another problem associated with the existence of the dilaton in particular is the need for it to settle into a minimum of the effective potential. If this occurs only with difficulty, there can be adverse effects on inflation. This problem will be discussed in section 9.3. On a slightly more positive note, the dilaton and moduli fields with their flat potentials before supersymmetry breaking are possible candidates for inflaton fields and this will be discussed in section 9.4.

The discussion up to this point in the chapter will assume that compactification to four dimensions has already taken place. However, there could be an era in the history of the universe during which all nine spatial dimensions are still of comparable size. The cosmology of this era and how it joins on to the era with just three large spatial dimensions will be discussed in section 9.5. In particular, the question of why only three spatial dimensions become large will be addressed.

All of this discussion is based on heterotic string theory. There are also promising candidate theories based on type IIA or type IIB string theory, containing extended solutions referred to as 'D-branes'. The cosmology of D-branes will be discussed in section 9.6.

Finally, in section 9.7 and section 9.8, we shall discuss two models for the universe which allow there to have been an evolution of the universe prior to the big bang. In the first of these models (pre-big-bang cosmology), the effect of the (weakly-coupled) heterotic-string dilaton on the cosmological field equations is exploited to obtain solutions with a growing positive Hubble parameter for $t < 0$ driving inflation *before* the big bang. In the second model (the 'ekpyrotic' universe), strongly coupled string theory is employed. Novel cosmology emerges from the existence of an 11th dimension in the dual M-theory which will be discussed in section 9.8.

9.2   Dilaton and moduli cosmology

Before discussing the cosmological implications of the existence of the dilaton and moduli fields and their supersymmetric partners, we give some arguments that allow the masses of these fields to be estimated [1]. The supergravity effective potential is given by (2.147). It is convenient here to rewrite this in terms of the F-term field

$$F^i = e^{G/2}(G^{-1})^i_j G^j \qquad (9.1)$$

and its adjoint

$$F_i \equiv (F^i)^* = e^{G/2}(G^{-1})^j_i G_j \qquad (9.2)$$

in the notation of (2.148). Then

$$V = F_i F^j G^i_j - 3e^G. \qquad (9.3)$$

The mass of the dilaton (or modulus) field, denoted by ϕ for the moment, is found by differentiating the relevant part of (9.3), namely

$$V = F_\phi F^\phi G^\phi_\phi - 3e^G + \cdots. \qquad (9.4)$$

This means that we need an estimate of F^ϕ at the absolute minimum of the effective potential to estimate these masses.

First, because the dilaton and moduli fields have only gravitational strength interactions, for any superpositions of the fields $G^i_j \sim 1$ in reduced Planck-scale units ($M_P = 1$). Because $V = 0$ at the absolute minimum (for zero cosmological constant), the square of the F-field responsible for supersymmetry breaking, assumed to be a superposition of the dilaton and moduli F-fields, is of order e^G. Moreover, the gravitino mass is given by

$$m^2_{3/2} = e^{G_0} \qquad (9.5)$$

where G_0 is the value of G at the absolute minimum. Thus,

$$F^2_i \sim m^2_{3/2} \qquad (9.6)$$

for the supersymmetry-breaking F-field superposition. If there is more than one superposition of the dilaton and moduli F-fields with a vacuum expectation value (VEV), we shall assume that (9.6) holds for each superposition or some are negligible. (With the usual definition of the scale of supersymmetry breaking m_{susy}, (9.6) is the statement

$$m^2_{3/2} \sim m^4_{SUSY} \qquad (9.7)$$

in reduced Planck-scale units.)

Returning to the dilaton and moduli masses,

$$m^2_\phi = \left\langle \frac{\partial^2 V}{\partial\phi\partial\phi^*} \right\rangle \qquad (9.8)$$

provided that the kinetic terms are minimal so that the ϕ-field does not need rescaling. With V given by (9.4) and F_ϕ of the order given by (9.6), m^2_ϕ is a sum of terms of order e^{G_0} and terms of order $F_\phi F^\phi$ both of which are of order $m^2_{3/2}$. Thus, we might expect that

$$m_\phi \sim m_{3/2} \qquad (9.9)$$

where ϕ denotes a dilaton or modulus field. Similar but somewhat more complicated arguments can be made for the dilatino and modulinos. Detailed calculations confirm these expectations [1].

The cosmology of dilatinos and modulinos, which are light fermions with masses of order $m_{3/2}$ with only gravitational strength interactions, resembles that of gravitinos. The dilatinos $\tilde{S}$ will have a decay rate

$$\Gamma_{\tilde{S}} \sim m^3_{\tilde{S}} M^{-2}_P \qquad (9.10)$$

where $m_{\tilde{S}}$ is the dilatino mass and a decay time

$$t_{\tilde{S}} \sim \Gamma^{-1}_{\tilde{S}}. \qquad (9.11)$$

To avoid ^{4}He and deuterium abundances being modified by the decay products of the dilatinos, we should insist on the temperature $T_{\tilde{S}}$ at the time of dilatino decay being larger than 1 MeV $\simeq 10^{-21} M_P$. If we assume a radiation-dominated universe during the period before the dilatons decay, then

$$T_{\tilde{S}} \sim t_{\tilde{S}}^{-1/2} \sim m_{\tilde{S}}^{3/2} M_P^{-1}. \tag{9.12}$$

Thus, we require that

$$m_{\tilde{S}} > 10^{-14} M_P \simeq 10 \text{ TeV}. \tag{9.13}$$

We expect the gravitino mass $m_{3/2}$, which controls the sizes of soft supersymmetry breaking masses for the matter fields, to be not greater than about 10 TeV to solve the hierarchy problem. This bound can just be satisfied, especially given that $m_{\tilde{S}} \sim m_{3/2}$ is only a rough estimate. The same discussion applies to the modulinos $\tilde{T}_i$.

In any case, the cosmological problems that arise from the decays of light, very weakly-interacting fermionic fields can be avoided if there is a period of inflation, additional to the main period of inflation, to dilute this fermionic field density. However, the reheat temperature after this period of inflation should not be too high in order to avoid regeneration of the dilatino and modulino densities. Thus, as discussed in section 8.5 for gravitinos, we should have a reheating temperature

$$T_R \lesssim 10^9 \text{ GeV} \tag{9.14}$$

so that any necessary regeneration of the baryon number density should occur through low-temperature baryogenesis.

The cosmology of the dilaton S and moduli T_i fields resembles that of the Polonyi scalar field discussed in section 8.6, and much of the calculation given there is unmodified. We shall focus on the dilaton field S but the discussion of the moduli fields T_i will be exactly similar. Consider first the case where the dilaton field has started to oscillate before the inflaton decays. With a dilaton field energy density at the end of inflation

$$\rho_S(t_f) \simeq \tfrac{1}{2} m_S^2 S^2 \tag{9.15}$$

and assuming for the moment that the dilaton field does not decay, we find, as in (8.115), that the requirement to avoid the dilaton field energy density dominating the energy density of the universe today and producing too large an expansion rate is

$$S \leq (10^{-15} - 10^{-16}) m_S^{-1/4}. \tag{9.16}$$

For

$$m_S \sim m_{3/2} \sim 100 \text{ GeV–10 TeV} \sim (10^{-16} - 10^{-14}) M_P \tag{9.17}$$

we get

$$S \leq 10^{-11} - 10^{-12} \tag{9.18}$$

in reduced Planck-scale units, which is much smaller than the value expected to be obtained when the effective potential for the dilaton is minimized in the presence of the inflaton field or when the VEV of the dilaton is shifted by quantum fluctuations during inflation. This value would instead be expected to be not many orders of magnitude less than unity, much as for the Polonyi field. (See (8.99).) Consequently, the dilaton field energy density would dominate the energy density of the universe today.

In practice, S would be expected to decay but the previous discussion suggests the possibility of excessive entropy production when the decay occurs. The reheat temperature T_{SR} when the dilaton field energy density decays is

$$T_{SR} \sim m_S^{3/2} \tag{9.19}$$

as a consequence of (7.72) with $\Gamma_S \sim m_S^3$. For successful nucleosynthesis to occur after reheating, we must have $T_{SR} \gtrsim 1$ MeV, which implies that

$$m_S \gtrsim 10^{-14} M_P = 10 \text{ TeV}. \tag{9.20}$$

There is then an entropy increase

$$\Delta = \left(\frac{T_{SR}}{T_{SD}}\right)^3 \tag{9.21}$$

where T_{SD} is the temperature at which the dilaton decay occurs. In analogy with (8.122),

$$T_{SD} \sim m_S^{11/6} S^{-2/3}. \tag{9.22}$$

Thus,

$$\Delta \sim \frac{S^2}{m_S}. \tag{9.23}$$

For $m_S \sim 10$ TeV,

$$\Delta \sim 10^{14} S^2 \tag{9.24}$$

in reduced Planck-scale units. When S is not much smaller than 1 in these units, this is very large and may dilute the baryon number density of the universe unacceptably. Then the reheat temperature T_{SR} needs to be high enough for regeneration of the baryon number density to be possible. This imposes the bound

$$m_S \gtrsim 10^{-10} - 10^{-11} \tag{9.25}$$

as in (8.131). Thus,

$$m_S \gtrsim 10^7 - 10^8 \text{ GeV} \tag{9.26}$$

is required. This is not consistent with a dilaton mass of order $m_{3/2}$. An exactly similar discussion applies for the moduli.

As in the case of the Polonyi field, the problem is less severe in the case that the inflaton vacuum energy density does not decay until after the dilaton field has already started to oscillate. If the dilaton field were not to decay then, following the discussion leading to (8.141), to avoid ρ_S dominating the energy density of the universe and producing an excessive expansion rate, we need $S \lesssim 10^{-5}$. This is not consistent with our expectation that S will not be very much less than unity in reduced Planck-scale units. Thus, ρ_S will dominate the energy density of the universe. This suggests that there might be a problem with entropy generation in the realistic case where S decays before $T = 2.7$ K is reached. Then, following the logic leading to (8.154), the increase in the entropy of the universe in the reheating after the S vacuum energy density decays is

$$\Delta \sim m_S^{-3/2} m_\phi^{3/2} S^2. \tag{9.27}$$

As discussed in section 8.6, electroweak baryogenesis can occur at the reheating after the inflaton vacuum energy density decays when

$$m_\phi \gtrsim 10^{-11} M_P \simeq 10^7 \text{ GeV} \tag{9.28}$$

and regeneration of the ^{4}He and deuterium densities can occur following reheating after the dilaton vacuum energy density decays if

$$m_S \gtrsim 10 \text{ TeV} \simeq 10^{-14} M_P. \tag{9.29}$$

If, for example, we take $m_\phi = 10^7$ GeV and $m_S = 10$ TeV, then even for $S \sim 1$, we get

$$\Delta \sim 10^5 \tag{9.30}$$

which may not be so large as to be inconsistent with sufficient baryon number density surviving. A similar discussion applies in the case of moduli fields.

As will be discussed next, even if too much entropy generation occurs when the dilaton or modulus vacuum energy density decays, it may be possible to regenerate the baryon number density (and ^{4}He and deuterium densities) in an extra stage of inflation referred to as 'thermal' inflation [2]. Such a period of inflation is produced by a scalar field α with mass m_α of order 100 GeV–1 TeV, an approximately flat potential, and a VEV $\langle \alpha \rangle$ which is large on the 100 GeV–1 TeV scale. If this vacuum expectation value is too large, then such a field will produce a Polonyi problem of its own and so we require an expectation value which is large on the previous scale, but not too close to the Planck scale. So-called 'thermal' inflation takes place while the field α is trapped in the metastable minimum at the origin by thermal effects. For this trapping to be possible, the temperature should satisfy $T \gtrsim m_\alpha$. Otherwise the field would sit at the zero-temperature minimum away from the origin. For inflation to occur, the vacuum energy density should dominate over the radiation energy density and so we should have

$$V_0 \gtrsim T^4 \tag{9.31}$$

where V_0 is the vacuum energy density of α in the minimum at the origin. It is then possible for inflation to occur when

$$m_\alpha \lesssim T \lesssim V_0^{1/4}. \tag{9.32}$$

Because RT is constant during inflation, the number N_e of e-folds $\ln(R_f/R_i)$ is just $\ln(T_i/T_f)$. Thus,

$$N_e = \ln\left(\frac{V_0^{1/4}}{m_\alpha}\right). \tag{9.33}$$

With $m_\alpha \sim 100$ GeV–1 TeV, the period of thermal inflation begins before the baryogenesis and nucleosynthesis that follow the earlier period of inflation. The thermal inflation may then sufficiently dilute the dilaton or modulus density to resolve problems with dilution of baryon number or ^{4}He or deuterium density caused by entropy generation when decay of the dilaton (or modulus) occurs. (Recall that the dilaton or modulus vacuum energy density is behaving like a gas of non-relativistic particles, as discussed before (7.64) in the case of the inflaton.)

9.3 Stabilization of the dilaton

As observed in section 9.1, the dilaton field has a flat effective potential before supersymmetry breaking. Let us assume that spontaneous symmetry breaking is due to non-perturbative 'gaugino condensation' in which a product of two gaugino fields develops a VEV. The effective potential of the dilaton including this effect may be calculated. It is then found that it is difficult for the dilaton to settle to a minimum of the effective potential. Before discussing this problem, it is necessary to review the form of the non-perturbative potential to be expected from gaugino condensation.

The six unobserved spatial dimensions are usually compactified on an 'orbifold' or a Calabi–Yau 3-fold which generally has three T-moduli determining the size of the compactified space in the three complex dimensions (as well as complex-structure moduli specifying its shape). For simplicity, assume that there is a single overall T-modulus field T (not to be confused with temperature).Thus, we take

$$T = T_1 = T_2 = T_3. \tag{9.34}$$

If the various factors in the hidden-sector gauge group are labelled by the index a, the non-perturbative gaugino condensate superpotential W_{np} is of the form (see, for example, [3] and references therein)

$$W_{\text{np}} = \sum_a W_{\text{np}}^a \tag{9.35}$$

with

$$W_{\text{np}}^a = d_a e^{24\pi^2 S/b_a} \eta(T)^{-6}. \tag{9.36}$$

In (9.36), $\eta(T)$ is the Dedekind eta function, b_a is the renormalization group coefficient for the ath factor of the hidden-sector gauge group including a contribution due to hidden-sector matter, d_a is a numerical cofficient and so-called

'Green-Schwarz terms' have been omitted. The Kähler potential for the dilaton and moduli fields is

$$K = \ln(S + \bar{S}) - 3\ln(T + \bar{T}). \tag{9.37}$$

Then, the effective potential V derived from (2.132) is given by

$$(S + \bar{S})(T + \bar{T})^3 V = \left| W_{np} - (S + \bar{S})\frac{\partial W_{np}}{\partial S} \right|^2 - 3|W_{np}|^2$$

$$+ 3\left| W_{np} - (T + \bar{T})\frac{\partial W_{np}}{\partial T} \right|^2. \tag{9.38}$$

The non-perturbative superpotential should be chosen such that

$$\text{Re } S \sim 2 \tag{9.39}$$

at the global minimum of the potential, because of the connection between Re S and the value g_{string} of the gauge coupling constant at the string scale

$$\text{Re } S = 2g_{\text{string}}^{-2}. \tag{9.40}$$

In particular, if the ath factor in the hidden-sector gauge group is $SU(N_a)$ and there are hidden-sector matter fields in M_a copies of $N_a + \bar{N}_a$ fundamental representations, then

$$b_a = -N_a + \tfrac{1}{3}M_a \tag{9.41}$$

and

$$d_a = \left(\frac{1}{3}M_a - N_a\right)(32\pi^2 e)^{3(M_a - N_a)/(3N_a - M_a)}\left(\frac{M_a}{3}\right)^{M_a/(3N_a - M_a)}. \tag{9.42}$$

It is not possible to satisfy (9.39) with a single condensate but with two condensates minimization of the effective potential gives

$$\text{Re } S \simeq 0.17\frac{N_2 M_1 - N_1 M_2}{3N_2 - M_2 - 3N_1 + M_1} \tag{9.43}$$

which allows (9.39) to be satisfied for many choices of the integer parameters, together with yielding a realistic value of the gravitino mass $m_{3/2}$ given by the value of $e^{G/2}$ at the minimum, as in section 9.5. There is also scope to tune the parameters to obtain $V = 0$ at this minimum and so zero cosmological constant.

The dilaton stabilization problem [4] is a result of the peculiar shape of the potential V illustrated schematically in figure 9.1. If Re S starts larger than 2, there is only a very small region of Re S which allows it to roll to the desired minimum at Re $S = 2$. If Re S starts smaller than 2, the very steep potential causes it to roll over the very low barrier, failing to be trapped at Re $S = 2$ unless it starts very close to Re $S = 2$. Thus, trapping the dilaton in the desired minimum requires fine tuning of the initial conditions. (Strictly, we should correct

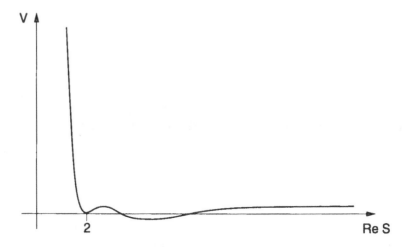

Figure 9.1. Representative two-condensate effective potential for ReS. The depths of the minima and height of the maximum are very small.

the discussion to allow for Re S having non-minimal kinetic terms. This makes little difference.)

However, the situation is drastically altered by several effects [5, 6] not included in the original treatment. First, the thermal energy density modifies the Hubble parameter and if the thermal energy density is large compared to the dilaton energy density, this can create sufficient 'friction' to allow the dilaton to roll into the Re $S = 2$ minimum of V for a considerable range of initial values of Re S. Second, the dilaton couples to the energy density of matter and gauge fields and, third, Re S couples to the axion field, which is the imaginary part of S. We shall focus here on the first of these effects.

For a homogeneous real scalar field ϕ with minimal kinetic terms and effective potential $V(\phi)$, as in (7.33),

$$\ddot{\phi} + 3H\dot{\phi} + V'(\phi) = 0. \tag{9.44}$$

We now want to allow for contributions to the Hubble parameter from the thermal energy density of matter fields or radiation. We shall take the field ϕ to have a potential of the form

$$V(\phi) = V_0 e^{-\lambda\phi} \tag{9.45}$$

in reduced Planck-scale units and shall discuss later how this relates to the dilaton field. The pressure p_ϵ and energy density ρ_ϵ of the matter fields or radiation satisfy an equation of state

$$p_\epsilon = (\epsilon - 1)\rho_\epsilon \tag{9.46}$$

where $\epsilon = 1$ for a matter-dominated universe and $\epsilon = 4/3$ for a radiation-dominated universe during the rolling of ϕ towards the minimum of its potential.

Then

$$H^2 = \tfrac{1}{3}(V + \tfrac{1}{2}\dot{\phi}^2 + \rho_\epsilon) \qquad (9.47)$$

is the generalization of (7.40) and (7.41), in reduced Planck-scale units. Using (9.47) and (9.44),

$$\dot{H} = -\frac{1}{2}\dot{\phi}^2 + \frac{1}{6H}\dot{\rho}_\epsilon. \qquad (9.48)$$

Also, from the energy–momentum conservation equation (1.45),

$$\dot{\rho}_\epsilon = -3H(p_\epsilon + \rho_\epsilon) \qquad (9.49)$$

so that

$$\dot{H} = -\tfrac{1}{2}(\dot{\phi}^2 + \rho_\epsilon + p_\epsilon). \qquad (9.50)$$

This ignores any interaction other than gravitational of ϕ with matter fields. These equations can be solved analytically for the asymptotic form of ϕ in the case of very steep potentials

$$\lambda^2 > 3\epsilon. \qquad (9.51)$$

The field ϕ increases at first but the 'friction' corresponding to the Hubble constant then freezes ϕ at a near-constant value $\tilde{\phi}_0$ for a time, where

$$\tilde{\phi}_0 = \phi_0 + \frac{\sqrt{6}}{3(2-\epsilon)} \ln\left(\frac{1+x_0}{1-x_0}\right) \qquad (9.52)$$

and

$$x_0 \equiv \frac{\dot{\phi}_0}{\sqrt{6}H_0} \qquad (9.53)$$

with ϕ_0, $\dot{\phi}_0$ and H_0 denoting initial values. Finally, ϕ approaches the asymptotic form

$$\phi(t) = \frac{1}{\lambda} \ln\left(\frac{2\lambda^2 V_0}{9H_0^2(2-\epsilon)\epsilon}\right) + \frac{3\epsilon}{\lambda} \ln R(t) \qquad (9.54)$$

where $R(t)$ is the scale factor ('radius') of the universe. The approach to the minimum is then slow and, after oscillations about the minimum with an exponentially damped amplitude, ϕ settles to its minimum.

Let us now apply these considerations to Re S, taken for the moment to have minimal kinetic terms. In the region Re $S < 2$ but not too close to Re $S = 2$ where the minimum has been produced by the balancing of two terms, the non-perturbative superpotential may be approximated by a single term W_{np}^a with S-dependence:

$$W_{np}^a \sim e^{-\Delta_a S} \qquad (9.55)$$

where

$$\Delta_a = \frac{24\pi^2}{N_a - \tfrac{1}{3}M_a} \qquad (9.56)$$

is positive and has the smaller value. Then, the parameter λ in (9.45) is given by

$$\lambda = 2\Delta_a. \tag{9.57}$$

A more careful treatment [5] with the correct non-minimal kinetic terms for Re S makes very little difference to (9.54), with ϕ replaced by Re S.

This discussion is valid provided that Re S is able to attain the asymptotic form of solution before passing through the minimum. Sufficient conditions [5] are to take the initial value of Re S, denoted by Re S_0, between the constant term in (9.54) and the value at the minimum Re $S_{min} \sim 2$, and also an initial velocity for Re S such that the constant value Re $\tilde{S}_0$, in the sense of (9.52), is also less than Re S_{min}. (In practice, the minimum we are trying to reach is at a smaller value of Re S than the adjacent maximum, and so we must start with Re $S <$ Re S_0 to have any chance of ending up in the minimum.) Thus, we require

$$\frac{1}{\lambda} \ln \left(\frac{2\lambda^2 V_0}{9H_0^2(2 - \epsilon)\epsilon} \right) < \text{Re } S_0 < \text{Re } S_{min} \tag{9.58}$$

and an appropriate initial velocity for Re S.

9.4 Dilaton or moduli as possible inflatons

A priori, the dilaton or moduli fields (for an orbifold or Calabi–Yau compactification) are good candidates for inflaton fields [7] because their potential is completely flat to all orders in string perturbation theory. Non-perturbative effects, such as gaugino condensation, can provide a non-trivial effective potential. As we shall see shortly, if the dilaton or modulus field is to be used as the inflaton, it is necessary to assume that the superpotential is the sum of two components. (See, for example, [8].) One of these components has a large scale and gives an effective potential with unbroken supersymmetry and zero cosmological constant at the global minimum when the other component is neglected. This large-scale component is responsible for driving inflation when the dilaton or modulus expectation value is in a flat region away from the minimum. The other component has a much smaller scale and is responsible for supersymmetry breaking in the low-energy world. It is the former component of the superpotential that we are interested in here. Neglecting the low-energy component of the superpotential, it is convenient to write the effective potential for the dilaton S in the form

$$V = \mu^4 F(S, \bar{S}) \tag{9.59}$$

where we are using reduced Planck-scale units, and $F(S, \bar{S})$ is of order 1. To obtain density perturbations consistent with the COBE data, we require

$$\mu \sim 10^{16}\text{–}10^{17} \text{ GeV} \tag{9.60}$$

as in (7.115). This contrasts with a scale of

$$(m_{3/2}M_P)^{1/2} \sim 10^{10}\text{--}10^{11} \text{ GeV} \tag{9.61}$$

for an effective potential due to the superpotential responsible for low-energy supersymmetry breaking with soft supersymmetry breaking masses on the scale of 10^2–10^3 GeV. This is the reason for assuming that the superpotential is the sum of two components. (It should be noted that the discussion of the previous section employed gaugino condensate superpotentials that had been designed to be responsible for low-energy supersymmetry breaking.) All of this discussion applies equally to the use of a modulus field as the inflaton.

Though it is a very attractive idea, it has proved difficult to find a realization of it in practice. Multiple gaugino condensate potentials for the dilaton, such as discussed in the previous section, tend to be too steep in the Re S direction for inflation to occur. If, instead, the overall modulus field T is employed as the inflaton, with a superpotential consistent with the modular invariance of an orbifold compactification, up to 20 e-folds of inflation can be obtained. However, this does not appear to be possible when we demand that the effective potential has a minimum with unbroken supersymmetry and zero cosmological constant [9].

9.5 Ten-dimensional string cosmology

Heterotic string theory begins as a ten-dimensional theory with the need for six dimensions to be compactified to provide us with the observed four-dimensional world (except in the case of direct constructions in four dimensions, such as the free-fermion construction). An attractive possibility is that the compactification of the six extra dimensions has a cosmological origin. In what follows we shall treat all spatial dimensions as being wrapped on a torus with all dimensions initially on the Planck scale. We shall study a mechanism, due to Brandenberger and Vafa [10], that naturally results in three of the spatial dimensions becoming very large, corresponding to a flat space, and the rest of the spatial dimensions remaining on the Planck scale. The dilaton plays a crucial part.

As discussed in section 9.3, we expect the dilaton to acquire a mass of the order of the electroweak scale, or one or two orders of magnitude larger, when supersymmetry breaking occurs. For consistent cosmology, it is crucial that the dilaton does acquire a mass, because it is a scalar field with only gravitational strength interactions, and a massless field of this kind is inconsistent with solar system observations. However, there is no *a priori* objection to the dilaton having been massless in the early stages of the universe before supersymmetry breaking at a temperature of around 100 GeV.

It will be assumed that the gravitational (metric) field and the dilaton field are slowly varying (adiabatic approximation) so that it is a good approximation to keep only the leading derivatives in the effective action. It will also be assumed that N spatial dimensions are large dimensions with time dependence, while the remaining N_c spatial dimensions are static compact dimensions. In the case of the heterotic string,

$$N + N_c = 9. \tag{9.62}$$

Eventually we want $N = 3$ but we shall leave N general for now. The effective action for the gravitational field and the dilaton is then (up to a multiplicative constant)

$$S_0 = \int d^{N+1}x \sqrt{|G|} e^{-2\phi} (\mathbb{R} + 4G^{AB} \partial_A \phi \partial_B \phi) \tag{9.63}$$

where G_{AB} is the metric tensor, $G = \det[G_{AB}]$, $\mathbb{R}$ is the curvature scalar for the $(N + 1)$-dimensional space with $A = 0, 1, \ldots, N$ and ϕ is the dilaton. In the case of $N = 3$, the connection with the value of the gauge coupling constant g_{string} at the string scale is $g_{\text{string}} = e^\phi$ and the connection with the dilaton field of section 9.3 is $\mathrm{Re}\, S = 2e^{-2\phi}$. (The antisymmetric tensor field in the supergravity multiplet is being ignored for simplicity.) All spatial dimensions will be taken toroidal with periodic length $a_i(t)$ for $i = 1, 2, \ldots, N$. Write

$$a_i(t) = e^{\lambda_i(t)}. \tag{9.64}$$

The metric is given by

$$ds^2 = dt^2 - \sum_{i=1}^N a_i^2(t)(dx^i)^2. \tag{9.65}$$

Keeping G_{00} general for the moment, and fixing it to 1 later so that we do not lose the field equation obtained by varying with respect to G_{00}, the curvature scalar (exercise 1) is

$$\mathbb{R} = -G^{00} \left[\sum_{i=1}^N (\dot\lambda_i)^2 + \left(\sum_{i=1}^N \dot\lambda_i \right)^2 + 2 \sum_{i=1}^N \ddot\lambda_i - \dot{G}^{00} \sum_{i=1}^N \dot\lambda_i \right]. \tag{9.66}$$

It is convenient to define

$$\Phi \equiv 2\phi - \sum_{i=1}^N \lambda_i \tag{9.67}$$

which absorbs a factor for the volume of the space because

$$e^{-2\phi} \sqrt{|G|} = \sqrt{G_{00}}\, e^{-\Phi}. \tag{9.68}$$

The action of (9.63) is then (exercise 2)

$$S_0 = -\left(\int d^N x \right) \int dt \, \sqrt{G_{00}}\, e^{-\Phi} G^{00} \left[\sum_{i=1}^N (\dot\lambda_i)^2 - (\dot\Phi)^2 \right] \tag{9.69}$$

In general, a thermal contribution S_T (from the gas of string modes in thermal equilibrium) should be added to the action. It has the form

$$S_T = \left(\int d^N x \right) \int dt \sqrt{G_{00}} F \left(\lambda_i, \beta \sqrt{G_{00}} \right) \tag{9.70}$$

where F is the free energy and $\beta = T^{-1}$ in units where the Boltzmann constant $k_B = 1$. The complete action is

$$S_{\text{tot}} = S_0 + S_T. \tag{9.71}$$

Varying this action with respect to Φ, λ_i and G^{00} gives (exercise 3), after combining field equations,

$$(\dot{\Phi})^2 - \sum_{i=1}^{N} (\dot{\lambda}_i)^2 = e^{\Phi} E \tag{9.72}$$

$$\ddot{\lambda}_i - \dot{\Phi} \dot{\lambda}_i = \tfrac{1}{2} e^{\Phi} P_i \tag{9.73}$$

$$\ddot{\Phi} - \sum_{i=1}^{N} (\dot{\lambda}_i)^2 = \tfrac{1}{2} e^{\Phi} E. \tag{9.74}$$

The thermodynamics that has been employed here is as follows. The free energy is

$$F = E - TS \tag{9.75}$$

where E is the energy and S is the entropy with

$$S = - \left(\frac{\partial F}{\partial T} \right)_V = \beta^2 \left(\frac{\partial F}{\partial \beta} \right)_V. \tag{9.76}$$

The pressure in the ith direction is

$$P_i = - \left(\frac{\partial F}{\partial \lambda_i} \right)_T. \tag{9.77}$$

As a consequence of (9.75) and (9.76),

$$E = F - T \left(\frac{\partial F}{\partial T} \right)_V = F + \beta \left(\frac{\partial F}{\partial \beta} \right)_V. \tag{9.78}$$

With this functional dependence of F,

$$E = F + 2 \frac{\partial F}{\partial G_{00}} \tag{9.79}$$

where we have set $G_{00} = 1$ at the end. From (9.72)–(9.74), we may deduce (exercise 4) that

$$\dot{E} + \sum_{i=1}^{N} \dot{\lambda}_i P_i = 0 \tag{9.80}$$

$$\beta^{-1} \dot{S} = \dot{E} - \sum_{i=1}^{N} \dot{\lambda}_i \frac{\partial F}{\partial \lambda_i} = \dot{E} + \sum_{i=1}^{N} \dot{\lambda}_i P_i. \tag{9.81}$$

Combining these, we have

$$\dot{S} = 0 \tag{9.82}$$

Thus, entropy is conserved.

In the first instance, S is a function of β and the λ_i. In principle, we can solve (9.82) to obtain β as a function of the λ_i. Then, E can be written as a function $E(\lambda)$ of the λ_i alone, where, for the moment, λ denotes the λ_i collectively. When the entropy is constant,

$$P_i = -\left(\frac{\partial E}{\partial \lambda_i}\right)_T. \tag{9.83}$$

To solve the equations (9.72)–(9.74), we need some knowledge of $E(\lambda)$. For the moment we ignore the contribution of winding modes. The properties of $E(\lambda)$, now assumed to be a function of a single λ when the λ_i have a common value, have been studied using the microcanonical ensemble [10]. It is T-duality symmetric, i.e. symmetric under the replacement of $a_i(t)$ by $a_i^{-1}(t)$ so that

$$E(\lambda) = E(-\lambda). \tag{9.84}$$

For $\lambda \sim 0$, the so-called 'Hagedorn region', $E(\lambda)$ is almost constant. For sufficiently large λ, only massless string modes contribute to the partition function, corresponding to a radiation-dominated universe. Then $E(\lambda)$ has the exponential behaviour

$$E(\lambda) \sim e^{-\lambda}. \tag{9.85}$$

Between these two limiting cases there is incomplete knowledge of $E(\lambda)$. However, it *is* known that E decreases with $|\lambda|$ for λ close to zero and this is believed to be correct for all λ.

This is enough information to see that a radiation-dominated era is approached as time increases if $\dot{\Phi}$ starts with a negative value. The argument is as follows. Because E is positive, (9.72) implies that $\dot{\Phi}$ can never become zero, so that $\dot{\Phi}$ can never change sign. Also, (9.74) implies that $\ddot{\Phi}$ is positive. Consequently $\dot{\Phi}$ increases and if it starts negative, it remains negative and approaches zero as $t \to \infty$. Now consider equation (9.73). Using (9.83), and assuming that $\lambda_i = \lambda$ for all i, (9.73) is

$$\ddot{\lambda} - \dot{\Phi}\dot{\lambda} = -\tfrac{1}{2} e^{\Phi} E'(\lambda) \tag{9.86}$$

(9.86) can be interpreted as the equation of motion for a particle with position λ moving in a potential $e^{\Phi} E(\lambda)$ with a damping term, because $\dot{\Phi} < 0$. Since the potential decreases as $|\lambda|$ increases (ignoring the λ dependence of Φ), λ slides towards increasing values of λ. The radius of the toroidal space is $a(t) = e^{\lambda(t)}$ and, hence, as λ increases a increases, when $\lambda > 0$. (There is a duality between large and small values of $a(t)$ and, hence, between positive and negative values of λ, as noted in (9.84).) Thus, we need only discuss $\lambda > 0$.) Since the spectrum of a string theory is known, the entropy can be calculated at a given temperature T in terms of the radius $a(t)$. We know that entropy is constant. Therefore, a relationship between $a(t)$ and T can be calculated numerically that ensures this. As $|\lambda|$ decreases, it is found that T increases towards a limit referred to as the 'Hagedorn temperature'. As $|\lambda|$ increases, T falls until the massive string modes go out of equilibrium and we enter a radiation-dominated era controlled by the massless string modes, as in the standard model of the universe.

We now ask the question: 'Why is the number N of large spatial dimensions equal to 3?' A possible explanation turns on the presence of winding modes in string theories compactified on a torus. (Remember that we have taken all spatial dimensions to be toroidal.) Because of the periodic nature of a torus, the closed-string boundary conditions for the spatial bosonic degrees of freedom $X^k(\tau, \sigma)$ can be satisfied when

$$X^k(\tau, \sigma + \pi) = X^k(\tau, \sigma) + 2\pi L^k \tag{9.87}$$

where centre-of-mass coordinates x^k on the torus have the identification

$$x^k \equiv x^k + 2\pi L^k \tag{9.88}$$

where L^k are referred to as 'winding numbers' and are proportional to the torus radius [11]. String modes with non-zero values of L^k are referred to as winding modes. If p^k is the centre-of-mass momentum of the string degree of freedom X^k, the mass-squared of a string state includes $(p^k + 2L^k)^2$ and $(p^k - 2L^k)^2$. As the radius of the torus increases, the squared mass of a winding mode increases.

The idea is that string winding modes, unlike other matter densities, will oppose expansion of the dimensions of the universe. The reason for this is that, in the presence of winding modes, the behaviour of $E(\lambda)$ is very different from that discussed earlier. As just discussed, the mass squared of any winding mode increases as the square of the winding number, for large values of the torus radius, and so as the square of the torus radius. This effect results in $E(\lambda)$ increasing as e^{λ}. Roughly, the growth of E with λ in (9.86) means that $\ddot{\lambda} < 0$ (up to a damping term) so that $\dot{\lambda}$ eventually becomes negative, λ starts to decrease and the radius of the universe starts to decrease. In this way, the winding modes first stop the expansion of the universe and then reverse it.

This argument is not quite correct because of the e^{Φ} term in (9.86). As discussed earlier, if $\dot{\Phi}$ starts negative, it remains negative so that Φ decreases with time. Thus, treating λ as the position of a particle, the strengthening of the potential with increasing λ as the universe expands can be offset by the behaviour of e^{Φ}. A more careful treatment shows that this does not affect the outcome.

Once the universe starts to contract, the momentum modes, which have behaviour dual to the winding modes, play a crucial role. They make a contribution to $E(\lambda)$ that increases as λ decreases and oppose the contraction. In this way, the universe is caused to oscillate between some minimum radius and some maximum radius within a few orders of magnitude of the Planck scale.

The question then is how the universe can ever expand to a large scale. The answer is that string winding modes can annihilate totally or partially into momentum states (with complete annihilation of winding number occurring between winding modes with equal and opposite winding number). In this way they are able to reach thermal equilibrium with other string states. In thermal equilibrium the number of winding modes becomes small as the torus radius increases and winding-mode masses increase. However, it is difficult to reach equilibrium if the winding modes find it difficult to collide to annihilate. A collision corresponds to the two-dimensional world surfaces of the two states intersecting. Generically, this does not occur when the dimensionality $N+1$ of the extended spacetime in which the winding modes move is greater than $2 + 2 = 4$. Thus, for $N + 1 > 4$, thermal equilibrium is not achieved and the winding modes stop the the universe expanding much beyond the Planck scale. However, for $N + 1 \leq 4$, the winding modes annihilate readily and thermal equilibrium is reached resulting in a low density of winding modes. The universe is then able to expand to a large scale.

This argument provides a partial explanation of the three-dimensional nature of our observed universe. If the universe starts to expand in some number $N > 3$ of (spatial) dimensions, then the expansion is stopped by the winding modes. It then oscillates for a while before expanding again in some number N of dimensions that may differ from the first time. This may happen many times until finally the universe starts to expand in some number N of dimensions with $N \leq 3$. Then, the expansion continues. Of course, this only explains why $N \leq 3$ and not why $N = 3$. Therafter, the discussion of the earlier part of this section, which neglected winding modes, applies and the universe evolves to a standard radiation-dominated universe.

9.6   D-brane inflation

The discussion so far in this chapter has been in the context of weakly coupled heterotic string theory. Alternative models of particle theory can be obtained from type II superstring theories because of the existence of extended so-called 'Dp-brane' solutions which occupy $p + 1$ dimensions of spacetime. (See, for example, [12] and references therein.) As well as closed strings, the theory contains open strings which are constrained to have their endpoints on Dp-branes. Chiral matter can be obtained from open strings whose endpoints are on Dp-branes located at an orbifold fixed point in the compact dimensions. We shall assume that this orbifold is a product of three two-dimensional tori with points on the tori identified under the action of a discrete Z_N or $Z_M \times Z_N$ group.

Models with the gauge fields of the standard model (up to some $U(1)$ factors) and the massless matter content of the standard model (up to some vector-like matter) have been obtained by employing D3-branes and D7-branes located at fixed points. The presence of D7-branes as well as D3-branes is necessary to satisfy certain 'twisted tadpole' cancellation conditions, which are required by a consistent string theory and, among other things, ensure non-Abelian gauge anomaly cancellation. The models also contain $\overline{D3}$-antibranes and $\overline{D7}$-antibranes which are needed to satisfy untwisted tadpole cancellation conditions. The chiral matter states are associated with open strings with their endpoints on D3-branes or with one endpoint on a D3-brane and the other on a D7-brane. In such theories, the string scale and the compactification scale do not necessarily coincide and low string scales are possible.

An important aspect of such constructions is that the Dp-branes and $\overline{Dp}$-antibranes can be prevented from moving from the fixed points in some models by the requirement that the twisted-tadpole conditions are always satisfied. Then, the brane–antibrane separations are fixed except to the extent that they share in the contraction or expansion of a toroidal space when the radius of space varies. Thus, a modulus field in the associated low-energy supergravity theory whose expectation value is a brane–antibrane separation is not a candidate inflaton. However, possible candidates are the moduli scalars T_i, $i = 1, 2, 3$, the real parts of which are associated with the radii R_i of the three tori in the form

$$t_i \equiv \mathrm{Re}\, T_i = e^{\phi} M_s^2 R_i^2 \tag{9.89}$$

where M_s is the string scale and ϕ is the ten-dimensional dilaton. The four-dimensional dilaton S is also a candidate. Models of inflation have been constructed [13] in which S or one of the T_i provides the inflaton while the other moduli (T_i or S) are frozen by some unidentified mechanism.

To discuss such models of inflation, we now require the form of the effective potential V as a function of the unfrozen modulus field. It is convenient to use T-duality with respect to all directions simultaneously:

$$R_i \rightarrow \frac{\alpha'}{R_i} \qquad i = 1, 2, 3 \tag{9.90}$$

where

$$\alpha' = M_s^{-2} \tag{9.91}$$

to map D3-branes into D9-branes and D7-branes into D5-branes. We need the potential due to the tension in the branes. This is proportional to the volume of the branes and, for a theory of D9-branes and D5-branes, is of the form

$$V = N_9 V_9 + \sum_{i=1}^{3} N_{5_i} V_{5_i} \tag{9.92}$$

where N_9 is the number of D9-branes, and N_{5_i} is the number of D5$_i$-branes, i.e. the number of D5-branes which wrap the ith torus (as well as living in four-dimensional spacetime). The potential V_9 due to the D9-branes is of the form

$$V_9 = \tau_9 R_1^2 R_2^2 R_3^2 \tag{9.93}$$

with the D9-brane tension

$$\tau_9 = \alpha_9 M_s^{10} e^{-\phi} \tag{9.94}$$

where α_9 is a dimensionless constant. Also, the potential due to the D5$_i$-branes is of the form

$$V_{5_i} = \tau_5 R_i^2 \tag{9.95}$$

with the D5-brane tension

$$\tau_5 = \alpha_5 M_s^6 e^{-\phi} \tag{9.96}$$

and α_5 a dimensionless constant. These potentials can be rewritten in terms of the real parts of the four-dimensional dilaton S and the T_i moduli fields:

$$s \equiv \text{Re } S = M_s^6 e^{-\phi} R_1^2 R_2^2 R_3^2 \tag{9.97}$$

$$t_i \equiv \text{Re } T_i = M_s^2 e^{-\phi} R_i^2 \qquad \text{for } i = 1, 2, 3. \tag{9.98}$$

Then,

$$V = M_s^4 \left[N_9 k_9 s + \sum_{i=1}^{3} N_{5_i} k_{5_i} t_i \right] \tag{9.99}$$

where k_9 and k_{5_i} are dimensionless constants of order 1. There is also a contribution to the potential from the exchange of massless bulk states, such as the graviton, which we are neglecting here. It can be shown that this is small compared to the retained terms when the moduli are large.

To apply this potential to the study of inflation, it is convenient to recast it in terms of fields with canonical kinetic terms. After Weyl rescaling to remove the factor of $e^{-2\phi}$ in front of the curvature scalar (displayed in (9.63)), the kinetic terms for moduli are

$$\mathcal{L}_{\text{kinetic}} = \frac{1}{4} M_P^2 \sqrt{|g|} g^{\mu\nu} \left(\partial_\mu \ln s \partial_\nu \ln s + \sum_{i=1}^{3} \partial_\mu \ln t_i \partial_\nu \ln t_i \right) \tag{9.100}$$

the Weyl rescaling being

$$g_{\mu\nu} \to \lambda g_{\mu\nu} \tag{9.101}$$

with

$$\lambda = \frac{M_P^2 e^{2\phi}}{M_s^8 R_1^2 R_2^2 R_3^2}. \tag{9.102}$$

These kinetic terms arise from the curvature scalar

$$\mathbb{R} \equiv G^{ab} R_{ab} \tag{9.103}$$

where $a, b = 0, 1, \ldots, 9$ and G_{ab} is of the form

$$G_{ab} = \mathrm{diag}[g_{\mu\nu}(x), -R_1^2(x)\delta_{mn}, -R_2^2(x)\delta_{rs}, -R_3^2(x)\delta_{uv}] \qquad (9.104)$$

where x denotes the four-dimensional spacetime coordinates. In the first instance, the kinetic terms are then given in terms of R_1, R_2 and R_3. After the Weyl rescaling (9.101), (9.102), these are recast in the form (9.100) in terms of s and t_i. The potential is then rescaled by a factor λ^2 to give

$$V = M_P^9 \left(\frac{k_9 N_9}{t_1 t_2 t_3} + \frac{k_{5_1} N_{5_1}}{s t_2 t_3} + \frac{k_{5_2} N_{5_2}}{s t_1 t_3} + \frac{k_{5_3} N_{5_3}}{s t_2 t_1} \right). \qquad (9.105)$$

If, for example, we now freeze the moduli t_1, t_2 and t_3 and treat s as the inflaton, then the inflaton field X with canonical kinetic terms is given by

$$s = \exp\left(\frac{\sqrt{2}X}{M_P} \right). \qquad (9.106)$$

The potential for the inflaton is of the form

$$V = \kappa_0 + \kappa_1 e^{-\sqrt{2}X/M_P} \qquad (9.107)$$

where κ_0 and κ_1 are constants

$$\kappa_0 = \frac{k_9 N_9 M_P^4}{t_1 t_2 t_3} \quad \text{and} \quad \kappa_1 = \frac{k_{5_1} N_{5_1}}{t_2 t_3} + \frac{k_{5_2} N_{5_2}}{t_1 t_3} + \frac{k_{5_3} N_{5_3}}{t_2 t_1}. \qquad (9.108)$$

We can now study slow-roll inflation with this potential in the usual way. The slow-roll parameters ϵ and η are defined in (7.184) and (7.185). In the present case,

$$\epsilon \simeq \frac{\kappa_1^2}{\kappa_0^2} e^{-2\sqrt{2}X/M_P} \quad \text{and} \quad \eta \simeq \frac{2\kappa_1}{\kappa_0} e^{-\sqrt{2}X/M_P} \qquad (9.109)$$

when $X \gg M_P$. Whenever $X \gg M_P$, the parameters ϵ and η are small and slow roll ocurs. Thus, slow roll is generic for large values of the modulus which is playing the role of the inflaton. This approximation also allows us to be in the low-energy field theory limit which requires weak coupling $e^\phi \ll 1$.

The next question to be addressed is when inflation ends. For $\kappa_1 < 0$, the potential is such that X (which is certainly positive in this approximation) will decrease with time and, eventually, the slow-roll conditions will no longer be satisfied if $|\kappa_1/\kappa_0| \gtrsim 1$. Conversely, for $\kappa_1 > 0$, X increases and there is no end to slow roll. However, as X grows, s grows, and either ϕ diminishes or one of the R_i diminishes. If the latter, then at least one of the R_i can become smaller than the string scale and low-energy field theory and the validity of the low-energy potential V break down. In this case, there is a striking mechanism which could end inflation. There can be a critical value of a radius at which a tachyon appears in the spectrum. Then, inflation ends in a version of hybrid inflation. In some orbifold models a brane–antibrane pair then annihilates to produce a single brane with one dimension more than the original brane, wrapped around this extra dimension, which is stable at a smaller radius. The reheating is then controlled by the difference in tension between the brane–antibrane pair and the single brane.

As to the number of e-folds of inflation, it does not appear possible [13] in the models described earlier to obtain a sufficient number of e-folds, except for large values of the N_{5_i}. In that case, the approximation of retaining only the tension term in the potential breaks down and even including exchange of single bulk states (e.g. the graviton) in the brane–antibrane interaction may not be sufficient. However, this problem may be overcome [13] when the compactified dimensions form an orientifold rather than an orbifold. Then, there still remains the difficulty that we have had to freeze all but one of the T_i moduli or dilaton fields arbitrarily.

An alternative type of brane model (see, for example, [14] and references therein) is based on intersecting D-branes with chiral matter living on (some of) the intersections, in the sense that it is associated with open strings that begin and end at a particular intersection of two D-branes. This type of model also provides a satisfactory model of inflation [15] up to a point but with the same difficulty of having to freeze all but one of the moduli discussed earlier.

9.7 Pre-big-bang cosmology

The presence of the dilaton in the heterotic string theory action allows for a possible alternative origin for inflation in a period of evolution of the universe before the big bang [16]. The basic idea is that it may be possible to join together two solutions of the cosmological field equations, one for $t < 0$ and one for $t > 0$, with the following properties. The $t < 0$ solution is chosen to have the dilaton ϕ growing to produce a growing Hubble parameter H, so that the universe expands rapidly (dilaton-driven inflation). This is an even more rapid expansion than the more familiar inflation driven by an approximately constant H. The $t > 0$ solution, conversely, has $|\phi|$ decreasing and ϕ rolling into a minimum of its potential. Thereafter, there is FRW cosmology (with a constant dilaton), possibly higher-dimensional, with compactification to three spatial dimensions to follow.

As $t = 0$ is approached from $t < 0$, we shall see that it is possible for the value of ϕ to be positive and diverge logarithmically, which pushes the universe into a strongly-coupled regime because $e^{2\phi}$ controls the strength of the gauge and gravitational interactions in heterotic string theory. At some point, weakly-coupled string theory breaks down and non-perturbative effects may allow the transition between the $t < 0$ and $t > 0$ weakly-coupled solutions. We shall now fill in a little of the detail of this idea.

The effective action for the gravitational field and dilaton is as in (9.63). We shall leave the number of spatial dimensions N arbitrary, allowing for the possibility that the cosmology starts higher-dimensional with compactification to three dimensions to follow, e.g. cosmology could start with the full nine spatial dimensions of the heterotic string. With the ansatz of (9.65) and (9.64), the action is given by (9.69) leading to the field equations (9.72) to (9.74). Making the simple assumption that the universe starts empty, the thermal contributions may be dropped and the field equations simplify to

$$(\dot{\Phi})^2 = \sum_{i=1}^{N} (\dot{\lambda}_i)^2 \tag{9.110}$$

$$\ddot{\lambda}_i = \dot{\Phi}\dot{\lambda}_i \tag{9.111}$$

$$\ddot{\Phi} = \sum_{i=1}^{N} (\dot{\lambda}_i)^2 \tag{9.112}$$

where λ_i is defined in (9.64) and Φ in (9.67).

For $t \neq 0$, there are solutions of (9.110)–(9.112) with (exercise 5)

$$\Phi(t) = \Phi_0 - \ln |t| \quad \text{and} \quad \lambda(t) = \lambda_0 \pm \frac{1}{\sqrt{N}} \ln |t| \tag{9.113}$$

in the isotropic case $\lambda_i = \lambda$ for all i. Equivalently, there are solutions

$$2\phi(t) = 2\phi_0 + (\pm\sqrt{N} - 1) \ln |t| \quad \text{and} \quad a(t) = a_0 |t|^{\pm 1/\sqrt{N}} \tag{9.114}$$

in the isotropic case $a_i = a$ for all i. The corresponding Hubble parameter is

$$H = \frac{\dot{a}}{a} = \pm\frac{1}{\sqrt{N}t} \tag{9.115}$$

which diverges as $t \to 0$. The two solutions for λ arise because of the T-duality symmetry of the equations (9.110)–(9.112) under the transformations

$$\lambda_i \to -\lambda_i \qquad \Phi \to \Phi \tag{9.116}$$

or, equivalently,

$$a_i \to \frac{1}{a_i} \qquad \phi \to \phi - \sum_{i=1}^{N} \ln a_i \tag{9.117}$$

which exchanges small and large scales accompanied by an obligatory action on the dilaton. The equations and their solutions also possess the usual $t \to -t$ symmetry of FRW cosmology. Because of the singularity at $t = 0$, in principle, we may pair either of the solutions for $t < 0$ with either of the solutions for $t > 0$. The hope is that non-perturbative effects will allow a smooth matching at $t = 0$. However, after the big bang, we require that the universe is expanding so, for positive t, we must choose the solution with $a(t) \propto |t|^{+1/\sqrt{N}}$. Now consider the

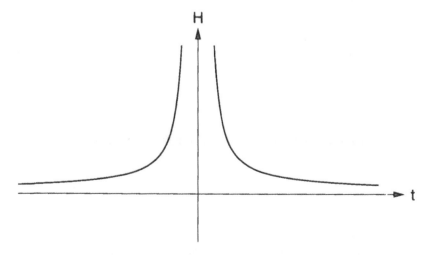

Figure 9.2. Hubble parameter for a possible solution linking the pre-big-bang era to the post-big-bang era.

behaviour of the solution obtained by pairing this positive-time solution with the negative-time solution in which the universe is also expanding. That is

$$2\phi(t) - 2\phi_0 = \begin{cases} (\sqrt{N} - 1)\ln t & t > 0 \\ -(\sqrt{N} + 1)\ln(-t) & t < 0 \end{cases}$$

$$a(t) \propto \begin{cases} t^{1/\sqrt{N}} & t > 0 \\ (-t)^{-1/\sqrt{N}} & t < 0 \end{cases} \tag{9.118}$$

and

$$H(t) = \frac{1}{\sqrt{N}|t|}. \tag{9.119}$$

As advertised earlier, (for $N > 1$) this gives $\phi(t)$ growing for all $t \neq 0$, whereas the Hubble parameter H, which is positive on both branches of the solution, is growing for $t < 0$ but decreasing for $t > 0$ (see figure 9.2): $\phi(t)$ is discontinuous and diverges logarithmically as $t \to 0$. This 'pre-big-bang' cosmology provides an alternative to inflation due to a scalar field rolling in a potential. There is no potential for the dilaton ϕ but nevertheless the universe inflates when $t < 0$.

The biggest difficulty, perhaps, is that it is not known whether the pre-big-bang and post-big-bang eras can be joined together smoothly (a form of the graceful exit problem) because the region close to $t = 0$ requires non-perturbative string theory. It has also been suggested that fine tuning [17] is involved in a successful pre-big-bang scenario. One aspect of this problem is that as a result of (9.118), when $a(t)$ increases by many orders of magnitude while $t < 0$ to solve the problems normally solved by inflation, $e^{\phi(t)}$ also increases by many orders of magnitude. However, as observed after (9.63), e^{ϕ} is g_{string}. This we expect to be of order unity in the post-big-bang universe and so the initial value of ϕ must be tuned to be very small. Moreover, the density perturbations due to the dilaton (playing the role of the inflaton here) are not consistent with the COBE data. However, it has been argued that this last problem can be overcome if the density perturbations required by COBE are due to an axion field in a non-perturbative potential. (See the third reference in [16].)

9.8   M-theory cosmology—the ekpyrotic universe

Another example of a cosmology in which there is a period of evolution of the universe prior to the big bang can be obtained by considering the collision of the so-called 'boundary branes' which occur in M-theory. We first review briefly the M-theory description of strongly-coupled heterotic string theory [18].

Strongly-coupled heterotic string theory in ten dimensions is known to be dual to a theory in 11 dimensions. If the strongly-coupled heterotic string theory has six dimensions compacified on a Calabi–Yau manifold or orbifold X, then the dual theory has seven dimensions compactified on $X \times S^1/Z_2$. Here S^1 is in the X^{10} direction with

$$-\pi\rho \le X^{10} \le \pi\rho \tag{9.120}$$

and the identification

$$X^{10} \sim X^{10} + 2\pi\rho. \tag{9.121}$$

The action of the Z_2 group is

$$Z_2 : X^{10} \to -X^{10}. \tag{9.122}$$

Consequently, there are two fixed points under the action of Z_2, namely

$$X^{10} = 0 \quad \text{and} \quad X^{10} = \pi\rho. \tag{9.123}$$

Because of the Z_2 symmetry (9.122), the quotiented circle is equivalent to a line segment of length $\pi\rho$.

The low-energy limit of this 11-dimensional theory is 11-dimensional supergravity. The E_8 gauge fields of the observable sector (and the observable chiral matter) live on one end of the line segment (on one four-dimensional boundary brane) while the E_8 gauge fields of the hidden sector live on the other end of the line segment (on the other boundary brane). The gravitational fields propagate in the 11-dimensional 'bulk'. The dilaton expectation value e^ϕ in the ten-dimensional theory is reinterpreted, up to a numerical factor, as the length of the line segment in the 11-dimensional theory in M_s^{-1} units.

The idea behind M-theory cosmology [19] (the so-called 'ekpyrotic universe—the universe being consumed by fire and reconstituted out of fire, as in Stoic philosophy) is that two four-dimensional boundary branes may move towards each other and collide before separating again. The size of the fifth dimension (which is the separation of the two four-dimensional boundary branes) goes through zero when the branes collide. Since the separation of the boundary branes is related to the dilaton in the dual description, the dilaton dynamics discussed in the previous section may be employed taking (the number of uncompactified spatial dimensions to be) $N = 4$ (and ignoring the six compactified dimensions.) As before, we require the universe to be expanding after the big bang but now we choose the alternative solution in which the universe is contracting before the big bang. Thus, the solution corresponds to the choice $\lambda = +(1/\sqrt{N}) \ln |t|$ on both branches. For $N = 4$, this gives

$$2\phi(t) - 2\phi_0 = \ln |t| \qquad a(t) \propto |t|^{1/2} \qquad \text{so that } H = \frac{1}{2t}. \qquad (9.124)$$

This choice is qualitatively different from the one made in pre-big-bang cosmology. There, before the big bang, $\phi(t)$ is positive and diverges as $t \to 0$, so that the gauge and gravitational interactions, whose strength is controlled by $e^{2\phi}$, become strong and the vicinity of $t = 0$ is in a strongly-coupled regime. Here, $\phi(t)$ is negative near $t = 0$ and diverges as $t \to 0$. Instead of strong coupling, the vicinity of $t = 0$ is in a weakly-coupled regime. When the branes collide, radiation modes are excited by the kinetic energy of the collision and a hot big bang is triggered.

The ekpyrotic universe allows a new solution to the horizon problem. The collision of the two boundary branes is a non-local event over a region much larger than the Hubble radius. It is this collision that generates the temperature of the hot big-bang universe. Thus, a large degree of homogeneity in the cosmic microwave background radiation is to be expected. However, a number of difficulties in trying to get the ekpyrotic universe to explain things normally explained by inflation has been pointed out. (See [20] and references therein.)

9.9 Exercises

1. Derive the curvature scalar of (9.66).
2. Derive the action (9.69).
3. Obtain the field equations (9.72)–(9.44) from the action (9.69).
4. Derive the finite-temperature equations (9.80) and (9.81).
5. Derive the solutions of the cosmological field equations for pre-big-bang cosmology (9.113).

9.10 General references

The books and review articles that we have found most useful in preparing this chapter are:

- Bailin D and Love A 1994 *Supersymmetric Gauge Field Theory and String Theory* (Bristol: IOP)

[1] de Carlos B, Casas J A, Quevedo F and Roulet E 1993 *Phys. Lett.* B **318** 447
[2] Lyth D H and Stewart G D 1995 *Phys. Rev. Lett.* **75** 201
     Lyth D H and Stewart G D 1996 *Phys. Rev.* D **53** 1784
[3] de Carlos B, Casas J A and Muñoz C 1993 *Nucl. Phys.* B **399** 623
[4] Brustein R and Steinhardt P J 1993 *Phys. Lett.* B **302** 196
[5] Barreiro T, de Carlos B and Copeland E J 1998 *Phys. Rev.* D **58** 083513, arXiv:hep-th/9805005
[6] Huey G, Steinhardt P J, Ovrut B A and Waldram D 2000 *Phys. Lett.* B **476** 379, arXiv:hep-th/0001112
[7] Binetruy P and Gaillard M K 1986 *Phys. Rev.* D **34** 3069
[8] Banks T, Berkooz M and Shenker S 1995 *Phys. Rev.* D **52** 3548, arXiv:hep-th/9503114
[9] Bailin D, Kraniotis G V and Love A 1998 *Phys. Lett.* B **443** 111, arXiv: hep-th/9808142
[10] Brandenberger R and Vafa C 1989 *Nucl. Phys.* B **316** 391
     Tseytlin A A and Vafa C 1992 *Nucl. Phys.* B **372** 443
[11] Bailin D and Love A 1994 *Supersymmetric Gauge Field Theory and String Theory* (Bristol: IOP)
[12] Aldazabal G, Ibáñez L E and Quevedo F 2000 *JHEP* **0001** 031, arXiv:hep-th/9909172
     Aldazabal G, Ibáñez L E, Quevedo F and Uranga A M 2000 *JHEP* **0008** 002, arXiv:hep-th/0005067
[13] Burgess C P 2002 *et al JHEP* **0203** 052, arXiv:hep-th/0111025
[14] Aldazabal G *et al* 2001 *JHEP* **0102** 047, arXiv:hep-ph/0011132
[15] Blumenhagen R, Körs B, Lüst D and Ott T 2002 *Nucl. Phys.* B **641** 235, arXiv:hep-th/0202124
     Jones N, Stoica H and Tye S H 2002 *JHEP* **207** 051, arXiv:hep-th/0203163
[16] Veneziano G 1991 *Phys. Lett.* B **265** 287
[17] Turner M S and Weinberg E J 1997 *Phys. Rev.* D **56** 4604
     Kaloper N, Linde A D and Bousso R 1999 *Phys. Rev.* D **59** 43508, arXiv:hep-th/9801073
[18] Horava P and Witten E 1996 *Nucl. Phys.* B **460** 506, arXiv:hep-th/9510209
     Witten E 1996 *Nucl. Phys.* B **471** 135, arXiv:hep-th/9602070
     Lukas A, Ovrut B A and Waldram D J 1998 *Nucl. Phys.* B **532** 43, arXiv:hep-th/9710208
[19] Khoury J *et al* 2002 *Phys. Rev.* D **65** 086007, arXiv:hep-th/0108107
     Khoury J, Ovrut B A, Steinhardt P J and Turok N 2001 *Phys. Rev.* D **64** 123522, arXiv:hep-th/0103239
[20] Linde A 2002 *Preprint* arXiv:hep-th/0205259

Chapter 10

Black holes in string theory

10.1 Introduction

Besides 'predicting' general relativity as the effective low-energy (classical) theory of gravitation, string theory also provides a (perturbative) *quantum* theory of gravity. Since it is the only such theory known, one might hope that string theory would offer insights into the quantum aspects of gravitation that are not available elsewhere. This is indeed the case.

In 1971, Hawking [1] showed that area of the event horizon of a black hole must increase with time. This prompted the observation that the area of the event horizon is analogous to the entropy of a thermodynamic system. It was subsequently shown that quantum mechanics requires that black holes besides absorbing radiation must also emit it and that black holes are indeed thermodynamic systems. Specifically, it was shown by Bekenstein [2] and Hawking [3] that the entropy is proportional to the area of the event horizon. Our experience with statistical mechanics as the microscopic theory underlying thermodynamics leads us to expect that this entropy is associated with the number of microstates of the (black-hole) system. It is precisely this aspect that is illuminated by string theory. As we shall see, for certain black-hole solutions of string theory, the number of microstates can be calculated, and the resulting multiplicity reproduces precisely the Bekenstein–Hawking formula for the entropy.

In the next section, we review the definition of the black-hole event horizon and outline the proof that the area of its two-dimensional section cannot decrease. In section 10.3, we show why quantum mechanics requires black holes to have a temperature that is determined by their 'surface gravity' and entropy proportional to the area of the (two-dimensional section of the) event horizon. The number of perturbative microstates in string theory is evaluated in section 10.4 and shown to be quite inadequate to explain the derived entropy of black holes. The special class of black holes for which string theory is able to provide a microscopic explanation of their entropy are (certain) 'extreme' black holes, which have both

275

mass and charge (or, more generally, charges). These and their generalization to five dimensions are discussed in the following section. To get black holes from string theory, we need to find analogous solutions to the underlying classical field theory. This is type II supergravity, which is described in section 10.6. It involves the field strengths of certain (antisymmetric) 'Ramond–Ramond' gauge form-fields. These lead to a generalized notion of electric charge, which, in turn, indicates the existence of (non-perturbative) extended objects called 'D-branes' which have non-zero Ramond–Ramond charges. This is described in section 10.7. In section 10.8 we construct the (five-dimensional, extreme) black-hole solutions, with three charges, that have an event horizon with non-zero area. If we use this area in the Bekenstein–Hawking formula, the entropy of the black hole is determined entirely by the (product of the) charges used. The non-zero charges have an immediate interpretation in terms of underlying microstates and the counting of these is done in section 10.9. The number of microstates obtained agrees precisely with that predicted from the calculated Bekenstein–Hawking entropy.

10.2   Black-hole event horizons

It is convenient to use mass units in which the Planck mass m_P and, hence, Newton's constant G_N, are unity: $m_P = G_N^{-1/2} = 1$. The most well-known black-hole solution of general relativity is the Schwarzchild solution which gives in spherical polar coordinates the line element outside of a spherical body of mass M:

$$ds^2 \equiv g_{\mu\nu} \, dx^\mu \, dx^\nu \tag{10.1}$$

$$= \left(1 - \frac{2M}{r}\right) dt^2 - \left(1 - \frac{2M}{r}\right)^{-1} dr^2 - r^2 \, d\Omega_2^2 \tag{10.2}$$

where

$$d\Omega_2^2 \equiv d\theta^2 + \sin^2\theta \, d\phi^2 \tag{10.3}$$

is the line element on the unit two-sphere S^2. The metric is singular at $r = 2M$ but this is merely a coordinate singularity. For example, a particle on a radial timelike geodesic $r = R(t)$ falls from its starting position at $r = R(0) > 2M$, through $R = 2M$, and reaches $R = 0$ in a finite proper time (exercise 1). On a radial null geodesic,

$$dt^2 = \left(1 - \frac{2M}{r}\right)^{-2} dr^2 \equiv (dr^*)^2 \tag{10.4}$$

where

$$r^* \equiv r + \ln\left|\frac{r - 2M}{2M}\right|. \tag{10.5}$$

As r ranges from $2M$ to ∞, r^* ranges from $-\infty$ to ∞. Thus, $d(t \pm r^*) = 0$ on radial null geodesics and the ingoing radial null (Eddington–Finkelstein) coordinate is defined as

$$v \equiv t + r^* = t + r + 2M \ln \left| \frac{r}{2M} - 1 \right| \qquad -\infty < v < \infty. \tag{10.6}$$

Using v as a coordinate instead of t gives

$$ds^2 = \left(1 - \frac{2M}{r}\right) dv^2 - 2dv \, dr - r^2 \, d\theta^2 - r^2 \sin^2 \theta \, d\phi^2. \tag{10.7}$$

This metric is defined initially for $r > 2M$, since the relation $v = r + r^*(r)$ is only defined for $r > 2M$. However, it can now be analytically continued to all $r > 0$ and, in these coordinates, there is no singularity at $r = 2M$. (There *is* a singularity at $r = 0$ where the curvature becomes infinite. Such a singularity cannot, of course, be removed by a coordinate transformation.) At large values of r, the light cones are almost Minkowskian and they allow a particle (or photon) to move outwards or inwards on a timelike (or null) worldline. However, as r decreases the lightcones gradually tilt over. When $r \le 2M$, $2dr \, dv \le 0$ for all non-spacelike (i.e. timelike or null) world lines. Since $dv > 0$ for future directed world lines, it follows that $dr \le 0$, with equality for radial null geodesics when $r = 2M$. When $r < 2M$, all non-spacelike curves necessarily move inwards and hit the singularity at $r = 0$. Thus, if the massive body emits light from its (spherical) surface at $r = r_B < 2M$, the light never escapes to an observer in the region $r > 2M$. Such an observer might infer the presence of the body from its gravitational field but she could not see it. The hypersurface traced out in spacetime by the spherical surface $r = 2M$ is called the 'event horizon' of the (Schwarzchild) black hole. The area of a two-dimensional section of the event horizon is

$$A_H = 4\pi (2M)^2 = 16\pi M^2. \tag{10.8}$$

Let $S(x)$ be a smooth function of the spacetime coordinates x^μ and consider a family of hypersurfaces $S(x) = $ constant. The vectors normal to the hypersurfaces are given by

$$l_\mu = \bar{f}(x) \frac{\partial S}{\partial x^\mu} \tag{10.9}$$

where $\bar{f}$ is an arbitrary non-zero function. If $l^2 = 0$ for a particular hypersurface $\mathcal{N}$ in the family, then $\mathcal{N}$ is said to be a 'null' hypersurface. So for the spherical hypersurfaces $S \equiv r = $ constant, with the black-hole metric (10.7),

$$l^2 = g^{\mu\nu} l_\mu l_\nu = g^{rr} \bar{f}^2 = -\left(1 - \frac{2M}{r}\right) \bar{f}^2. \tag{10.10}$$

Thus, the event horizon $r = 2M$ is a null hypersurface and (exercise 2)

$$l^\mu |_{r=2M} = g^{\mu\nu} l_\nu |_{r=2M} = -\bar{f} \delta^\mu_v. \tag{10.11}$$

We can think of the metric (10.2) or (10.7) as being that which arises following the spherically symmetric collapse of a star with mass $M \gtrsim 1.5$–$2m_\odot$. It is instructive to consider a series of light flashes emitted near the centre of the collapsing star, which is assumed to be made of transparent matter. In the early stages, the density of the star is low, the wavefront of the light will be approximately spherical and its area proportional to the square of the time elapsed since the emission of the flash. However, the gravitational attraction of the stellar matter through which the light is passing deflects neighbouring rays towards each other, reducing the rate at which they are diverging from each other. In other words, the gravitational effect of the matter is to focus the light and to reduce the area of the wavefront from what it otherwise would have been. In the early stages, the wavefront continues to increase in area, crossing the surface of the collapsing star and eventually reaching infinity. As the collapse continues, the matter density increases and so does the focusing effect, until a critical wavefront emerges from the surface of the star with zero divergence. Outside of the star, this wavefront will remain constant and will be the surface $r = 2M$ discussed earlier whose spacetime evolution is the event horizon. It is the boundary of the spacetime region from which it is not possible to escape to infinity. It is generated by null geodesics which have no future endpoint but which do have past endpoints (at the emission of the flash.) The divergence of these null geodesic generators is positive during the collapse phase and zero in the final time-independent state. The area of a two-dimensional section of the event horizon increases monotonically from zero to the final value (10.8). Subsequent flashes will be focused so much by the stronger gravitational focusing that their rays begin to converge and the area of the wavefront decreases.

Now consider what happens when a thin spherical shell of matter of mass δM collapses from infinity at some later time and hits the singularity at $r = 0$. During the collapse, the metric is spherically symmetric but, of course, time-dependent. Afterwards, it will have the form (10.2) or (10.7) but with M replaced by $M + \delta M$. Since δM is necessarily positive, the area of the two-dimensional section of the event horizon must *increase*:

$$\delta A_H = 32\pi M \delta M > 0. \tag{10.12}$$

These results illustrate general results for black holes that are true even without spherical symmetry. The focusing or converging effect that follows from the fact that the gravitational mass is always positive can be described quantitatively using the positive-definiteness of the energy density. Consider a set of null geodesics, and let $l^\mu = dx^\mu/dv$ be a null tangent vector to these geodesics, where v is an affine parameter for the geodesic. At each point, we can define two unit spacelike vectors a^μ and b^μ that are orthogonal to each other and to l^μ. It is convenient to define the complex vectors

$$m^\mu \equiv \frac{1}{\sqrt{2}}(a^\mu + ib^\mu) \quad \text{and} \quad \bar{m}^\mu \equiv \frac{1}{\sqrt{2}}(a^\mu - ib^\mu). \tag{10.13}$$

Then

$$m^\mu m_\mu = 0 = \overline{m}^\mu \overline{m}_\mu = l^\mu m_\mu = l^\mu \overline{m}_\mu \quad \text{and} \quad m^\mu \overline{m}_\mu = -1. \quad (10.14)$$

The fact that the curves of this set are geodesics requires that

$$l_{\mu;\nu} m^\mu l^\nu = 0 \quad (10.15)$$

where, as usual, the semi-colon indicates covariant differentiation. The average rate of convergence of nearby null geodesics is encoded by the quantity

$$\rho \equiv l_{\mu;\nu} m^\mu \overline{m}^\nu \quad (10.16)$$

which is real provided that the null geodesics lie in a three-dimensional null hypersurface, as we shall assume. Let $\mathcal{N}$ be the null hypersurface generated by null geodesics with tangent vectors l^μ and let ΔT be a small element of a spacelike two-dimensional surface in $\mathcal{N}$. We can move each point of ΔT a parameter distance δv up the null geodesics. Then the area A of ΔT changes by

$$\delta A = -2A\rho\delta v. \quad (10.17)$$

Thus, as we should expect, the area decreases if the convergence ρ is positive. The behaviour of ρ along the geodesics is determined from the Newman–Penrose equations [4] which for an affine parametrization (so $l^\nu l_{\mu;\nu} = 0$) give

$$\frac{d\rho}{dv} = \rho^2 + \sigma\bar{\sigma} + \phi_{00} \quad (10.18)$$

where

$$\sigma \equiv l_{\mu;\nu} m^\mu m^\nu \quad \text{and} \quad \phi_{00} \equiv \tfrac{1}{2} R_{\mu\nu} l^\mu l^\nu. \quad (10.19)$$

The Einstein field equations are

$$R_{\mu\nu} - \tfrac{1}{2} g_{\mu\nu} \mathbb{R} = 8\pi T_{\mu\nu} \quad (10.20)$$

where $T_{\mu\nu}$ is the energy–momentum tensor. Thus,

$$\phi_{00} = 4\pi T_{\mu\nu} l^\mu l^\nu. \quad (10.21)$$

The local energy density measured by an observer with velocity vector v^μ is $T_{\mu\nu} v^\mu v^\nu$ and it is reasonable to assume that this is always non-negative. Then, from continuity, the 'weak energy condition'

$$T_{\mu\nu} w^\mu w^\nu \geq 0 \quad (10.22)$$

follows for any null vector w^μ. With this assumption, (10.21) shows that $\phi_{00} \geq 0$ and then, from (10.18), that the effect of the matter is always to increase the average convergence, i.e. to focus the null geodesics.

For a general discussion of what can be seen from infinity and, therefore, of the event horizon, we need to determine the light cone structure of spacetime. For this purpose, it is useful to do a conformal transformation of the metric

$$g_{\mu\nu} \rightarrow \Omega^2 g_{\mu\nu}. \tag{10.23}$$

Such a transformation leaves the light-cone structure unaffected but can be chosen so as to compress everything near infinity and bring it to a finite distance. For Minkowski space, the line element is

$$ds^2 = dt^2 - dr^2 - r^2 d\Omega_2^2. \tag{10.24}$$

In light-cone coordinates $u \equiv t - r$ and $v \equiv t + r$, this becomes

$$ds^2 = du\, dv - \tfrac{1}{4}(v - u)^2 d\Omega_2^2. \tag{10.25}$$

Now define new coordinates p and q such that $\tan p = v$ and $\tan q = u$ with $p - q \geq 0$. Then

$$ds^2 = \sec^2 p \sec^2 q[dp\, dq - \tfrac{1}{4} \sin^2(p - q)\, d\Omega_2^2] \tag{10.26}$$

which shows that the Minkowski metric is conformally related to the metric whose line element $d\bar{s}^2$ is in the square brackets. With a further coordinate transformation $p \equiv t' + r', q \equiv t' - r'$ the line element becomes

$$d\bar{s}^2 = dt'^2 - dr'^2 - \tfrac{1}{4} \sin^2(2r')\, d\Omega_2^2. \tag{10.27}$$

Thus, Minkowski space is conformal to the region bounded by the null surfaces $\mathcal{I}^+ \equiv \{t' + r' = \pi/2\}$ and $\mathcal{I}^- \equiv \{t' - r' = -\pi/2\}$. $\mathcal{I}^+$ is the past light cone of the point $i^+ = \{r' = 0, t' = \pi/2\}$ and $\mathcal{I}^-$ is the future light cone of the point $i^- = \{r' = 0, t' = -\pi/2\}$. All timelike geodesics start at i^-, representing past timelike infinity, and end at i^+, representing future timelike infinity. Null geodesics start at some point on the surface $\mathcal{I}^-$ and end at some point on $\mathcal{I}^+$. We are interested in (black-hole) spacetimes that are asymptotically flat. This means they must be 'like' Minkowski space near infinity, and so should have a similar conformal structure at infinity. In fact, the conformal metric is, in general, singular at the points i^+ and i^- but regular on the null surfaces $\mathcal{I}^+$ and $\mathcal{I}^-$.

Consider the set $J^-(S)$ consisting of a set S of spacetime points plus all points from which S can be reached by future-directed non-spacelike curves. The region of spacetime from which one can escape to infinity along a future directed non-spacelike curve is, therefore, $J^-(\mathcal{I}^+)$, the causal past of future null infinity. The boundary of this region $\dot{J}^-(\mathcal{I}^+)$ is the general definition of the event horizon. It is generated by null geodesics segments which may have past endpoints but can have no future endpoints. Now, using the positivity of ϕ_{00}, it follows from (10.18) that

$$\frac{d\rho}{dv} \geq \rho^2. \tag{10.28}$$

Suppose that the convergence ρ of neighbouring generators has a positive value $\rho_0 > 0$ at some point q on a generator of $\dot{J}^-(\mathcal{I}^+)$, then ρ increases to infinity within a finite affine distance $\Delta v \leq 1/\rho_0$ to the future of q. The point r at which ρ becomes infinite is a focal point at which neighbouring null geodesics intersect. In other words, if the generators of $\dot{J}^-(\mathcal{I}^+)$ ever start converging, they are destined to have future endpoints within a finite affine distance. This contradicts the previously stated property that such generators have *no* future endpoints. It follows [1] that $\rho \leq 0$ everywhere on the event horizon and, therefore, from (10.17), that the area of the two-dimensional cross section *cannot decrease* with time[1]. As already noted, this prompted the observation that this area is analogous to the entropy of a thermodynamic system.

10.3   Entropy of black holes

In fact, the area of the two-dimensional section of the event horizon will remain constant only if the black hole is in a stationary state. If the black hole interacts with anything else the area always increases. In this respect, the area behaves similarly to the entropy of a thermodynamic system. In favourable circumstances, one can arrange that the increase in area can be made arbitrarily small, which corresponds to nearly reversible transformations in thermodynamics.

During black-hole formation in the collapse of a star, the metric is strongly time-dependent and a complete classification of all solutions has not been found. However, the possible final *stationary* states have been identified. An asymptotically flat metric is called stationary if there exists a Killing vector[2] k that is timelike near infinity (where it may be normalized such that $k^2 = 1$). In other words, outside of the horizon $k = \frac{\partial}{\partial t}$, where t is a time coordinate. In these coordinates, the general stationary metric has a line element of the form

$$ds^2 = g_{00}(x)\,dt^2 + 2g_{0i}(x)\,dt\,dx^i + g_{ij}(x)\,dx^i\,dx^j. \tag{10.29}$$

A stationary metric is called *static* if it is also invariant under time-reversal, at least near infinity. Thus, the general static metric has $g_{0i} = 0$ and the line element takes the form

$$ds^2 = g_{00}(x)\,dt^2 + g_{ij}(x)\,dx^i\,dx^j. \tag{10.30}$$

---

[1] It would be possible to escape this conclusion if the generators were prevented from reaching the finite affine distance to the endpoint because of an intervening singularity. However, Hawking [5] has shown, using the general requirements of asymptotic predictability, that this does not occur.

[2] A general coordinate transformation $x \rightarrow x'$ is called an 'isometry' if the transformed metric $g'_{\mu\nu}(x')$ is the same function of its argument x'^μ as the original metric $g_{\mu\nu}(x)$ was of its argument x^μ. The generators of such transformations may be found by considering an infinitesimal transformation in which $x'^\mu = x^\mu + \epsilon\xi^\mu$ with $\epsilon \ll 1$. This is an isometry if ξ satisfies $\xi_{\mu;\nu} + \xi_{\nu;\mu} = 0$, and any vector satisfying this is called a 'Killing' vector. We may equivalently write the Killing vector as $\xi = \xi^\mu \frac{\partial}{\partial x^\mu}$.

With modest assumptions about causality, it can be shown that the only *static* (single) black-hole solution of the Einstein field equations is the (spherically symmetric) Schwarzchild solution given in (10.2).

This result can be generalized to black holes with electric charge Q by solving the Einstein–Maxwell equations. They are derived from the action

$$S_{EM} = \frac{1}{16\pi G_N} \int d^4x \, (-g)^{1/2} [\mathbb{R} - G_N F_{\mu\nu} F^{\mu\nu}]. \tag{10.31}$$

where $F_{\mu\nu}$ is the electromagnetic field strength. Then the Einstein field equations (10.20) have the energy–momentum tensor

$$T_{\mu\nu} = \frac{1}{4\pi} \left(F^\gamma_\mu F_{\nu\gamma} - \frac{1}{2} g_{\mu\nu} F_{\gamma\delta} F^{\gamma\delta} \right). \tag{10.32}$$

The only static solution is the Reissner–Nordström (RN) solution with line element

$$ds^2 = \left(1 - \frac{2M}{r} + \frac{Q^2}{r^2} \right) dt^2 - \left(1 - \frac{2M}{r} + \frac{Q^2}{r^2} \right)^{-1} dr^2 - r^2 \, d\Omega_2^2 \tag{10.33}$$

and gauge potential 1-form

$$A \equiv A_\mu \, dx^\mu = \frac{Q}{r} \, dt. \tag{10.34}$$

Like the Schwarzchild metric, this has a curvature singularity at $r = 0$. For $M \geq |Q|$, there are coordinate (but not curvature) singularities at $r = r_\pm \equiv M \pm \sqrt{M^2 - Q^2}$. This assumes that $M \geq |Q|$, since otherwise there are no horizons and the curvature singularity at $r = 0$ is 'naked'. More generally, these results can be extended to stationary black-hole solutions. The stationary solutions of the Einstein equations are the (axially-symmetric) Kerr solutions, classified by two parameters, the mass M and the angular momentum J. Generalizing to solutions of the Einstein–Maxwell equations leads to the three-parameter Kerr–Newman metrics:

$$ds^2 = \frac{\Delta - a^2 \sin^2\theta}{\Sigma} dt^2 + 2a \sin^2\theta \frac{r^2 + a^2 - \Delta}{\Sigma} dt \, d\phi$$
$$- \frac{(r^2 + a^2)^2 - \Delta a^2 \sin^2\theta}{\Sigma} \sin^2\theta \, d\phi^2 - \frac{\Sigma}{\Delta} dr^2 - \Sigma \, d\theta^2 \tag{10.35}$$

where

$$\Sigma \equiv r^2 + a^2 \cos^2\theta \quad \text{and} \quad \Delta \equiv r^2 - 2Mr + a^2 + Q^2. \tag{10.36}$$

The three parameters are M, a and Q. a is related to the total angular momentum J by

$$a = \frac{J}{M}. \tag{10.37}$$

If we allow magnetic charge P as well as the electric charge Q, we replace Q by

$$e \equiv \sqrt{Q^2 + P^2}. \tag{10.38}$$

The Maxwell 1-form is

$$A = \frac{Qr}{\Sigma}(dt - a\sin^2\theta \, d\phi) - \frac{P\cos\theta}{\Sigma}[a \, dt - (r^2 + a^2) \, d\phi]. \tag{10.39}$$

The future and past event horizons are at

$$r = r_{\pm} \equiv M \pm (M^2 - Q^2 - a^2)^{1/2} \tag{10.40}$$

so that the future event horizon has area A_H given by

$$A_H = 4\pi[2M^2 - Q^2 + 2(M^4 - M^2Q^2 - J^2)^{1/2}]. \tag{10.41}$$

This can be rewritten as

$$M^2 = \frac{A_H}{16\pi} + \frac{4\pi J^2}{A_H} + \frac{\pi Q^4}{A_H} + \frac{Q^2}{2}. \tag{10.42}$$

The first term on the right-hand side is the 'irreducible' part of M^2 that is irretrievably lost down the black hole. The second term is the contribution from the rotational energy of the black hole and the third and fourth terms arise from the electrostatic energy. The mass M, as opposed to M^2, can also be written in an elegant form due to Smarr [6]:

$$M = \frac{\kappa A_H}{4\pi} + 2\Omega_H J + \Phi_H Q \tag{10.43}$$

where, on the future event horizon, the surface gravity (i.e. the acceleration of a static particle as measured at spatial infinity) is κ, the angular velocity is Ω_H and the co-rotating electrostatic potential is Φ_H; all of these are constant on the horizon. If such a black hole is perturbed and settles down to another stationary black hole with parameters $M + dM$, $J + dJ$ and $Q + dQ$, then (exercise 4)

$$dM = \frac{\kappa}{8\pi} dA_H + \Omega_H \, dJ + \Phi_H \, dQ. \tag{10.44}$$

Comparing this with the thermodynamical (first law) formula

$$dU = T \, dS + p \, dV + \mu \, dN \tag{10.45}$$

we see that if some multiple of the area A_H of a section of the event horizon is analogous to entropy, then some multiple of the surface gravity κ on the horizon is analogous to the temperature. Bekenstein [7] suggested that these are not merely analogues but, in some sense, actually *are* the entropy and temperature of the black hole.

However, if the black hole has a temperature, it should radiate a black-body spectrum, thereby contradicting the defining property of a black hole that it can absorb particles or radiation but not emit them. Hawking's resolution of this paradox noted that this absorption-only property is a feature of the *classical* theory of gravitation. He showed that *quantum* mechanical effects cause black holes to create and emit particles as if they were hot bodies with a temperature T_{bh} given by

$$T_{bh} = \frac{\kappa}{2\pi}. \tag{10.46}$$

Then, from (10.44) the (Bekenstein–Hawking formula for the) entropy S_{bh} is

$$S_{bh} = \tfrac{1}{4} A_H. \tag{10.47}$$

The result is obtained by treating the spacetime metric classically but the matter fields to which it is coupled are treated quantum mechanically. We should expect this semi-classical approximation to be excellent except near a spacetime singularity. In flat Minkowski spacetime, a massless real scalar field ϕ, for example, satisfies the field equation $\eta^{\mu\nu}\phi_{;\mu\nu} = 0$ and ϕ can be expanded in terms of annihilation and creation operators $\hat{a}_i$ and $\hat{a}_i^\dagger$ as

$$\phi = \sum_i (\hat{a}_i f_i + \hat{a}_i^\dagger f_i^*) \tag{10.48}$$

where the $\{f_i\}$ are a complete orthonormal set of positive-frequency (complex) solutions of the wave equation $\eta^{\mu\nu} f_{i;\mu\nu} = 0$: the positive frequency is defined with respect to the usual Minkowski time coordinate. The vacuum $|0\rangle$ is then defined to satisfy

$$a_i |0\rangle = 0 \quad \forall i. \tag{10.49}$$

In a curved spacetime with metric $g_{\mu\nu}$, the field equation becomes $g^{\mu\nu}\phi_{;\mu\nu} = 0$ with the semi-colon indicating covariant differentiation. However, in general, positive and negative frequencies have no invariant meaning in a curved spacetime and the expansion of ϕ in annihilation and creation operators is not defined. In a region of spacetime which was flat or asymptotically flat such an expansion can be made, but if we have a spacetime with an initial flat region (1), followed by a region of curvature (2), and then another flat region (3), the initial vacuum $|0_1\rangle$ will not be the same as the final vacuum $|0_3\rangle$. This will lead to the interpretation that the time-dependent metric in (2) has led to the creation of a number of particles of the scalar field ϕ. This is what happens in the core of a black hole [8], hidden from outside observers by the event horizon. When the radius of curvature of spacetime is smaller than the Compton wavelength of a given species, there is an indeterminacy in the particle number, that is to say, particle creation. Although these effects are negligible locally, they can have a significant influence on the black hole over the lifetime of the universe.

The most elegant derivation of the quantum result utilizes Feynman's path-integral formulation of quantum field theory [9]. In this, the generating function $W[J]$ for the Green functions of the quantum field theory is a functional integral over classical fields ϕ

$$W[J] = \int \mathcal{D}\phi \exp\{iS[\phi, J]\} \tag{10.50}$$

where

$$S[\phi, J] = \int d^4x \, [\mathcal{L}(\phi, \partial_\mu\phi) + J\phi] \tag{10.51}$$

is the action integral with $\mathcal{L}$ the Lagrangian density, and J a source current. In the present context, the functional integral is over both matter fields ϕ and metrics $g_{\mu\nu}$. The latter should include both metrics that can be continuously deformed to the flat metric as well as homotopically disconnected metrics such as those of black holes. The evaluation of the action integral is problematic for black-hole metrics because of the spacetime singularities they contain. However, this difficulty can be surmounted by complexifying the metric and evaluating the integral over a contour that avoids the singularities [10].

As an example, we take the Schwarzschild metric (10.2) which, as already noted, has a coordinate singularity at $r = 2M$ and a curvature singularity at $r = 0$. The former can be removed by transforming to Kruskal coordinates in which the line element has the form

$$ds^2 = 32M^3 \frac{e^{-r/2M}}{r}(dz^2 - dy^2) - r^2 \, d\Omega_2^2 \tag{10.52}$$

where

$$-z^2 + y^2 = \left(\frac{r}{2M} - 1\right) e^{r/2M} \tag{10.53}$$

$$\frac{y+z}{y-z} = e^{t/2M}. \tag{10.54}$$

The singularity at $r = 0$ is now on the surface $z^2 - y^2 = 1$ but it can be avoided by defining a new coordinate $\zeta = iz$. Then the metric has the Euclidean form

$$-ds^2 = 32M^3 \frac{e^{-r/2M}}{r}(d\zeta^2 + dy^2) + r^2 \, d\Omega_2^2 \tag{10.55}$$

where now

$$\zeta^2 + y^2 = \left(\frac{r}{2M} - 1\right) e^{r/2M} \tag{10.56}$$

$$\frac{y - i\zeta}{y + i\zeta} = e^{t/2M}. \tag{10.57}$$

Thus, on the contour where ζ and y are real, r is real and $r > 2M$. Further, on this contour we define an imaginary time τ by $\tau = it$ and then (10.57) shows that $\tau =$

$4M \arg(y + i\zeta)$ is an angular coordinate with period $\beta \equiv 8\pi M$. The functional integral defining the generating function should, therefore, be over matter fields and metrics with this periodicity in τ. But this is just the partition function Z for a canonical ensemble of the fields at temperature $T = \beta^{-1} = (8\pi M)^{-1}$. The surface gravity κ for the Schwarzschild black hole is $\kappa = (4M)^{-1}$, so that the temperature is $T = \kappa/2\pi$ in accordance with (10.46). (For the generalization of this result to the Reissner–Nordström black hole, see exercise 7.) With this established, it is clear that the black hole must have the entropy given by (10.47). Nevertheless, it is an interesting excercise to verify this directly by evaluating the action.

In order to obtain a finite result, it is necessary not only to compute the (Euclidean) action by integrating over the previous imaginary time coordinate τ but also over a finite region of space. In general, for a finite region $\mathcal{M}$ of D-dimensional spacetime, the Einstein–Hilbert action must be supplemented by a contribution evaluated on the boundary $\partial\mathcal{M}$ which allows variations of the metric that vanish on $\partial\mathcal{M}$ but which might have non-vanishing derivatives normal to it. In units where $G_N = 1$, the action can be written as

$$S[g, 0] = \frac{1}{16\pi} \int_{\mathcal{M}} \sqrt{-g}\,\mathbb{R}(g)\,\mathrm{d}^D x + \frac{1}{8\pi} \int_{\partial\mathcal{M}} \sqrt{-h}\,B\,\mathrm{d}^{D-1}\xi \qquad (10.58)$$

where h_{ab} is the induced metric on $\partial\mathcal{M}$

$$h_{ab} = \frac{\partial x^\mu}{\partial\xi^a}\frac{\partial x^\nu}{\partial\xi^b} g_{\mu\nu} \qquad (10.59)$$

ξ^a are $D - 1$ coordinates on $\partial\mathcal{M}$, which is specified by an equation of the form $f(x^\mu(\xi^a)) = 0$. Up to a term C that depends only on the induced metric h_{ab}, $B = K + C$ is just the trace K of the extrinsic curvature (the second fundamental form) K_{ab} of the boundary $\partial\mathcal{M}$:

$$K_{ab} = \frac{\partial x^\mu}{\partial\xi^a}\frac{\partial x^\nu}{\partial\xi^b} n_{\mu;\nu} \qquad (10.60)$$

where n_μ is the unit outgoing normal to $\partial\mathcal{M}$ and the semi-colon denotes a covariant derivative. Then

$$n_\mu = \pm \left| g^{\lambda\nu} \frac{\partial f}{\partial x^\lambda}\frac{\partial f}{\partial x^\nu} \right|^{-1/2} \frac{\partial f}{\partial x^\mu}. \qquad (10.61)$$

For asymptotically flat metrics in $D = 4$ dimensions, where $\partial\mathcal{M}$ can be chosen to be the product of the (imaginary) time axis with a 2-sphere of large radius R, it is natural to choose the constant C so that the action is zero for the flat Minkowski space metric $\eta_{\mu\nu}$. Then

$$B = K(g) - K(\eta). \qquad (10.62)$$

Since $\mathbb{R}(g) = 0$, the action for the Schwarzschild black hole derives entirely from the surface term in (10.58). In the case of a spherical surface, $f(x) \equiv r - R = 0$

and we may use the remaining coordinates for the ξ^a. Then the unit outgoing normal is

$$n^\mu = \frac{1}{\sqrt{|g_{rr}|}} \delta^\mu_r \qquad (10.63)$$

and

$$K_{ab} = \frac{1}{2} n^\mu \frac{\partial g_{ab}}{\partial x^\mu}. \qquad (10.64)$$

For the Schwarzschild metric (10.2), this gives (exercise 8)

$$K(g) = \frac{2R - 3M}{R^2(1 - 2M/R)^{1/2}} \qquad \text{and} \qquad \sqrt{-h} = -i\left(1 - \frac{2M}{R}\right)^{1/2} R^2 \sin\theta \qquad (10.65)$$

and then

$$\int_{\partial M} \sqrt{-h}[K(g) - K(\eta)] \, d^3x = -i4\pi\beta\left[2R - 3M - 2R\left(1 - \frac{2M}{R}\right)^{1/2}\right]. \qquad (10.66)$$

Thus, in the limit $R \to \infty$, the Schwarzschild action is

$$S[g_{(S)}, 0] = \frac{1}{2} i\beta M = \frac{i\pi}{\kappa} M. \qquad (10.67)$$

The general result [10] for Kerr–Newman metrics $g_{(KN)}$ of the form (10.35) is that the action integral has the value

$$S[g_{(KN)}, 0] = \frac{i\pi}{\kappa}(M - \Phi_H Q). \qquad (10.68)$$

(The rotation does not affect the evaluation of the action.) The dominant contribution to the path integral (10.50) comes from fields (in our case metrics g) with the correct periodicity that minimize the action. Such fields are solutions of the classical equations of motion and, in the present context, are the Kerr–Newman metrics $g_{(KN)}$. Thus,

$$\ln W[0] = \ln Z \simeq iS[g_{(KN)}, 0]. \qquad (10.69)$$

In a thermodynamic system, the partition function Z for a grand canonical ensemble at temperature $T = \beta^{-1}$ with chemical potentials μ_i associated with conserved charges N_i is defined as

$$Z = \mathrm{Tr} \exp\left[-\beta\left(H - \sum_i \mu_i N_i\right)\right] \qquad (10.70)$$

and its logarithm is related to the free energy F by

$$\ln Z = -\beta F = -\beta\left(E - TS - \sum_i \mu_i N_i\right). \qquad (10.71)$$

For the case we are considering in which $\beta = 2\pi/\kappa$, this gives

$$\ln Z = -\frac{2\pi}{\kappa}(M - TS - \Phi_H Q - \Omega_H J) = -\frac{\pi}{\kappa}(M - Q\Phi_H) \qquad (10.72)$$

using (10.68) and (10.69) for the second equation. Thus

$$\tfrac{1}{2}M = TS + \tfrac{1}{2}\Phi_H Q + \Omega_H J. \qquad (10.73)$$

From Smarr's formula (10.43), we also have that

$$\frac{1}{2}M = \frac{\kappa}{8\pi}A_H + \frac{1}{2}\Phi_H Q + \Omega_H J. \qquad (10.74)$$

Then, comparing the two expressions, restoring the factors of $\hbar, c, G_N$ and k_B, and using

$$T = \frac{\hbar\kappa}{2\pi k_B} = \frac{\hbar c^3}{8\pi G_N M k_B}$$

we deduce that the entropy is given by the Bekenstein–Hawking formula

$$S = \frac{c^3 A_H}{4 G_N \hbar} \qquad (10.75)$$

as anticipated in (10.47).

It is worthwhile pausing for a moment to reflect upon this extraordinary result. We are accustomed to extensive quantities, such as the entropy of a thermodynamic system, being proportional to the volume of the system. Yet here the entropy is scaling as the surface area. The result generalizes to spacetimes with dimension D as

$$S = \frac{c^3 A_D}{4 G_D \hbar} \qquad (10.76)$$

where A_D is the $(D - 2)$-dimensional 'area' of the event horizon and G_D is the D-dimensional Newton constant. The entropy is thus essentially the horizon area measured in Planck units. This area dependence is an example of the 'holographic' principle, and suggests that the fundamental degrees of freedom describing the system may be characterized by a quantum field theory with one fewer space dimensions and with an ultraviolet cut-off at the Planck scale [11]. We shall not pursue this intriguing suggestion further.

The identification of the Bekenstein–Hawking entropy with the physical entropy of the black hole leads to two important puzzles. The first is the so-called 'information problem'. Hawking [12] showed that the outgoing radiation from the radiating black hole is purely thermal and depends only on the conserved charges coupled to long-range fields. This clearly entails a loss of information, since two different, macroscopic objects having the same mass, a graduate student and her supervisor, for example, falling into the black hole would, according to an observer outside of the horizon, generate the same Hawking radiation.

String theory provides a unitary, quantum theory of gravity, so information cannot be lost. Thus, the information loss entailed in Hawking's derivation must be an artefact of the semi-classical approximation used. However, it is not clear precisely where or how the approximation does breakdown nor, if it does, how the information is returned. We shall not pursue this topic further either. Instead, we turn to the second puzzle thrown up by the Bekenstein–Hawking result and the one on which string theory has been able to shed some light, namely whether there is an explanation of black-hole entropy in terms of microstates.

10.4   Perturbative microstates in string theory

Thermodynamics is only an approximation to a more fundamental description based on the statistical properties of the microstates of the system. So the fact that black holes have entropy suggests that this too should be understood microscopically. Roughly speaking, a thermodynamic system with entropy S is associated with a number e^S of microstates of the system. For the Schwarzchild black hole, with a horizon area given by (10.8), we see from (10.75) that the entropy is $S_{bh} = 4\pi G_N M^2/\hbar c$. For the moment, the only important feature is that the associated number of states grows like e^{M^2}. We might have hoped that this approximates the number of perturbative string states with mass M. However, this is *not* the case, as we shall now demonstrate.

It suffices to consider the open bosonic string. This has mode expansion [13]

$$X^\mu = x^\mu + l_s^2 p^\mu \tau + i l_s \sum_{n \neq 0} \frac{1}{n} \alpha_n^\mu \cos(n\sigma) \qquad (10.77)$$

where the oscillator coefficients α_n^μ are creation (annihilation) operators for $n < 0$ ($n > 0$). In units where the string length scale $l_s \equiv 1/\sqrt{\pi T}$ is 1 (T is the string tension), the mass eigenvalues are given by the eigenvalues of

$$\tfrac{1}{2} M^2 = N - 1 \qquad (10.78)$$

where

$$N = \sum_{n=1}^\infty \alpha_{-n}^i \alpha_n^i \qquad (10.79)$$

is the number operator; the sum over i is over the $D_T = 24$ transverse dimensions of the bosonic string. We wish to estimate the number of (degenerate) states d_n that have number-eigenvalue n. It is convenient to define a generating function

$$G(w) \equiv \operatorname{tr} w^N = \sum_{n=0}^\infty d_n w^n \qquad (10.80)$$

with $|w| < 1$. Now

$$\mathrm{tr}\, w^N = \prod_{m=1}^{\infty} \mathrm{tr}\, w^{\alpha^i_{-m}\alpha^i_m} = \prod_{m=1}^{\infty} \left(\frac{1}{1-w^m}\right)^{24}. \tag{10.81}$$

The function

$$f(w) \equiv \prod_{m=1}^{\infty}(1 - w^m) \tag{10.82}$$

can be written as follows:

$$\ln f(w) = \sum_{m=1}^{\infty} \ln(1 - w^m) = -\sum_{m,p=1}^{\infty} \frac{w^{mp}}{p} = -\sum_{p=1}^{\infty} \frac{w^p}{p(1-w^p)}. \tag{10.83}$$

When $w \sim 1$, we can expand $w^p \simeq 1 + p(w-1) + \cdots$, so we can approximate $\ln f(w)$ by

$$\ln f(w) \simeq \frac{-1}{1-w} \sum_{p=1}^{\infty} p^{-2} = -\frac{\pi^2}{6(1-w)}. \tag{10.84}$$

Thus,

$$G(w) \sim \exp\left(\frac{4\pi^2}{1-w}\right) \qquad \text{when } w \sim 1. \tag{10.85}$$

The required degeneracy d_n may be obtained from $G(w)$ by performing a contour integral

$$d_n = \frac{1}{2\pi i} \oint_C \frac{G(w)}{w^{n+1}}\, dw \tag{10.86}$$

where C is a closed loop around $w = 0$. The integrand vanishes rapidly as $w \to 1$ and, when n is large, w^{n+1} is small near $w = 0$. Thus, for large n, there is a saddle point near $w = 1$. In fact, the integrand is stationary when

$$1 - w \simeq \frac{2\pi}{\sqrt{n+1}} \simeq -\ln w. \tag{10.87}$$

It follows that

$$d_n \propto \exp(4\pi\sqrt{n}) \qquad \text{as } n \to \infty. \tag{10.88}$$

As a function of the mass M given in (10.78), the number of states $\rho(M)$, therefore, increases as

$$\rho(M) \propto \exp(\sqrt{2}\pi M) \tag{10.89}$$

quite inadequate for the black-hole entropy requirement that the number of states increases as

$$\rho_{\mathrm{bh}}(M) \propto \exp(4\pi G_N M^2). \tag{10.90}$$

Nevertheless, using a similar treatment, we shall see later that string theory *can* account for the entropy but the associated microstates are *non*-perturbative. Even then, the accounting has so far only been successful for certain special types of black hole called 'extreme' black holes. We therefore first describe the features of extreme black holes that are important for understanding the microscopic origin of their entropy.

10.5   Extreme black holes

It might, in any case, be objected that associating perturbative string states with black holes is absurd. The former are obtained by quantizing the string in a flat background spacetime. How could they be equivalent to a black hole? Nevertheless, certain perturbative states *can* be associated with 'extreme' black holes. We have already noted the constraint $M \geq |Q|$ for the Reissner–Nordström (RN) metrics (10.33). Metrics saturating this constraint are called 'extreme' (RN) black holes. The usual (supersymmetric) perturbative states in string theory satisfy $M \geq |Q|$ and states saturating this inequality are called 'BPS states' after Bogomolnyi [14] and Prasad and Sommerfield [15]. They have the crucial property that their mass cannot receive quantum corrections and it is this that allows their association with extreme black holes. As the string coupling strength g_s increases, the mass M of the perturbative state is unaltered, since it is independent of g_s classically, and because of supersymmetry, there are no quantum corrections. However, the gravitational field of the state is determined by $G_N M$ (in four dimensions) and G_N is proportional to g_s^2. Thus, as g_s increases, the gravitational field increases, there is a back reaction on the perturbative state and, eventually, it may be described by a curved spacetime with large curvature. Thus, in principle at least, it might be possible to associate extreme black-hole spacetimes with perturbative states.

The RN black hole looks like a natural starting place in the search for black holes in string theory. We can think of the Einstein–Maxwell action (10.31) as the bosonic part of the $\mathcal{N} = 2$ supergravity theory in four dimensions. The (massless) gravity supermultiplet contains the graviton, two (fermionic) gravitinos and a vector boson called the 'graviphoton'. The supersymmetry algebra is [13]

$$\{Q_\alpha^A, \overline{Q}_{B\dot{\beta}}\} = 2\delta_B^A \sigma_{\alpha\dot{\beta}}^\mu P_\mu$$

$$\{Q_\alpha^A, Q_\beta^B\} = 2\epsilon_{\alpha\beta} Z^{AB} \tag{10.91}$$

where Q_α^A, with $\alpha = 1, 2$ a (Weyl) spinor index and $A = 1, 2$, are the two supersymmetry generators; the $\overline{Q}$s are defined by $\overline{Q}_{A\dot{\alpha}} \equiv (Q_\alpha^A)^\dagger$; the 2×2 matrices σ^μ with $\mu = 0, 1, 2, 3$ are defined by $\sigma^\mu = (I_2, \sigma^i)$, with σ^i ($i = 1, 2, 3$) the standard Pauli matrices; $\epsilon_{\alpha\beta}$ and the 'central charge' Z^{AB} are antisymmetric with $\epsilon_{12} = +1$ and $Z^{12} \equiv Z$ and, without loss of generality, we may choose $Z \geq 0$. (The graviphoton is a $U(1)$ gauge boson coupled to a charge which is, in fact, the central charge.) For massive representations, we may work in the rest frame where $P_\mu = (M, \mathbf{0})$. Then it is easy to see (exercise 9) that we may form two linear combinations a_α and b_α of the generators that satisfy

$$\{a_\alpha, a_\beta^\dagger\} = 4(M + Z)\delta_{\alpha\beta}$$

$$\{b_\alpha, b_\beta^\dagger\} = 4(M - Z)\delta_{\alpha\beta}$$

$$\{a_\alpha, b_\beta^\dagger\} = 0 = \{a_\alpha, b_\beta\} \tag{10.92}$$

Up to a normalization factor, these are just the anticommutation relations obeyed by two independent sets of fermion annihilation and creation operators. In general, starting with a state $|\psi\rangle$ that is annihilated by a_α and b_α, we can construct a total of 16 states using the creation operators $a_\alpha^\dagger$ and $b_\alpha^\dagger$. Since $\langle\phi|\{a_\alpha, a_\alpha^\dagger\}|\phi\rangle \geq 0$ for any state $|\phi\rangle$ and, similarly, for b_α, the 'BPS bound'

$$M \geq |Z| \tag{10.93}$$

follows. The inequality is saturated by representations (BPS states) for which $|\psi\rangle$ is also annihilated by one set of creation operators ($b_\alpha^\dagger$ if $Z > 0$). Thus, such states are 'short' massive representations of the supersymmetry algebra, since they are constructed using only the $a_\beta^\dagger$ creation operators. They are invariant under half of the supersymmetry algebra.

The extreme RN black hole, obtained from (10.33) and (10.34) by setting $M = Q$, *is* part of such a short hypermultiplet [16]. In this case, the two horizons are both at $r = M = Q$ and, defining $\rho \equiv r - Q$, we may write the solution in the 'isotropic" form in which the spatial part of the metric is conformal to flat space:

$$ds^2 = H^{-2} dt^2 - H^2(d\rho^2 + \rho^2 d\Omega_2^2) \tag{10.94}$$

$$A = (1 - H^{-1}) dt \tag{10.95}$$

where

$$H \equiv 1 + \frac{Q}{\rho}. \tag{10.96}$$

Both the temporal and spatial 'warp' factors (the factors multiplying the two parts of the metric), as well as the electromagnetic vector-potential 1-form, are determined by a single (harmonic) function H. Extreme solutions have the important property that they are easily generalized to a case representing N extreme black holes with charges $q_i = m_i$ (with $i = 1, 2, \ldots, N$) by replacing the function H given in (10.96) by

$$H = 1 + \sum_{i=1}^{N} \frac{q_i}{|\mathbf{r} - \mathbf{r}_i|}. \tag{10.97}$$

By Gauss' law, the total charge is $Q = \sum_{i=1}^{N} q_i$ which, by the BPS bound, is also the total mass

$$M = \sum_{i=1}^{N} m_i = \sum_{i=1}^{N} q_i = Q. \tag{10.98}$$

There is, thus, no binding energy between the individual black holes: the gravitational binding is precisely cancelled by the electrostatic repulsion. Near the horizon $|\rho| \ll Q$, the metric is approximated by $AdS_2 \times S^2$ (exercise 11). Thus, the extreme black hole may also be regarded as a soliton that interpolates between the Minkowski vacuum when $\rho \gg Q$ and $AdS_2 \times S^2$ when $|\rho| \ll Q$. Generalizations of this special form, involving extra dimensions, extended objects and other charges, are important in what follows.

In particular, the simplest example of a string-theory black hole for which a microscopic description can be found is provided by a five-dimensional analogue of the RN solution (10.33) and (10.34). The static, solution of the five-dimensional Einstein–Maxwell equations outside of a (S^3)-spherically symmetric body of mass M and charge Q is

$$ds^2 = \left(1 - \frac{2M}{r^2} + \frac{Q^2}{r^4}\right) dt^2 - \left(1 - \frac{2M}{r^2} + \frac{Q^2}{r^4}\right)^{-1} dr^2 - r^2 d\Omega_3^2 \quad (10.99)$$

and the gauge potential 1-form

$$A = \frac{Q}{r^2} dt. \qquad (10.100)$$

(The r^{-2} dependence of the potential arises, of course, because there are now four spatial dimensions.) As before, the extreme case where $M = Q$, with the horizon at $r = \sqrt{Q}$, may be cast in an isotropic form by using the coordinate $\rho \equiv \sqrt{r^2 - Q}$. Then the horizon is at $\rho = 0$ and

$$ds^2 = H^{-2} dt^2 - H(d\rho^2 + \rho^2 d\Omega_3^2) \qquad (10.101)$$
$$A = (1 - H^{-1}) dt \qquad (10.102)$$

where now the harmonic function is

$$H = 1 + \frac{Q}{\rho^2}. \qquad (10.103)$$

To connect this black hole with string theory, we need to consider the effective field theory describing string theory in the low-energy limit. This is a generalization of Einstein–Maxwell theory both with respect to the number of spacetime dimensions (ten) and the fields involved. The fields involved are just the massless modes that arise in (perturbative) string theory and the field theory that desribes them is type II supergravity.

10.6 Type II supergravity

The massless states that arise in superstring theory include both bosons and fermions. Black holes are, of course, solutions of the classical bosonic field equations, so we shall only be concerned with the bosonic degrees of freedom. All closed superstring theories have massless (bosonic) modes associated with the graviton field $G_{\mu\nu}$ (a symmetric, traceless, rank-2 tensor), the Kalb–Ramond field $B_{\mu\nu}$ (an antisymmetric, rank-2 tensor) and the dilaton field ϕ. (The expectation value of the dilaton fixes the string coupling constant g_s via $g_s = \langle e^\phi \rangle$.) In type II superstring theories, these states arise in the NS–NS sector, i.e. they are states which are constructed using the Neveu–Schwarz (half-integer-moded) world-sheet fermion creation operators for both left- and right-movers. Massless bosonic states also arise in the R–R sector: these are constructed using the Ramond (integer-moded) world-sheet fermion creation operators for both left- and right-movers. In type IIA (in the light-cone gauge), the R–R states transform as $[1] = 8_v$ and $[3] = 56$ representations of the (transverse) $SO(8)$ group, where $[n]$ denotes the totally antisymmetric rank n tensor. They may, therefore, conveniently be represented by form fields $C_{(1)}$ and $C_{(3)}$ where

$$C_{(n)} \equiv \frac{1}{n!} C_{\mu_1\mu_2\ldots\mu_n} \, \mathrm{d}x^{\mu_1} \wedge \mathrm{d}x^{\mu_2} \wedge \ldots \wedge \mathrm{d}x^{\mu_n}. \tag{10.104}$$

The representations $[n]$ and $[8-n]$ of $SO(8)$ are the same, since they related by the eight-dimensional ϵ-tensor, so we could as well represent these fields by the forms $C_{(7)}$ and $C_{(5)}$ respectively. The tree-level Weyl invariance of the string world-sheet action is preserved in the quantum string theory provided that the (renormalization-group) beta functions associated with these fields all vanish. The resulting equations amount to spacetime field equations for the background fields that would arise from the effective action:

$$
\begin{aligned}
S_{\mathrm{IIA}} = \frac{1}{2\kappa_{10}^2} \Bigg\{ & \int \mathrm{d}^{10}x \, (-G)^{1/2} e^{-2\phi} \mathrm{R}(G) \\
& + \int [e^{-2\phi} (4\mathrm{d}\phi \wedge {}^*\mathrm{d}\phi - \tfrac{1}{2} H_{(3)} \wedge {}^*H_{(3)}) - \tfrac{1}{2} F_{(2)} \wedge {}^*F_{(2)} \\
& - \tfrac{1}{2} \tilde{F}_{(4)} \wedge {}^*\tilde{F}_{(4)} - \tfrac{1}{2} B_{(2)} \wedge F_{(4)} \wedge F_{(4)}] \Bigg\}
\end{aligned}
$$

$$\tag{10.105}$$

where $G \equiv \det[G_{\mu\nu}]$,

$$2\kappa_{10}^2 = (2\pi)^7 \alpha'^4 g_s^2 \tag{10.106}$$

is related to the ten-dimensional Newton constant by $2\kappa_{10}^2 = 16\pi G_{10}$; α' is related to the string tension T by $\alpha' = \frac{1}{2\pi T} = \frac{1}{2} l_s^2$ (where l_s is the string length scale); the dilaton 1-form is $\mathrm{d}\phi = \partial_\mu \phi \, \mathrm{d}x^\mu$, and the field-strength forms are related to the potentials by

$$
\begin{aligned}
H_{(3)} &= \mathrm{d}B_{(2)} \qquad F_{(2)} = \mathrm{d}C_{(1)} \qquad F_{(4)} = \mathrm{d}C_{(3)} \\
\tilde{F}_{(4)} &= F_{(4)} + C_{(1)} \wedge H_{(3)}
\end{aligned}
\tag{10.107}
$$

$B_{(2)}$ is the 2-form associated with the Kalb–Ramond field and the (10-dimensional) Hodge dual $^*C_{(p)}$ is

$$^*C_{(p)} \equiv \frac{(-G)^{1/2}}{p!(10-p)!} \epsilon_{\nu_1 \nu_2 \ldots \nu_{10-p} \mu_1 \mu_2 \ldots \mu_p} C^{\mu_1 \mu_2 \ldots \mu_p} \, dx^{\nu_1} \wedge dx^{\nu_2} \wedge \ldots \wedge dx^{\nu_{10-p}}.$$

$$(10.108)$$

It is important to bear in mind that the effective action (10.105) is an approximation that is good in the low-energy limit $\alpha' \to 0$. Higher-order terms in the curvature are negligible provided $\mathbb{R}(G)\alpha' \ll 1$. Roughly speaking, we may say that the metric obtained by solving the lowest-order field equations is only well defined on length scales $l \gg l_s$. Note that all terms from the NS–NS sector are multiplied by $e^{-2\phi}$, while terms from the R–R sector are not coupled to the dilaton. This is a feature of the 'string frame' in which the action (10.105) is written. To remove the dilaton factor from the curvature term, as in the conventional 'Einstein frame' in which we have worked hitherto, we perform the following field redefinition

$$G_{\mu\nu} = e^{\phi/2} g_{\mu\nu}. \qquad (10.109)$$

Then the effective bosonic action in the Einstein frame is

$$S_{\text{IIA}} = \frac{1}{2\kappa_{10}^2} \left\{ \int d^{10}x \, (-g)^{1/2} \mathbb{R}(g) - \frac{1}{2} \int [d\phi \wedge {}^*d\phi + e^{-\phi} \tfrac{1}{2} H_{(3)} \wedge {}^*H_{(3)} \right.$$
$$\left. + e^{3\phi/2} F_{(2)} \wedge {}^*F_{(2)} + e^{\phi/2} \tilde{F}_{(4)} \wedge {}^*\tilde{F}_{(4)} + B_{(2)} \wedge F_{(4)} \wedge F_{(4)}.] \right\}$$

$$(10.110)$$

Type II string theory has $\mathcal{N} = 2$ supersymmetry which, in 10 dimensions, is realized by two Majorana–Weyl spinors Q_α and $\tilde{Q}_\alpha$, each having have 16 real components, so there is a total of 32 supersymmetry charges; Q acts on the right-movers, and $\tilde{Q}$ on the left-movers. In type IIA theory, the two spinors have opposite chirality, while type IIB both spinors have the same chirality. Thus, type IIB is a chiral theory and type IIA non-chiral. The R–R states in type IIB are also represented by form fields but now with components transforming as even-ranked tensors $C_{(0)}$, $C_{(2)}$ and $C_{(4)}$—the only subtlety is that the $[4]_+ = 35$ of $SO(8)$ is self-dual. Analogous forms of the actions (10.105) and (10.110) may also be written in terms of the field strengths $F_{(1)}$, $F_{(3)}$ and $F_{(5)}$ derived from these R–R sector fields. Otherwise, the structure of the two forms of the action is very similar to the type IIA case and we shall not reproduce them here. The self-duality constraint (of the 5-form field strength) must be applied as an extra condition on the solution of the field equations. The important point is that we may consistently truncate the type IIA and type IIB effective actions to include only the graviton, dilaton plus *one* field-strength tensor $F_{(n)}$ (or $H_{(3)}$). This is non-trivial because it must be verified that the (local) supersymmetry variation of the (zero) fermion fields is constantly zero. In the Einstein frame, this gives an action of the form

$$S_n = \frac{1}{2\kappa_{10}^2} \left\{ \int d^{10}x \, (-g)^{1/2} R(g) - \frac{1}{2} \int [d\phi \wedge {}^*d\phi + e^{-a\phi} F_{(n)} \wedge {}^*F_{(n)}] \right\}.$$

$$(10.111)$$

The value of a determines the coupling to the dilaton and may be read off from the type IIA or IIB action. If the chosen field strength is $H_{(3)}$ deriving from the NS–NS form $B_{(2)}$, then $a = 1$, whereas if the chosen field strength derives from an R–R field, then $a = (n-5)/2$. In the latter case, $n = 2, 4, \ldots$ corresponds to type IIA, and $n = 1, 3, 5, \ldots$ corresponds to type IIB. The fact that the type II string theory effective action (10.110) involves various field strengths $F_{(n)}$ suggests that there should be some objects in the underlying string theory that couple directly to the associated gauge form fields, just as an electron is coupled by its charge to the Maxwell gauge potential A_μ. We shall see in the next section that these objects are extended objects, p-branes, generally having p spatial dimensions.

10.7   Form fields and D-branes

There is a geometric aspect of the antisymmetric forms which gives important insights. A gauge field A_μ is coupled naturally to the world line $X^\mu(\tau)$ of a charged particle by a term in the action of the form

$$S \sim \int A_\mu \frac{dX^\mu}{d\tau} \, d\tau.$$

$$(10.112)$$

Under a $(U(1))$ gauge transformation, the vector potential 1-form

$$A_\mu \, dx^\mu \equiv A_{(1)} \to A_{(1)} + d\Lambda_{(0)}$$

$$(10.113)$$

where $\Lambda_{(0)}$ is a 0-form, i.e. a function. The field strength 2-form $F_{(2)} = dA_{(1)}$ is gauge invariant and satisfies

$$dF_{(2)} = 0$$

$$(10.114)$$

which in four dimensions are two of Maxwell's equations. The other two are

$$d \, {}^*F_{(2)} = {}^*J_{(1)}$$

$$(10.115)$$

where $J_{(1)} \equiv J_\mu \, dx^\mu$ is the current 1-form. ${}^*F_{(2)}$ and ${}^*J_{(1)}$ are the Hodge duals. They are defined in ten dimensions in (10.108) but have an obvious generalization to any dimensionality. In four dimensions, ${}^*J_{(1)}$ is a 3-form, and we may use Gauss' theorem to find the electric charge Q in some spatial volume V_3 enclosed by a surface S_2:

$$Q = \int_{V_3} {}^*J_{(1)} = \int_{S_2} {}^*F.$$

$$(10.116)$$

The string world-sheet $X^\mu(\tau, \sigma)$ has an analogous coupling to the Kalb–Ramond field $B_{\mu\nu}$:

$$S \sim \int B_{\mu\nu} \epsilon^{\alpha\beta} \partial_\alpha X^\mu \partial_\beta X^\nu \, d^2\xi \tag{10.117}$$

where $\xi_{0,1}$ are the two world-sheet coordinates (τ, σ) respectively, $\epsilon^{\alpha\beta}$ is the antisymmetric tensor with $\epsilon^{01} = -\epsilon^{10} = -1$. The gauge transformation

$$B_{(2)} \rightarrow B_{(2)} + d\Lambda_{(1)} \tag{10.118}$$

leaves the 3-form field strength $H_{(3)} = dB_{(2)}$ invariant and, analogously to (10.114),

$$dH_{(3)} = 0 \tag{10.119}$$

Evidently, the string, and (some of) its excitations, have a non-zero value of the NS–NS 'electric' charge associated with the $B_{\mu\nu}$ gauge field. The generalization of (10.116) is that the electric charge Q_1 of the (one-dimensional) string, associated with the 2-form gauge potential that is enclosed by the seven-dimensional hypersurface S_7, is given by

$$Q_1 = \int_{S_7} {}^*H_{(3)}. \tag{10.120}$$

For a general antisymmetric $(p + 1)$-dimensional tensor gauge field, the generalization of (10.112) and (10.117) is a term in the action of the form

$$S \sim \int C_{\mu_1,\mu_2,\ldots,\mu_{p+1}} \epsilon^{\alpha_0 \alpha_1 \ldots \alpha_p} \partial_{\alpha_0} X^{\mu_1} \partial_{\alpha_1} X^{\mu_2} \ldots \partial_{\alpha_p} X^{\mu_{p+1}} \, d^{p+1}\xi. \tag{10.121}$$

This describes the coupling of the $C_{(p+1)}$ form gauge field to the $(p + 1)$-dimensional world volume of an extended object having p spatial dimensions (a p-brane) that has a non-zero value for the NS–NS or R–R electric charge associated with the NS–NS or R–R sector gauge form fields described in the previous section. Under a gauge transformation, a gauge form field transforms as

$$C_{(p+1)} \rightarrow C_{(p+1)} + d\Lambda_{(p)} \tag{10.122}$$

and the field strength $F_{(p+2)} = dC_{(p+1)}$ is invariant. The electric charge of the p-brane associated with the gauge potential, that is enclosed by the hypersurface S_{8-p} is, therefore, generally given by

$$Q_p = \int_{S_{8-p}} {}^*F_{(p+2)}. \tag{10.123}$$

In addition to the field strength $H_{(3)}$ deriving from the Kalb–Ramond NS–NS form field $B_{(2)}$, which we have already noted is coupled 'electrically' to the string world sheet, the effective action for type IIA superstrings (10.105) or (10.110) also involves the field strengths $F_{(n)}$ (with n even) that derive from the R–R fields $C_{(n-1)}$. However, unlike the NS–NS field $B_{(2)}$, these fields are *not* coupled electrically to the string world sheet or its excitations. In fact, the vertex operator for the emission of an R–R state involves the corresponding field strength rather than the gauge field. It therefore vanishes at zero momentum and, in consequence, the pertubative string states are electrically neutral with respect to the charge associated with the R–R gauge fields. The foregoing discussion suggests that the R–R field strengths $F_{(n)}$ are naturally associated with branes having $p = n - 2$ dimensions. Thus, type IIA superstring theory is also associated with 0-branes (i.e. point particles), 2-branes (membranes) and 4-branes having non-zero values of the associated R–R charges. Similarly, type IIB superstring theory is associated with 1-branes (R–R strings), 3-branes and 5-branes. Since the perturbative string states are neutral with respect to the charges associated with the R–R fields, type II (A or B) string theory has to be augmented with *non*-perturbative dynamical objects (*p*-branes) that *do* have R–R charges, i.e. electric charges associated with the R–R gauge form fields. The branes are called 'Dirichlet *p*-branes' or 'D*p*-branes', because besides the closed-string sector, there has to be an open-string sector in which the open string ends on these *p*-dimensional hyperplanes [17, 18]. Hence, the open-string world sheet has Dirichlet boundary conditions in the directions perpendicular to the branes. It turns out that these D*p*-branes *do* couple electrically to the associated (closed-string) R–R fields, just as the fundamental string is coupled electrically to the NS–NS Kalb–Ramond field $B_{\mu\nu}$. (More precisely, D-branes act as a source for the associated R–R gauge fields.) It is this enlargement of (the theory formerly known as) string theory that has led to the understanding of black-hole entropy in terms of the associated microstates. We now turn to the construction of the explicit black-hole solutions of the type II supergravity field equations whose entropy we shall eventually be able to explain in terms of D*p*-branes.

10.8   Black holes in string theory

As we have just noted, besides fluctuations of the string, string theory also has various non-perturbative solitons. These are static, finite-energy solutions of the classical field equations, just as RN black holes are static finite-energy solutions of the classical Einstein–Maxwell field equations. It follows from the previous discussion that a ('black brane') solution of the field equations deriving from the action (10.111) for $g_{\mu\nu}$, with non-zero $F_{(n)}$, will give the gravitational field associated with an $(n - 2)$-brane having the associated NS–NS or R–R charge. The field equations may be simplified by looking for solutions that have Poincaré invariance in the $p + 1 = n - 1$ dimensions associated with the *p*-brane world volume and rotational invariance in the remaining transverse directions. The coordinates x^μ are, therefore, split into longitudinal ones, denoted x^a with $a = 0, 1, \ldots, p$, and transverse ones, denoted y^i with $i = (p + 1), \ldots, 9$. The metric is then assumed to have the 'warped' form reminiscent of that encountered earlier in (10.94) and (10.101):

$$ds^2 = e^{2A(r)} \eta_{ab} \, dx^a \, dx^b - e^{2B(r)} \delta_{ij} \, dy^i \, dy^j \tag{10.124}$$

where A and B are functions of the radial coordinate $r = \sqrt{y^i y^i}$. When the p spatial dimensions $x^1, \ldots, x^p$ are compactified the metric could then describe a higher-dimensional black hole. For the dilaton, the ansatz is

$$\phi = f(r) \tag{10.125}$$

and, for the antisymmetric tensor, the assumption is

$$C_{012\ldots p} = e^{C(r)} - 1. \tag{10.126}$$

Then it turns out [19] that the classical field equations following from (10.111) have a solution in which all of the functions A, B, C, f are determined by a single harmonic function $H(r)$:

$$ds^2 = H_p(r)^{-(7-p)/8} \eta_{ab} \, dx^a \, dx^b - H_p(r)^{(p+1)/8} \delta_{ij} \, dy^i \, dy^j \tag{10.127}$$

$$e^\phi = H_p(r)^{(3-p)/4} \tag{10.128}$$

$$C_{(p+1)} = [H_p(r)^{-1} - 1] \, dx^0 \wedge dx^1 \ldots \wedge dx^p \tag{10.129}$$

with

$$H_p(r) = 1 + \frac{L_p^{7-p}}{r^{7-p}}. \tag{10.130}$$

The length L_p is defined by

$$L_p^{7-p} \equiv \frac{2\kappa_{10}\sqrt{\pi} q_p}{(7-p)\Omega_{8-p}} (2\pi \sqrt{\alpha'})^{3-p} \tag{10.131}$$

where q_p is an integer and

$$\Omega_n = \frac{2\pi^{(n+1)/2}}{\Gamma((n+1)/2)} \tag{10.132}$$

is the volume of the unit n-sphere S^n. Using (10.109), the line element in the string frame is

$$ds^2 = H_p(r)^{-1/2} \eta_{ab} \, dx^a \, dx^b - H_p(r)^{1/2} \delta_{ij} \, dy^i \, dy^j \tag{10.133}$$

with all other fields unaltered. In fact, these solutions are extreme solitons. As in the case of the four-and five-dimensional black-hole solutions (10.33) and (10.99), the mass M_p can be read off from the warp factor $H_p(r)$. It is the coefficient of $2\kappa_{10}^2/(7-p)\Omega_{8-p}$ that plays the rôle of Newton's constant in this case. Thus,

$$M_p = q_p \frac{\sqrt{\pi}}{\kappa_{10}} \left(2\pi \sqrt{\alpha'}\right)^{3-p} = q_p \frac{2\pi}{g_s} \left(2\pi \sqrt{\alpha'}\right)^{-(1+p)}. \tag{10.134}$$

As expected of Dp-branes, the solitons we have found are non-perturbative objects whose mass diverges as $g_s \to 0$. However, the fact that the mass scales as g_s^{-1} rather than g_s^{-2} shows that they are unconventional solitons, quite unlike the sphalerons whose action is given in (4.172), for example. The electric charge Q_p associated with the $C_{(p+1)}$ form gauge potential is given in (10.123):

$$Q_p = \frac{1}{2\kappa_{10}^2} \int_{S_{8-p}} {}^*F_{(p+2)} = q_p \frac{\sqrt{\pi}}{\kappa_{10}} \left(2\pi\sqrt{\alpha'}\right)^{3-p}. \tag{10.135}$$

Thus, the solitons have q_p units of the fundamental Dp-brane R–R 'electric' charge $\mu_p \equiv (2\pi)^{-p}(\alpha')^{-(p+1)/2}/g_s$. Further, $M_p = Q_p$, so, as claimed, these are extreme states. There is an exact cancellation between the attractive forces from NS–NS fields due to the mass of the soliton and the repulsive Coulomblike electrical forces from the R–R fields due to its charge. As in the four-dimensional case, this signals the fact that one-half of the supersymmetry charges are preserved by the solution, which, in this case, is invariant under the 16 supersymmetry charges

$$Q_\alpha + P\tilde{Q}_\alpha \tag{10.136}$$

where Q and $\tilde{Q}$ act, respectively, on the right- and left-movers and P represents the operator that reflects the directions y^i transverse to the brane.

The first question then is: do any of these solutions give us the metric of a black hole? If, with the benefit of hindsight, we try to obtain the extreme five-dimensional RN solution given in (10.101), (10.102) and (10.103), then the obvious first attempt is to choose $p = 5$. (We are, therefore, considering type IIB superstring theory, since p is odd.) This gives four (transverse) spatial dimensions y^6, y^7, y^8, y^9,

$$r^2 \equiv (y^6)^2 + (y^7)^2 + (y^8)^2 + (y^9)^2 \tag{10.137}$$

and the harmonic function $H_5(r)$ varies as r^{-2}. As previously noted, if the five longitudinal coordinates x^1, x^2, x^3, x^4, x^5 are compactified (on a 5-torus T^5 say), the metric resembles the five-dimensional RN black hole (10.101). However, the warp factors in both the longitudinal and transverse directions are wrong and, further, the dilaton is singular on the event horizon at $r = r_H = 0$. (The quantum states associated with a soliton are found by identifying the zero modes (or collective coordinates) and quantizing them. This is not possible if the soliton is singular.) Now, because the solitons are extreme, we may combine solutions with different values of p, provided that (some) supersymmetry is preserved. This requires that some of the supersymmetry charges (10.136) preserved by one solution are also preserved by the other. Thus, if P_1 and P_2 represent, respectively, reflection of the coordinates transverse to the p_1- and p_2-solitons, the preserved supersymmetry charges satisfy

$$Q_\alpha + P_1\tilde{Q}_\alpha = Q_\alpha + P_2\tilde{Q}_\alpha = Q_\alpha + P_1(P_1^{-1}P_2)\tilde{Q}_\alpha \tag{10.138}$$

and we see that the unbroken supersymmetries correspond to $+1$ eigenvalues of $P_1^{-1}P_2$. In general, this requires that the number of directions that are transverse to one brane and parallel to the other is a multiple of four and that a $(p-2)$-brane can lie within a p-brane [20]. In the present context, therefore, we may combine the $p = 5$ solitons described earlier with the $p = 1$ solitons, which we take to have longitudinal coordinates x^0, x^1—the remaining four x^a coordinates are transverse to the 1-brane but parallel to the 5-brane. Then again just half of the supersymmetry charges are preserved leaving just eight, corresponding to $\mathcal{N} = 2$ supersymmetry in four dimensions.

In the first instance, the harmonic function H_1 is a function of $\tilde{r}$ where

$$\tilde{r}^2 \equiv (x^2)^2 + (x^3)^2 + (x^4)^2 + (x^5)^2 + r^2 \tag{10.139}$$

and varies as $\tilde{r}^{-6}$. However, when the four directions x^2, x^3, x^4, x^5 are compactified on $T^4 \subset T^5$ of volume V_4, this has the effect of 'smearing' the D1-branes. For example, compactifying the coordinate x^2 on a circle of radius R, the system of D1-branes is replaced by an infinite array of parallel branes a distance $2\pi R$ apart. Because they are extreme states, the effective harmonic function $\tilde{H}_1$ for the array is easily written down and, for $r \gg R$, this function varies as $[\tilde{r}^2 - (x^2)^2]^{-5/2}$, as if the D1-branes also wrapped x^2. The effective number of D2-branes is $q_1\sqrt{\alpha'}/R$. This result obviously generalizes to the case in which all four coordinates are compactified on T^4. Then $\tilde{H}_1$ varies as r^{-2}, just like $H_5(r)$, and is given by (10.130) with $p = 5$ but with the effective number $\tilde{q}_5$ of D5-branes given by

$$\tilde{q}_5 = \frac{(2\pi\sqrt{\alpha'})^4}{V_4}q_1. \tag{10.140}$$

The composite system has warp factors that are given by the 'harmonic function sum rule' [21]. With the proviso that the harmonic function H_1 for the D1-branes is replaced by the smeared version $\tilde{H}_1$ just described, so that both of the harmonic functions are functions of the overall transverse distance r, the warp factors of the composite system are just the product of the separate warp factors in each of the three sectors, namely parallel to both, transverse to both and transverse to the 1-brane but parallel to the 5-brane. The dilaton is also given by the product. Thus, for the combined system,

$$ds^2 = H_5^{-1/4}\tilde{H}_1^{-3/4}[(dx^0)^2 - (dx^1)^2] - H_5^{-1/4}\tilde{H}_1^{1/4}[(dx^2)^2 + \cdots + (dx^5)^2]$$
$$- H_5^{3/4}\tilde{H}_1^{1/4}[(dy^6)^2 + \cdots + (dy^9)^2] \tag{10.141}$$

$$e^\phi = H_5^{-1/2}\tilde{H}_1^{1/2} \tag{10.142}$$

$$C_{(5)} = [H_5^{-1} - 1]dx^0 \wedge dx^1 \ldots \wedge dx^5 \qquad C_{(1)} = [\tilde{H}_1^{-1} - 1]dx^0 \wedge dx^1. \tag{10.143}$$

Then the dilaton is, as required, finite at the event horizon. (We note, incidentally, that this is only possible when at least two 'charges' are activated.) In addition, the warp factor associated with the y^i directions now approximates the H warping in (10.101). Thus, ignoring the directions x^2, x^3, x^4, x^5 for the moment, all that is needed to complete the resemblance to (10.101) is a further warping approximating H^{-1} in the x^0-direction. In any case, in its present form, the metric does not have a finite horizon 'area', because g_{11} vanishes at the horizon: it requires a further warping approximating H. In fact, the missing ingredient is supplied by adding momentum in the x^1-direction. Since the extreme solution is boost invariant, we cannot add momentum simply by boosting it. The generalization arises [21] because the solutions have a null-hypersurface orthogonal isometry which permits the replacement

$$(dx^0)^2 - (dx^1)^2 \rightarrow (dx^0)^2 - (dx^1)^2 + H_P(r)(dx^0 - dx^1)^2 \qquad (10.144)$$

for an arbitrary harmonic function $H_P(r)$. This is exactly what happens if we do a Kaluza–Klein reduction along the x^1-direction but with a non-zero off-diagonal component in the metric $g_{01}/g_{11} = H_P(r)^{-1}$ (see exercise 7). Indeed, the resulting metric can be obtained from a non-extreme metric by this means: the dilaton and gauge potential form fields are unaffected. The final form of the metric is

$$ds^2 = H_5^{-1/4} \tilde{H}_1^{-3/4} [(dx^0)^2 - (dx^1)^2 + H_P(dx^0 - dx^1)^2]$$
$$- H_5^{-1/4} \tilde{H}_1^{1/4} [(dx^2)^2 + \cdots + (dx^5)^2] - H_5^{3/4} \tilde{H}_1^{1/4} [dr^2 + r^2 \, d\Omega_3^2].$$
$$(10.145)$$

We may write the harmonic functions in the form

$$\tilde{H}_1 = 1 + \frac{r_1^2}{r^2} \qquad H_5 = 1 + \frac{L_5^2}{r^2} \qquad H_P = \frac{r_P^2}{r^2} \qquad (10.146)$$

where, using (10.106), (10.131) and (10.140),

$$r_1^2 = q_1 g_s \alpha' \frac{(2\pi \sqrt{\alpha'})^4}{V_4} \qquad L_5^2 = q_5 g_s \alpha' \qquad r_P^2 = n g_s^2 \alpha' \frac{(2\pi \sqrt{\alpha'})^4}{V_4} \frac{\alpha'}{R^2} \qquad (10.147)$$

with R the radius of the circle upon which the x^1-coordinate is compactified, and n an integer specifying the (right-moving) momentum $P = n/R$ on the circle. Adding this momentum breaks a further half of the supersymmetries leaving a total of four preserved supersymmetry charges. This corresponds to $\mathcal{N} = 1$ supersymmetry in five dimensions. The 'area' A_H of the event horizon is defined as the (eight-dimensional) volume of the time slice at the horizon. Taking the limit $r \rightarrow 0$ from above, this gives a product of factors, one from each of the three disjoint pieces of the metric:

$$A_H = [(r_1^{-3/4} L_5^{-1/4} r_P) 2\pi R][(r_1^{1/4} L_5^{-1/4})^4 V_4][(r_1^{1/4} L_5^{3/4})^3 2\pi^2]$$
$$= 4\pi^3 R V_4 r_1 L_5 r_P. \qquad (10.148)$$

The surface gravity and, hence, the black-hole temperature is zero, as might be expected from an extreme black hole (see exercise 7). Using the (generalized)

Bekenstein–Hawking formula (10.75), the entropy S_{bh} associated with this solution is

$$S_{bh} = \frac{A_H}{4G_{10}} = 2\pi\sqrt{q_1 q_5 n} \qquad (10.149)$$

independently of the size of the compact dimensions and of the string coupling g_s. Thus, the entropy is determined entirely by the integers q_1, q_5 and n and it is this feature that allows the identification of the associated microstates. The fact that the entropy of a black hole at zero temperature is non-zero does not imply a violation of the third law of thermodymamics, if indeed the analogy between black-hole dynamics and thermodynamics is exact. The version of the third law that suggests that there is, is a statement about the equations of state for ordinary matter.

10.9 Counting the microstates

We have already argued that Dp-branes are sources of R–R charge and we have also shown in (10.135) that the soliton solution (10.129) has q_p units of p-brane R–R charge. Thus, the obvious interpretation of the black-hole solution that we have just constructed is that it is a bound state made out of q_5 D5-branes and q_1 D1-branes (D-strings) with some momentum n/R. However, Dp-branes are defined as pointlike objects in their transverse dimensions in an otherwise flat spacetime. The R–R solitons that we have derived are only asymptotically flat, so why do we believe that they are made of D-branes? We have noted previously that the effective action from which these solitons derive is a good approximation so long as the curvature is small $\mathbb{R}(G)\alpha' \ll 1$. The length scale defined in (10.131), associated with the solution (10.129), is given by $L_p \sim (g_s q_p)^{1/(7-p)}\sqrt{\alpha'}$. Thus, when $g_s q_p > 1$, the curvature is small and the soliton solutions are valid. In fact, the effective supergravity equations were derived using string perturbation theory, which is valid only when $g_s < 1$. Consequently the soliton solutions apply only when q_p is large, so that the curvature is small. When the string coupling g_s is very small, the R–R solitons are very massive, as is apparent from (10.134). However, their gravitational field is proportional to $G_{10}M_p$ and, since $G_{10} \propto g_s^2$, the associated spacetime becomes flat as $g_s \to 0$. Also, the horizon area $A_H = 4G_{10}S_{bh}$ approaches zero in this limit. When it is smaller than the string scale $l_s^2 \equiv 4\pi^2\alpha'$, the higher-order curvature terms become important and the flat space description becomes valid. Provided that $g_s q_1 \ll 1$ and that $g_s q_5 \ll 1$, we are considering weakly-coupled D-branes in a flat spacetime. In this case, it is straightforward to count the number of configurations. We shall return shortly to the case where $g_s q_p > 1$ in which our black-hole solutions are valid.

The configuration in question (10.145) breaks the 10-dimensional Lorentz symmetry $SO(1,9) \to SO(1,1) \times SO(4)_\parallel \times SO(4)_\perp$. The first factor acts on the D-string world sheet (x^0, x^1), the second factor on the rest of the D5-brane world volume (x^2, x^3, x^4, x^5), and the final factor on the remining dimensions (y^6, y^7, y^8, y^9) that are transverse to both. This symmetry forbids rigid branes from carrying linear or angular momentum. So the question arises as to what degrees of freedom do carry the momentum n/R. An obvious possibility is the massless states of the open strings that begin and end on D-branes. (The massive excitations of the D-branes have masses proportional to g_s^{-1} and, hence, do not play a role when the coupling is weak.) The 1–1 states, in which the open string begins on a D1-brane and ends on a D1-brane, generate a vector supermultiplet in the adjoint representation of the $U(q_1)$ gauge group, similarly for the 5–5 states with gauge group $U(q_5)$. In geometrical terms, the VEVs of the scalar fields in these supermultiplets correspond to separations of the individual D1- and D5-branes from each other. This takes us away from the black-hole state, which has maximal degeneracy. So instead we consider the 1–5 and 5–1 states. These generate hypermultiplets in the $(q_1, \bar{q}_5) + (\bar{q}_1, q_5)$ of $U(q_1) \times U(q_5)$, which gives a total of $4q_1q_5$ (scalar) bosons and an equal number of (Weyl) fermions. The vacuum expectation values of the scalars are associated with the 4 coordinates (transverse to the D1-branes) of each D1-brane giving its position relative to each D5-brane. This configuration must be made to carry $P = n/R$ momentum in the x^1-direction. With the four coordinates x^2, x^3, x^4, x^5 compactified on a torus whose size is small compared to that of the circle on which x^1 is compactified $(V_4^{1/4} \ll R)$, we effectively have a two-dimensional field theory on the world volume (x^0, x^1) of the D-string. The Hamiltonian is $H = n/R$ and this has to be distributed among the $4q_1q_5$ bosons and fermions. Apart from the (minor) complication introduced by having fermions, this is precisely the problem discussed in section 10.4 in which we estimated the number of (bosonic) string states having mass n (in string units).

As before, the problem may be solved using a generating function (the partition function)

$$Z(w) \equiv \operatorname{tr} w^N = \sum_{n=0}^{\infty} d_n w^n \tag{10.150}$$

where $N \equiv PR$ and d_n is the (required) number of states having eigenvalue n of N (and, therefore, right-moving momentum n/R). Then

$$\operatorname{tr} w^N = \prod_{m=1}^{\infty} \left(\frac{1 + w^m}{1 - w^m} \right)^{4q_1q_5}. \tag{10.151}$$

The fermions give the terms in the numerator and the bosonic contribution in the denominator is derived precisely as in (10.81). As in section 10.4, we may estimate d_n for large values of n (exercise 12) with the result [22]

$$d_n \sim \exp\left(2\pi \sqrt{q_1 q_5 n}\right). \tag{10.152}$$

Hence,

$$\ln d_n \sim 2\pi \sqrt{q_1 q_5 n} = S_{\text{bh}} \tag{10.153}$$

using (10.149). This is a spectacular result. We have recovered the Bekenstein–Hawking formula (10.149) for the entropy of the extreme black hole (10.145) from counting the microstates that make up this black hole. We assumed that $V_4^{1/4} \ll R$ in order to simplify the counting of states but this is not essential. Since the number of states (10.152) does not change when the radii are continuously varied, we know that the entropy is given by (10.153) for all values of R. Also, the states were counted in the limit $g_s q_1 \ll 1$ and $g_s q_5 \ll 1$ where we have D-branes in flat space, whereas the black-hole solution (10.145) is valid only when $g_s q_1 > 1$ and $g_s q_5 > 1$ so that higher-order curvature corrections are negligible. However, because these are extreme solutions, they are protected by their supersymmetry and we may assume that the calculated degeneracy will not undergo renormalization by quantum effects.

This five-dimensional RN black hole utilizes three non-zero $U(1)$ charges—the two R–R charges q_1 and q_5, and the momentum n/R in the internal x^1-direction—and this is the minimum number needed to get a finite area with a regular horizon. In four dimensions, a minimum of four non-zero charges is needed. The result can be generalized to near-extreme black holes, in which case the entropy becomes a function of the mass of the black hole as well as its four charges. We shall not pursue this further. The interested reader is referred to one of the excellent reviews [23–25] in the literature.

10.10 Problems

1. Show that a particle on a radial timelike geodesic $r = R(t)$ in a Schwarzchild spacetime falls from rest at $r = R(0) > 2M$ to $R = 0$ in proper time
$$\tau = \pi M \left(1 - \sqrt{1 - 2MR(0)} \right)^{-3/2}.$$

2. Verify that the vector l^μ normal to the event horizon of a Schwarzchild black hole has components (10.11) in Eddington–Finkelstein coordinates.

3. Verify that the Kerr–Newman metric reduces to the Reissner–Nordström metric when the angular momentum $J = 0$.

4. Verify Smarr's formula and (10.44) for the Kerr–Newman metric where the surface gravity is $\kappa = (r_+ - r_-)/2(r_+^2 + a^2)$ and the co-rotating electrostatic potential is $\Phi_H = \frac{Qr_+}{(r_+^2 + a^2)}$.

5. The surface gravity κ can be calculated directly using the formula
$$\kappa^2 = -\tfrac{1}{2} \chi^{\mu;\nu} \chi_{\mu;\nu}|_{r=r_H}$$
where χ is a timelike Killing vector normal to the horizon (and normalized so that $\chi^2 = 1$ at spacelike infinity) and the semi-colon indicates the covariant derivative. Using the metric (10.7) and the Killing vector $\chi^\mu = \delta^\mu_v$, verify that χ is a unit timelike Killing vector and that the previous formula is satisfied by $\kappa = 1/4M$, as required.

6. Calculate the Schwarzchild radius for a black hole having mass $M = M_\odot$. Calculate also its temperature and entropy.

7. By choosing new coordinates $\rho = \sqrt{r - r_+}$ and $\tau = it$, show that near the horizon the line element for the Reissner–Nordström black hole may be written in the form

$$-ds^2 \simeq A\rho^2 d\tau^2 + 4A^{-1} d\rho^2 + r^2 d\Omega_2^2 \qquad (10.154)$$

where $A \equiv (r_+ - r_-)/r_+^2$. Hence, show that τ is an angular coordinate with period $4\pi A^{-1}$ and, therefore, that the temperature of the extreme black hole is zero.

8. Evaluate the extrinsic curvature K_{ab} of the Schwarzschild metric on a spherical shell of radius R and verify equations (10.65) and (10.66).

9. Show that the $\mathcal{N} = 2$ supersymmetry algebra (10.91) can be written in the form (10.92).

10. Express the function $f(w)$ given in (10.82) in terms of the Dedekind eta function

$$\eta(\tau) \equiv e^{i\pi\tau/12} \prod_{m=1}^{\infty} (1 - e^{2\pi im\tau}).$$

Using the property

$$\eta(-1/\tau) = (-i\tau)^{1/2}\eta(\tau)$$

show that

$$f(w) = \left(\frac{-2\pi}{\ln w}\right)^{1/2} w^{-1/24} \exp\left(\frac{\pi^2}{6\ln w}\right) f\left(\frac{4\pi^2}{\ln w}\right).$$

Hence, show that

$$f(w) \sim A(1 - w)^{-1/2} \exp\left(-\frac{\pi^2}{6(1 - w)}\right) \qquad \text{for } w \sim 1$$

and find the power-law correction to the exponential dependence given in (10.89).

11. For the extreme Reissner–Nordström black hole with the metric (10.94), show that near the horizon the metric approximates $AdS_2 \times S^2$.

12. Show that the partition function $Z(w)$ defined in (10.150) and (10.151) has the asymptotic behaviour

$$Z(w) \sim \exp\left(\frac{q_1 q_5 \pi^2}{1 - w}\right) \qquad \text{for } w \sim 1$$

and, hence, verify the degeneracy (10.152) of states having momentum n/R.

10.11   General references

The books and review articles that we have found most useful in preparing this chapter are:

- Hawking S W 1973 The Event Horizon *Black Holes* Cours de l'Ecole d'été de Physique Theorique, Les Houches (New York: Gordon and Breach) 1
- Townsend P K Cambridge University lecture notes on *Black Holes*, arXiv:gr-qc/9707012
- Horowitz G T 1996 The origin of black hole entropy in string theory *Seoul 1996, Gravitation and Cosmology* 46, arXiv:gr-qc/9604051
- Johnson C V 2003 *D-Branes* (Cambridge: Cambridge University Press)

Bibliography

[1] Hawking S W 1971 *Phys. Rev. Lett.* **26** 1344
[2] Bekenstein J 1972 *Lett. Nuovo Cimento* **4** 737
[3] Hawking S W 1974 *Nature* **248** 30
[4] Newman E T and Penrose R 1962 *J. Math. Phys.* **3** 566
[5] Hawking S W 1973 The event horizon in black holes *Cours de l'Ecole d'été de Physique Theorique, Les Houches* (New York: Gordon and Breach) p 1
[6] Smarr L 1973 *Phys. Rev. Lett.* **30** 71
    Smarr L 1973 *Phys. Rev. Lett.* **30** 521(E)
[7] Bekenstein J 1973 *Phys. Rev.* D **7** 2333
    Bekenstein J 1974 *Phys. Rev.* D **9** 3392
[8] Hawking S W 1975 *Commun. Math. Phys.* **43** 199
[9] See, for example, Bailin D and Love A 1986 *Introduction to Gauge Field Theory* (London: IOP)
[10] Gibbons G W and Hawking S W 1977 *Phys. Rev.* D **15** 2752
[11] 't Hooft G 1990 *Nucl. Phys.* B **335** 138
[12] Hawking S W 1976 *Phys. Rev.* D **14** 2460
[13] See, for example, Bailin D and Love A 1994 *Supersymmetric Gauge Field Theory and String Theory* (Bristol: IOP)
[14] Bogomolnyi E B 1976 *Yad. Fiz.* **24** 861 (Engl. transl. *Sov. J. Nucl. Phys.* **24** 449)
[15] Prasad M and Sommerfield C 1975 *Phys. Rev. Lett.* **35** 760
[16] Gibbons G W and Hull C M 1982 *Phys. Lett.* B **109** 190
[17] Dai J, Leigh R G and Polchinski J 1989 *Mod. Phys. Lett.* A **4** 2073
    Leigh R G 1989 *Mod. Phys. Lett.* A **4** 2767
[18] Polchinski J 1994 *Phys. Rev.* D **50** 6041
    Polchinski J 1995 *Phys. Rev. Lett.* **75** 4724
[19] Horowitz G T and Strominger A 1991 *Nucl. Phys.* B **360** 197
[20] Polchinski J, Chaudhuri S and Johnson C V *Notes on D-branes*, arXiv:hep-th/9602052
[21] Tseytlin A A 1996 *Nucl. Phys.* B **475** 149, arXiv:hep-th/9604035
[22] Strominger A and Vafa C 1996 *Phys. Lett.* B **379** 99, arXiv:hep-th:9601029
[23] Mandal G 1999 *Lectures Presented at the ICTP Spring School, April 1999* arXiv:hep-th/0002184

[24] Wadia S R 1999 *Lectures Given at Advanced School on Supersymmetry in the Theories of Fields, Strings and Branes, Santiago de Compostela, Spain, 26–31 July 1999 and at Workshop in String Theory, Allahabad, India, 18 Oct–18 Nov 1999 and at ISFAHAN STRING SCHOOL AND WORKSHOP, Isfahan, Iran, 1–14 May 2000* arXiv:hep-th/0006190

[25] Peet A W 1999 Theoretical Advanced Study Institute in Elementary Particle Physics (TASI 99) Lectures: *Strings, Branes, and Gravity, Boulder* pp 353–433, arXiv:hep-th/0008241

Index

For Product Sales, Contact and Ordering please contact or
its representative GRS Publishing GmbH & Co. KG or its Reader
Verlag GmbH, Kaufingerstraße 21, 80331 München, Germany